INTERNET and WEB TECHNOLOGY

A. RAVICHANDRAN

KHANNA BOOK PUBLISHING CO. (P) LTD.

Publisher of Engineering and Computer Books

4C/4344, Ansari Road, Darya Ganj, New Delhi-110002

Phone : 011-23244447-48 **Mobile:** +91-9910909320

E-mail : contact@khannabooks.com

Website : www.khannabooks.com

Dear Readers,

To prevent the piracy, this book is secured with HIGH SECURITY HOLOGRAM on the front title cover. In case you don't find the hologram on the front cover title, please write us to at contact@khannabooks.com or whatsapp us at +91-99109 09320 and avail special gift voucher for yourself.

Specimen of Hologram on front Cover title:

Moreover, there is a SPECIAL DISCOUNT COUPON for you with EVERY HOLOGRAM.

How to avail this SPECIAL DISCOUNT:

Step 1: Scratch the hologram
Step 2: Under the scratch area, your "coupon code" is available
Step 3: Logon to www.khannabooks.com
Step 4: Use your "coupon code" in the shopping cart and get your copy at a special discount
Step 5: Enjoy your reading!

Price: ₹ 650.00

ISBN: 978-93-80016-269

Internet and Web Technology
by A. Ravichandran

First Edition: 2011
Revised Edition: 2022
Reprint Edition: 2025

To view complete list of books, Please scan the QR Code:

Published by:
Khanna Book Publishing Co. (P) Ltd.
CIN: U22110DL1998PTC095547

Visit us at: **www.khannabooks.com**
Write us at: **contact@khannabooks.com**

Printed in Bharat.

Contents

4. Extensible Markup Language (XML) 105

1 Introduction to Web Development Strategies

Introduction and Web Development Strategies

global hypertext system that uses the Internet as its transport mechanism is called World Wide Web. In a hypertext system, you navigate by clicking hyperlinks, which display another document (which also contains hyperlinks). What makes the Web such an exciting and useful medium is that the next document you see could be housed on a computer next door-or halfway around the world. The Web makes the Internet easy to use.

Created in 1989 at the European Laboratory for Particle Physics (CERN), a research institute in Switzerland, the Web relies upon the HyperTExt Transport Protocol (HTTP), an Internet standard the specifies how an application can locate and acquire resources (such as document, sound or graphic) stored on another computer on the Internet. HTTP provides transparent, easy-to-use access to Web documents, FTP (File transfer Protocol) file archives, Gopher menus, and even UseNet newsgroups, users of Web navigation software, called Web browsers, simply click on an underlined word or phrase, and HTTP takes care of locating and downloading the desired documents. Most Web documents are created using the HyperText markup language (HTML), a markup language that is easy to learn and will soon be supplanted by automated tools. Incorporating hypermedia (graphics, sound, animations, and video), the Web has become the ideal medium for publishing information on the Internet.

A key 1995 agreement on security protocols ensures that Web will quickly become an important commercial medium, in which consumers can browse on-line catalog and place orders without worrying about anyone intercepting the supplied credit card information.

History of Web

In 60s a project was undertaken by the U.S. Defense Advanced Research Projects Agency (DARPA). It was in fact looking for some technology that could enable it to maintain its strategic military-based communication worldwide in case of a nuclear attack. This can be said as the main conception of the Internet.

Later, these developments led to the establishment of the Advanced Research Projects Agency Net (ARPANet). The main interest of this was looking for a technology that could link computers in various locations by using a new technology called Packet Switching Technology. This new technology enabled several users to simultaneously share a single communication line. It also allowed the creation of nets that could automatically route data around downed circuits or computers. This technology was then used by National Science Foundation to create to its own net, and called it NSDnet. It met with a large of success though its main users were universities and schools for information available over it.

Since the users were mostly scientists and researchers, the demand went on increasing endlessly, NSF found itself unable to cope with it. As a result, a decision was taken to open the Net for use by private organizations as well. This allowed anyone with a modem and a computer to access the Net. The beginning was made. Thus, NSFnet became the backbone of communication service for the

Internet and continues to do so even now. At present, NSFNet comprises a set of highspeed data connections that join the major network all over the world.

This lead to several users using the net for information and then came the phase when its potential for doing business was exploited and companies started marketing services and information over the net. Today in fact, anybody with a computer and modem with dedicated phone line can access Internet with the help of some service provided by local Internet Service Providers.

Protocols governing Web

Networks provide computes the basic ability of transferring bits from one computer to another. In order to use networks, we need a set of rules, which all the members of a network agree upon. This is called a protocol. A communication protocol is a standard that is designed to specify how computers interact and exchange messages. A protocol usually specifies the format of messages and how to handle errors.

Protocol Suite

To simplify the design and implementation of protocols, a set of protocols is designed so that each one has different responsibilities, instead of a single protocol being responsible for handling all forms of communication. The set of such protocols is called a Protocol Suite, which covers all forms of communication needs.

PPP (Point-to-Point Protocol)

It is a communication protocol, used over serial lines to support Internet connectivity. This protocol encompasses both synchronous as well as asynchronous systems and covers connections, both from a host to a network and between two networks. Used to linked personal computers (PCs) to the Internet through modems and telephone lines, it also offers a built-in security mechanism.

FTP (File Transfer Protocol)

It is used to transfer files from one computer to another. It is a very commonly used procedure to download and upload files over the Internet. With FTP, you can login to another Internet site and transfer files. Some sites have public file archives that you can access by using FTP with the account name 'anonymous' and your e-mail address as the password. This type of access is called anonymous FTP.

Telnet: Terminal Emulation Protocol

This protocol is used over TCP connections. It enables users to login to remove hosts and use their resources, as if, these resources exist on their (users') own computers.

HTTP (Hyper TextTransfer Protocol)

It is used as the communication protocol to the transport hypertext documents over the Internet. It tells the server what to send to the client, so that the client can view Web pages, FTP sites or other areas on the Net.

Transmission Control Protocol (TCP)

Transmission Control Protocol (TCP) is widely used protocol which enables a computer to send data to a remote computer. TCP is a reliable protocol that sees to it that the packets of a message (each with its own address and checksum) reach their destination safely in the correct order.

Internet Protocol (IP)

IP allows precise data transmission by laying down the procedures that identify each machine on the

Internet by imparting each machine a unique IP address. In fact, each IP address is made up of a unique set of numbers.

Address Resolution Protocol (ARP)

ARP is a method for finding a host's hardware address from it's Internet address. The sender uses the ARP to put forth a message seeking to delineate the IP address from the hardware address and waits to receive the latter. Each host maintains a cache of address translations to reduce delay and loading. ARP allows the Internet address to be independent of the hardware address but it only works if all hosts support it.

Network News Transport Protocol (NNTP)

A protocol used mainly to run a bulletin board service used by Newsgroups worldwide.

Creating Websites for Individual and Corporates World

Designing an effective web site requires more than just gathering relevant information and posting it on the web. Like a good paper or research presentation, a quality web project demands as much attention to the selection, organization, and presentation of material as to the underlying research itself.

You should strive, above all, to be both clear and engaging in every aspect of site design. Without the first, you will quickly lose your audience. Without the second, you'll never catch their attention in the first place. Whenever we design consumer oriented sites like Bazee.com, sifi.com, the customer can avail end products that starts from buying a notebook having cost Rs. 5 as well as buying PC having cost Rs. 15000. Here are some concrete suggestions for making your site a winner: Before you begin following are the some design strategies:

- Consider your audience and your goals. You should have a clear sense of who will be using your site (mostly college undergraduates) and what kind of experience you are hoping to provide. What exactly are you trying to accomplish here? Why is this important.
- Plan your site on paper first. You can draw a "family tree" of pages with arrows indicating links. Or you can make a hierarchical outline. Either way, it is essential to organize your information and layout the architecture of your site before attempting to implement your vision.

Cyber Laws

So much is available on the computers that the eagerness to sabotage is high. To avoid these so called hackers, who spoil the show, we have the Cyber laws. It is a very wide term that includes:

- computers
- computer networks
- the Internet
- data
- software etc.

Cyber law includes laws relating to:

1. Electronic and Digital Signatures
2. Computer Crime
3. Intellectual Property
4. Data Protection and Privacy
5. Telecommunications Laws

Let us discuss each of them in brief:

Electronic and Digital Signature Laws: Electronic Signatures especially Digitally Signatures are fast becoming the de-facto standard for authentication of electronic records, Electronic Data Interchange, Emails etc. Comprehensive laws are required so that uniform standards and procedures can be established.

These laws relating to Electronic Signatures e.g. the Electronic Signatures in Global and National Commerce Act of the USA are part of cyber law.

Computer Crime Law: Our governing dependence on computers and the Internet has made us all potential victims of Internet threats.

Some countries have enacted legislations that especially deal with computer crime and yet others have adapted their existing laws to make computer crime an offence under existing statutes.

Intellectual Property Law: Cyber law covers the intellectual property laws that relate to cyber space and its constituents.

This includes:

- Copyright law in relation to computer software, computer source code etc.
- Trademark law in relation to domain names
- Semiconductor law which relates to the protection of Semiconductor Design and Layouts
- Patent Law in relation to computer hardware and software.

Data Protection and Privacy laws: many nations have enacted legislation relating to data protection and privacy within their jurisdictions. It is pertinent to note that due to the nature of the Internet and the amount of information that may be accessed through it, such legislation is critical to protect the fundamental rights to privacy of an individual.

These laws would probably play a vital role, as the dependence of insecure networks such as the Internet grows further.

Telecommunication Laws: As has been seen above cyberspace does not only include the Internet and computes, but is present where two or more cables or wires etc. meet. Telecommunications systems also call within the ambit of cyber space and therefore would form an integral part of cyber laws. The word cyber and its relative dot.com pare probably the most commonly used terminologies of the modern era.

In the information age the rapid development of computers, telecommunications and other technologies has led to the evolution of new forms of trans-national crimes known as "cyber crimes". They may be defined as "any crime with the help of computer and telecommunication technology", with the purpose of influencing the functioning of computer or the computer systems.

The extent of loss involved worldwide of cyber crimes is tremendous as it is estimated that about 500 million people who use the Internet can be affected by the emergence of cyber crimes. Cyber crimes are a very serious threat for the times to come and pose one of the most difficult challenges before the law enforcement machinery.

Most cyber crimes do not involve violence but rather greed, pride or play on some character weakness of the victims. It is difficult to identify the culprit, as the Net can be a vicious web of deceit and can be accessed from any part of the globe. For these reasons, cyber crimes are considered as "white-collar crimes". To understand cyber crime as a significantly new phenomenon, with potentially profoundly new consequences, it is necessary to recognize it as a constituent aspect of the wider political, social and economic reconstructing currently effecting counties worldwide.

This new technology not only provides opportunities for the profitable development of an international information market but has also raised the specter of new criminal activities to exploit them. The very technology that enables multinationals to do business more effectively and challenge the individual controls and regulations of nation states, also offers the prospect of globally organized criminal networks.

Moreover the free flow of uncensored information on electronic networks and web-sites is an attractive Just as crimes have changed with the growth of information technology so have the categories of criminals who engage in such crimes. There are three basic categories of criminals who engage in such crimes, ranging from hackers, information merchants and mercenaries, to terrorists, extremists and deviants.

Web Applications

It is not that the Internet is used only for sending and receiving information. It has its commercial aspect too. Internet today offers a number of standardized services to all its users. In fact, it is like a multipurpose tool to communicate with one another, collect information and enhance your knowledge about the subjects that you like, conduct research, play games, invest in shares or just look around (usually called browsing) in search of recipes or even jokes or the latest technological developments. The description that follows will acquaint you with some of the well-known ways in which Internet is used worldwide. Some of the famous usages of Internet are:

Information Retrieval: One of the most commonly used services on the Net comprises retrieval of information about various topics that interest you. This retrieval (getting) of information is made possible through the World Wide Web (popularly called WWW or W3).

Internet and e-mail: The Internet facility used to the maximum is e-mail. It involves writing messages on a computer and transmitting them to another computer so that the addressee can read them, thus saving paper, time, energy and cost. In fact, it can be said that e-mail is the essence of all communications on the Internet. Practically, everyone with Internet access does have an e-mail account.

Newsgroups (Usenet): The Internet provides you a major avenue to communicate with large groups. Popularly, referred to as 'Newsgroups', this service is technically called Usenet and comprises a distributed bulletin board system. The information (news) shared by Usenet groups can be 'unmoderated' (unedited) or 'moderated' (edited.).

Accessing People: Internet would help you in locating your long lost friend, who was probably lost in school or college. That person must have an e-mail account. Rest is easy. For this purpose, you can visit a commonly used site such as whowhere (use your search engine to locate it by typing http://whowhere.lycos.com). Now, to find the e-Mail address, type the name of the person in the name field. Thereafter, type the e-mail server domain name (like hotmail.com) in the domain field. Once you click the Go Get It button, the result will be displayed on your computer's screen. You will also find that the on-screen display serves as a link, wherefrom, you can get the person's e-mail address.

Telnet: It is a text-based Internet service that connects you to a remote host (server). Using a special protocol known as the Network Terminal Protocol, it enables you to log onto another computer on the Internet and use its resources as if they existed on your own machine. To use this service, you need to provide your valid login and user password.

Chat: Everybody likes to chat. Even if it on the Internet. You can in fact chat with totally unknown people. Technically known as Internet Relay Chat (IRC), it is multi-user and multi-channel chatting net that allows users to communicate in real time. Remember! Chatting through IRC is in the written form, i.e., while communicating with some one, you type your message and the receiver responds (types back) with his/her comments. In short, it is form of instant talking, almost like a telephone conversation but in a written form. Residing on many systems on the Net, it is a software, which provides access to a series of interactive services. By availing these services, one can chat or play on-line games with people on the Internet.

FTP (File Transfer Protocol): It connotes an Internet service that transfers files from one computer to another. It is a common procedure to download and upload files over the Internet. With FTP you can login to another Internet site and transfer (meaning send or receive) files. FTP works on the client/server principle. A client program enables the user to interact with a server in order to access information and services on the server computer.

Anonymous FTP: Some sites have public file archives that you can access by using FTP with the account name 'anonymous' and your e-Mail address as the password. This type of access is called anonymous FTP.

e-commerce: e-commerce means doing business online. It refers to any manner of conducting business on line by an individual/organization. As part of e-commerce, large organizations also send data from the Internet to conduct research and plan their marketing strategies globally. With credit cards becoming more popular along with computerized banking services, payment for services through Internet is becoming very easy.

Employment Generation: Job placement agencies and employers have started to increasingly use the Internet as a source of recruitment by advertising on the Net to fill up vacancies. Side by side, people seeking job or better employment opportunities also sue the Internet.

Medicare: Doctors now increasingly use the Internet to know the latest treatment techniques to benefit their patients. Hospitals sometimes use video conferencing to provide on-line guidance for conducting complicated operations.

Shopping Online: Some department stores offer their products, both new and old to Internet users to payment, which is usually made in advance through a credit card. While in the case of software, developers often post the software they have written – a process similar to posting a message. You can also download and use these programs free of cost and give a feedback to the programmer about its usefulness.

Entertainment: Internet now gives you unlimited opportunities to watch latest films, TV programs and listen to music. You can download your favorite movies and music from different sources.

World Wide Web (WWW): The World Wide Web is a collection of million of files stored in thousand of computers (called Web server) all over world. Using WWW a user can download files, listen to sounds, view video files and jump to other documents on or Net sites by using hypertext links.

Educational Opportunities: Suppose you wish to study in America after passing the Senior Secondary examination. Just surf the Web and you will find that almost every university in the U.S. maintains a web site. Each of these sites provides extensive information, ranging from courses available to credit prices, course fees, etc., details of programs leading to various degrees and career planning services

Lastly I can say that with the help of computer and Internet, you can see the World on your desktop only. And that too anytime of the day.

Writing Web Projects

Any person having skills can write a Web project. A web project consists of various web pages which are uploaded to a Web site. The various steps involved in designing a Web project are:

1. Project goal
2. Project objectives
3. User targets

4. Scope of the site
5. Budget of the site
6. Planning issues

Identification of Objects and Target Users

Designing an effective web site requires more than just gathering relevant information and posting it on the web. Like a good paper or research presentation, a quality web project demands as much attention to the selection, organization, and presentation of material as to the underlying research itself.

You should strive, above all, to be both clear and engaging in every aspect of site design. Without the first, you will quickly lose your audience. Without the second, you'll never catch their attention in the first place. Whenever we design consumer oriented sites like Bazee.com, sifi.com, the customer can avail end products that starts from buying a notebook having cost Rs. 5 as well as buying PC having cost Rs. 15000. Here are some concrete suggestions for making your site a winner: Before you begin following are the some design strategies:

- Consider your audience and your goals. You should have a clear sense of who will be using your site (mostly college undergraduates) and what kind of experience you are hoping to provide. What exactly are you trying to accomplish here? Why is this important.
- Plan your site on paper first. You can draw a "family tree" of pages with arrows indicating links. Or you can make a hierarchical outline. Either way, it is essential to organize your information and layout the architecture of your site before attempting to implement your vision.

Web Team

Web team as the name suggests, is the team which develops the web site. To start with the small time programmers were doing these jobs. Now-a-days a full time team works on it.

There are basically two types of web teams that are:

a) Server-side: Server side web teams are web teams, hired by a company to develop a website.
b) Client-side: Client side web teams are part of the company that is putting together the website. It is entirely possible that you put together Web teams comprised of both these subteams.

Planning and Process Development

Terms Required for Developing Website

There are several terms which are important for developing website which are as follows:

1. Planning and Process development: Strive for consistency. You want your project to have an identity, so all the pages in your project should have a common feel: there should be consistency among backgrounds, color schemes, navigational tools, and tone of voice. This is especially important if you are dividing up your site's pages among several team members. Otherwise, your project will seem like two or three separate projects lumped together, rather than a single, unified whole. Provide a rich set of links within your site. Ideally, there should be multiple ways for your user to navigate your pages. You should consider including a prominent 'home' link or icon on each page, a menu or table of contents, and highlighted links within textual material to related information elsewhere on the site.

   Feel free, of course, also to provide links to other pages within ECE or elsewhere on the web. Don't hide important information. Users don't like to click too many times to find the information they want — if information is particularly important, make it accessible up front. On any given page, remember that as with a newspaper, the top left comer is the. most prominent.

2. Early planning: Much of page layout on the Web has to be accomplished within tables because HTML has no method to set margins or columns. Once designed, the HTML document can serve as a template for subsequent documents on your site. When creating the template for your site:
   1. Keep lines short, approximately 40-60 characters or about 11 words per line.
   2. Keep paragraphs four to eight lines.
   3. Use wider margins to increase white space for readability online.
   4. Use headlines, bullets, pull quotes, and sidebars.
   5. Use darker or brighter contrast in small areas.
   6. Give all pages a common look and feel.
3. Contents planning: Give your site a descriptive title. Your title should convey the content of your site in a concise but engaging manner. Remember, the title is how your site will be identified on the ECE home pages. Ideally, it should pique the curiosity of users and prompt them to explore your project pages.

   Include a brief introduction. This should be part of your site's home page, and should explain the scope and purpose of the site. Once users have noticed your title and followed a link to your site, they will expect quickly to find a further elaboration of your title, a brief paragraph or two describing what the site is all about and what makes it interesting. You've caught the user's eye with your title; the introduction is your chance to heighten their interest and persuade them to actually stick around and explore.

   Make your site's home page as useful a starting point as possible. The viewer should be able to see at a glance what your site is about, how it is laid out, and what kinds of resources and features it includes. Ideally, all of this information (along with your site title and introduction) should be visible on a single fast-loading screen that requires a minimum of scrolling.
4. Technical planning: If you have spent any time on the Web, you can probably remember at some point being lost in cyberspace. The ability to link virtually anywhere from a page can be confusing. A well-designed site offers good navigational tools to give the reader a clear, quick, and consistent method to move within the site. Books tend to be linear, where one starts at the beginning and proceeds to the end. Many Web site designs begin with a hierarchical diagram with the home page at the top and the main branches linked below.
5. Production planning: Thoroughly test your site. Ask a friend to sit down and explore your site. .Ask them to think out loud, and watch them navigate the site. Do they get lost? Do they have trouble finding links? Do they have trouble understanding your labels? Do they understand your prose? Ideally, you should elicit and incorporate feedback about your site in the course of developing it as well as when its nearing complet:on. Be sure to test your site both of the most commonly used browsers, Netscape and Internet Explorer.

   Proofread carefully! ECE is a public resource sponsored by the University of Michigan and all of its pages should maintain a high level of professionalism. Check carefully for spelling and grammatical *errors* before posting *your* written materials to the web.

Communication

It is not that the Personal Computers are for personal use only. They can be used for sharing of information too with others having similar type of equipments. Logically it sometimes is fisible to have a number of PCs then having a bigger computer which can give the same thoughput as the number of

PCs. But, overall it becomes a expensive proposition and sometimes is not fisible, specially for individuals and small businesses or small companies.

Another problem arises for those who are physically placed at a distance. They would need to send messages and receive messages, or use the same data at, or share program at, many different locations. One answer to this problem is to keep separate data files at each location, but this approach is very costly and potentially a dangerous solution. It is very expensive to maintain these redundant files of data and difficult to ensure the continuing accuracy of data at any one location. The solution of both, the need for more power and the ability to receive, send messages and share data, are found in data communication technology.

Communicating Data

When you send data from one location to another using any electronic means, it is called Data communication. Linking one computer to another permits the power and resources of that computer to be tapped. It also makes possible the updating and sharing of data at different location. Computer that are located in the same room office building, can communicate data through a direct cable link. For example, in a multiuser system, a single processing unit is linked by several dumb terminals through direct cables. Since multi-user systems link inexpensive terminals instead of the more expensive personal computers used with other systems, multi-user systems provide cost savings for small companies purchasing a complete system or adding users to a system currently in operation. In general, multi-user systems are most appropriate for companies requiring a shared data base for extensive data input and output. Computer located far away from the processor use a special form of data communication – telecommunication or telegraph. For example, teller and point-of-sale terminals are spotted in location away from computers: airlines, car-rentals, railway reservation systems and hotel reservation systems have thousands of terminals located many miles from the processors. In these cases, the process of using communication facilities such as the telephone system and microwave relays to send data between computers is a form of data communication, sometimes referred to as telecommunication, or tele-processing.

Dr. George Stibitz was the first to link computer and computing devices and send data way back in 1940. He used telegraph lines to send data files from Dartmouth College in New Hampshire, USA to a Bell Laboratories calculator in New York City, USA. An early large-scale business application was the Sabre passenger reservation system, developed in late 1950s and early 1960s by American Airlines. Hundreds of terminals were linked to the central processing unit. Communication usage has grown steadily since then.

Now-a-days minis and larger machines are in constant communication with outlying terminals. With available attachments, computers can use telecommunications channels to link up with information retrieval services, banks and electronic bulletin boards. An electronic-mail message can be sent from one place to another and an answer can be returned by electronic mail as soon as it is received. This service eliminates the hassle of continually trying to return a call, or long mail delays.

All communication between computers are managed by data communication software. The precise procedures used depend on the particular software, but the following account outlines the general procedures.

Communication Between Two Microcomputers

A small business that uses microcomputers has accumulated information that it wants to send to a customer who also uses microcomputers.

The procedure used to send the information is as follows:

1. Both the business and the customer start the communication program, which gives them a menu of options.

2. Both the business and the customer choose the same options, including the number of start and stop bits, the type of parity checking, the communication rate, full-duplex or half-duplex transmission and so on.
3. The communication link is established. The users decide who will originate the telephone call and the originator enters the telephone number. The communication program then instructs the modem to detail the number and reports to the users if a connection has been made.
4. After the connection has been made, the users can have an on-line conversation. They can type messages back and forth to each other and can record their conversation.
5. When the business is ready to transfer the information, it sets its computer to send; the customer sets its computer to receive. The business starts, the transfer. The communication program takes over by reading the information from the disk and sending it through the communication link. The receiving computer stores the incoming information on a disk. The transfer proceeds without any further human attention.
6. When the transfer has been completed, the business and the customer say "goodbye" and terminate the connection.

Multi-Department and Large Scale Sites

The main issues which need to be considered while developing a multi-departmental and large scale sites.

1. Getting to know your audience: If you do not know your audience, you will not be able to build a successful web site, because you would not understand what the web site must contain in order to meet your client's goal.
2. Balancing the needs of different units: To do so, first we need to have an overall corporate objectives, or mission statement, for the entire site. Having this statement will help you formulate your interview questions and help you prioritize the need of different units.
3. Corporate politics: Being aware of any corporate politics in your internal organization and your client's organization is very important. Being aware and alert, yet objective and professional, is the way to handle corporate politics, especially when they are put directly in front of you.
4. Site wide branching: To make a large site branded approximately is a tough issue at a corporate level. Each department or business unit will naturally want to have its own personality. However, if each department, or unit receives different branding, the site wil be in danger of becoming non-usuable. The information architect has to ensure that visual design allows to have both features included.
5. Quality Issues: In a large scale website, where there are multiple content authors, servers and production systems, you face big challenges in ensuring that all subsections of the site are up to an acceptable level of quality. Configuration management also needs to be taken care of. One even needs to make the site.

Quality Assurance

In the World of web development, both cost and time induce the quality of a product. Quality Assurance means to achieve a standard of quality that matches the objective of your project and that is on time and fits the budget. The client or the strategy team wants you to deliver a product that meets the obectives of the scope document. The scope document is transformed into a functional and technical specifications. We can measure the success of website and hence its quality with the help of metrics and requirments. Common websites metrics include browser, operating system configuration file and application size, connection speed, assets like audio, media, moving or static images. Before publishing a site, the Internet or intranet, it needs to be tested. Tesing are of two types:

1. Alpha Testing: This the testing process that takes place during the development of the site itself. The developer will constantly test the page on the browser and change the design as and when required.
2. Beta Testing: This testing is done once the stie is developed by a different group of employees or it can also be tested by the people for whom this site is being developed

Technical Advances for Web Team

There are different technological advances considered and their impact on the working web teams. The web has many audiences, all of whom come looking for information, goods and services on demand. It is an active medium; users for most part, seek information at their own level of desire. This kind of information on demand has been the most prevelent kind of web application.

However, as people begin to use the web for more and more consumer purposes, such as comparision shopping or advertising, the web will become more like Television or Radio, where information is pushed towards the consumer through channels, stations and yet to be developed means.

1. **Web as Interactive Television:** One of the biggest is the broadband internet access high will make. Web to provide rich media like audio, video ad virtual reality. Web can come from more accessibility television sets.
2. **The web as information store:** The Web is becoming central repository of information for national and international organizations like postal service, libraries and universities. Hence such software are being developed to search information accurately in both textual and binary formats.
3. **Web and E-commerce:** The web is being used vastly now for carrying out commercial activity like buying, and selling of products, services, information, etc. For this security concepts, database and network features are being incorporated.

Search Engines

They are programs that locate needed information in a database, but especially an Internet accessible search service that enables you to search for information on the Internet. To use a search engine, you type one or more key words; the result is a list of documents or files that contain one or more of these words in their titles, descriptions, or text. The databases of most Internet search engines contain World Wide Web (WWW) documents; some also contain items found in Gopher menus and File Transfer Protocol (FTP) file archives.

Compiling the database requires an automated search routine called a spider, forms filled out by Web authors, or a search of other databases of Internet documents. The effectiveness of search engine can be measured in terms of Indexing and Term specificity. Since the Internet is a vast collection of information, it is difficult to find specific information you actually need. Therefore, search engines are more than a boom.

E-Commerce

It is nothing but adding electronic to the commerce, i.e., doing business on the net. It can be buying and selling of stocks and bonds, selling a product, buying goods and services for your company's operations or purchasing a fancy dress, or any other such commercial transaction. E-commerce covers the vast of amount of business-to-consumer and business-to-business transactions taking place on the Web (with the help of internet). It is not Just Web Shopping, it is more than the definition.

Despite popular belief, there is a lot more to e-commerce than buying goods online. Every day, millions of transactions are conducted via the Web, all of which fall under the E-commerce. E-Commerce is to conduct business by using the IT (Information Technology, i.e., Computer technologies and the electronic communications). In other words it is the buying and selling of items of goods and services on the Web. E-Commerce is the process of doing business electronic. It involves automation of a variety of B2B (Business to Business) and B2C (Business to Consumer) transaction through reliable and secure connections.

The use of electronic transmission medium (tele-communication, internet) to engage in the exchange of business document including buying and selling products and services with physically or digitally from location to another using computer as a component and internet as a media. E-Commerce is a composite of technologies process and business strategies that foster the instant exchange of information within between organization. E-Commerce strengthens relationship with buyers make it easier to attract new customer improves customer responsiveness and open new markets on a global scale.

E-Commerce is the application of various communication technologies to provide the automated exchange of business information with internal and external customer, suppliers and financial institutions. E-Commerce is about setting your business on the Internet, allowing visitors to access your website and go through virtual catalog or your product/services online. When the visitor wants to buy something he/she likes, they merely, add it to their virtual shopping basket, items in the virtual shopping basket can be added or deleted and when you are all set to checkout . You head to the virtual checkout counter, which has your complete total and will ask you for your name, address etc. and method/way of payment (usually credit card). Once all the information is entered the buying and selling procedure is complete and you have to just wait for delivery.

Traditional Commerce

The major difference between E-Commerce and Traditional Commerce is that with E-Commerce, these exchanges of goods and services are carried out over the web instead of the traditional physical act of going to a trader for goods and services. It wasn't until the accessibility of the Internet reached a global level that E-Commerce took off. Now that such a large number of people have access to the Internet it is no surprise that individuals and organizations have seen the opportunity to make some money out of it. Judging by the current fingers that represent the E-commerce industry, it looks as if it is here to stay. With 60% of home Internet users buying over the Internet, market researches predict that by the end of this year, Internet sales to consumers will be worth Rs. 60 crores.

Successful E-Commerce strategies allow organizations distinct advantages in terms of both costs and revenues – the two fundamentals of all business. This is because costs can be cut immediately as retail outlets are not required. However, as attractive as E-Commerce may look to the prospective E-Entrepreneurs, the fact that a badly conceived or badly run business will not succeed still needs to be observed. With the most popular purchases over the Internet being computer products, books and music, the prospective E-business needs to also access how E-Commerce works in their chosen sector of industry. Much of the cost associated with traditional high capital business is eliminated and or transformed into profit in the Internet environment. Throw away all previous paradigms of business; that mentality won't kill your business immediately but without beginning the Internet transition soon it will be impossible to grow with the rest of the business world.

Difference between E-Commerce & Traditional System

Reduction of Data Error

E-commerce — It does not involve data at multi points. With E-commerce data goes directly from one to another computer without involving human being.

Traditional commerce	The buyer and seller create purchase order on their system print it or fare it or mail to receiver or seller or trading partner. The receiver/seller then re-enter the same information on the computer. This will create the error
Reduction in Cost	
E-commerce	Initial cost of E-Commerce is very high as compared to paper process but over a long period of time it is very effective.
Traditional commerce	As time is money, time is directly linked to saving the money. In it there is a repetition of same work at every level. So it involves a lot of waste of time and if the error is arisen that will lead to more wastage money
Reducing paper Work	
E-commerce	E-Commerce data in the electronic form makes it easy to share it across the organization.
Traditional commerce	It requires re-entry of data at each level and requires lot of time. So the peak time is wasted in re-entering and printing of the reports
Reducing Processing Cycle Time	
E-commerce	E-Commerce reduces the processing cycle time of complete cycles as the data is entered the system it is simultaneously processed.
Traditional commerce	When the buyer order in a paper format, the data is re-entered into the seller's computer and then only processing can take place which is a time consuming and full fervor

Designing Strategies

Increasing the accessibility of your Web site can improve information clarity for all visitors to your pages. Web accessibility tools can be used to check to make sure images include *alt tags* and there are no incorrect or missing HTML tags. The Web Accessibility Initiative (W AI) provides information on creating Web sites which provide equal access to information for students with disabilities.

As you develop your pages, make sure they can meet the guidelines below. Otherwise, you will be denying access to students with disabilities who are viewing the pages using standard assistive computer technologies.

Graphics : Make sure graphics support the message. Using a graphic repeatedly throughout the site reduces download time because once an image is downloaded it remains in the computer's cache. During the browser session, the image is called from cache instead of going out to the Web site. Avoid overusing graphics and animations because they draw away from the message and increase download time.

Highly graphic pages, however, can be extremely slow loading when using a modem or the graphics can distract the reader. Web pages need a mixture of visual sensation, text information, and interactive hypermedia links. Types of Media on the Web contains further information on graphic types and size. Dense text documents without the contrast and visual relief offered by graphics and careful page layout are difficult to read. The Yale Style Guide recommends chunking information so information can be scanned and located quickly because:

- Few Web users spend time reading long passages of text on-screen
- Concise segments of information fit better on the computer screen.
- Long pages tend to be printed instead of read online.

What to Avoid : Large graphics with slow download time that delay page display. Bandwidth is a I primary restriction of the Web and large graphic, audio, and video files that are slow loading distract the reader. If the load time is not worth the wait, the reader becomes frustrated with the site. Buttons or links that is hard to use or identify. Logic behind the way screens are linked and the navigation elements need to be clear.

Long pages without white space or relevant graphics. Unless the page is meant to be printed instead of read on the screen.

Backgrounds that result in unreadable text or are distracting. Avoid dark colors combined with dark text fonts or distracting images that overshadow the text. To begin, stay with a white background.

Broken links are a sign of an outdated site or one with a low quality control

Overuse of fonts, colors, and animations.

Writing for the Web: Readers seek clarity, order, and accuracy in information sources, whether they are traditional paper documents or Web pages. The spatial organization of graphics and text on a Web page can engage the student, direct the student's attention, prioritize information, and make the student's interactions with your online course material more. enjoyable and more effective.

Write in an inverted-pyramid style with the conclusion first. Use lists when possible.

Skill set needed for Web Team

No matter what kind of website you are producing, there are some core responsibilities and roles. for other kind of sites, generally larger and more complex sites, there are extended team member, these are people whose skills might not always be necessary or who might have cross-functional roles. Special team people are brought into to do work that is not part of your core or extended team but may become so. Example of core team could be project manager/producer, technical lead, web production specialist, creative lead, designer, etc., where as account manager, programmer, network engineer, information architect, etc., are extended members. On the other hand, security expert, audio engineer, video engineer, 3-D modeler, etc., are special member. When a team is put together, we must make sure the following skills are covered.

- Project management skill
- Information design/architecture design skill
- Graphic design skill
- Graphic production skill
- Content development skill
- Programming skill
- Technical/Network infrastucture skill

Database to Web Applications

Database management deals with the storage, retrieval and editing of data within the database. Important functions of Database management are the security and validations of data items in Database. Visual Basic makes it simple to handle database by supplying a variety of tools and methods for data management. Visual Basic version 6.0 has a full-fledged database engine called **Jet**. This engine contains virtually all the database functions that many applications would need.

Visual Basic also provides a powerful and easy-to-use front-end development environments for database management. Using this, you can manipulate database in various formats. It also includes an in-built support for **Open Database Connectivity** (ODBC) – an industry standard for data access from multiple database formats.

A multi-user database system, if implemented on PC, generally falls into one of the two categories:

- **File-server system**
- **Client/Server system**

In a file-server system, no intelligent process is active at the server level. All intelligence resides on the individual workstation. When your application needs data, it is the workstations software responsibility to determine which file should be read and how to access the network drive. Because all data in the database file must be sent from the server to the workstation, network traffic is increased.

A multi-user database system within a file-server environment is made up of the following:

- A database that resides on a network file server running an operating system such as **Microsoft Windows 2000 Server** or **Novell NetWare**.
- One or more users accessing the database using a workstation's application program.

In a Client/Server system, the server is responsible for intelligently processing requests for data. The workstation does not request data at the file level. It sends a request to the server to execute a specific query and gets the results. The primary advantage of this technique is that network traffic is reduced because server sends the result to the workstation.

A Client/Server system typically has a back-end database residing on the server controlled and maintained by the server software, such as **Microsoft SQL Server**.

Revision

1. A global hypertext system that uses the Internet as its transport mechanism is called World Wide Web.
2. Web allows you to skip from one page to another using hyperlink.
3. Web is independent of the platform.
4. With Web you can get information from the user too.
5. Web browsers allow you to see the web pages.
6. Most popular Web browser is Microsoft Internet Explorer.
7. Web server manages the communications protocols and the related software required to create a Web site on the Internet.
8. HTTP is used as the communication protocol to transport hypertext documents over the Internet.
9. A URL describes the location and a method of accessing a resource on the Internet.
10. Search engines are used to search for different information on the net.
11. Yahoo! is the popular search engine.
12. Cyber laws applicable in India.
13. Internet Protocol is responsible for the addressing and sending data from one computer to another computer.
14. Internet Service Provider is one gateway to the Internet.
15. At the heart of the Web is the ability to display multimedia information, such as images, audio, video, animation, and other multimedia data types.
16. Transmission Control Protocol uses a set of rules to exchange messages with other Internet points at the information packet level.
17. Web Browser is a software application that resides on your PC and can display text, images, and multimedia data found on different Web pages.
18. A Web server refers to a location (computer) on the Internet that contains information in the form of Web pages.
19. A Web site comprises of a collection of Web pages.
20. A Web page refers to a document on the Web.
21. Internet can be used for various purposes.
22. E-commerce refers to electronic commerce.

23. The major difference between E-Commerce and Traditional Commerce is that with E-Commerce, these exchanges of goods and services are carried out over the web instead of the traditional physical act of going to a trader for goods and services.

Multiple Choice Questions

1. The browser is which is available free:
   (a) Internet Explorer
   (b) Netscape Navigator
   (c) Lynx
2. Which of the following is not a search engine:
   (a) Yahoo!
   (b) Internet Explorer
   (c) Lycos
3. ________ is a set of rules, that is used to communicate applications to each other.
   (a) HTML
   (b) Website
   (c) Protocol
4. Types of web team are :
   (a) Two
   (b) Three
   (c) Four
5. ________ is the primary protocol used to distribute information on the web.
   (a) TCP/IP
   (b) FTP
   (c) HTTP

Answers

1. (c) 2. (b) 3. (c) 4. (a) 5. (c)

Subjective Questions

1. What is World Wide Web?
2. What are the basic features of World Wide Web?
3. Describe the various popular web browsers.
4. What is the use of a Web server?
5. Describe the various protocols used.
6. What is a Search Engine?
7. Which are the cyber laws applicable in India?
8. What does the Web do?
9. Describe the various Multi-Department and Large Scale Sites.
10. How would you assure the Quality Assurance of the site?
11. Describe the various technical advances taking place in terms of web team.
12. What is E-commerce?
13. How does e-commerce differ with traditional commerce?
14. What skills are needed by the Web team?
15. How is database linked with web sites?

2 Introduction to HTML

Introduction

HTML stands for Hypertext Markup Language. This is text oriented language used for writing programs to create web sites. There is no need for compiling or doing anything else as you would do in any other programming language. However, there are few things which you have to take care of as you do in your text related documents.

You can have Headings, Paragraphs, Bulleted text, etc. You can even have tables too. However, there is no fix page layout as such. HTML does not support it. You can have the text in Bold, Italics, etc. But, each one of the above set of text, has to be tagged accordingly, e.g., Heading, for the text to be taken as headings. In fact Markup means this only. But, remember the file has to be text and not a word processing document. A simple ASCII text file.

As said earlier, HTML is used to create web sites. Since we now know that the text files are created using HTML but how are these so-called text files going to be used. These text files are used with programs called HTML browsers, take these HTML files and create web sites/

HTML browsers

As said above a browser is used to create web document. It is important that your computer has more than one browser in it. It is important since it lets you view the document through more than one browser. It gives your document more versatility and makes you sure that your document would run all the browsers.

The two most popular browsers are Netscape Communicator (used to be called Navigator) and Microsoft Internet Explorer. Both are quite popular and can be both used in your computer.

Netscape Communicator

More web designer use this software than any other. It is also called Netscape. Developed by some of the same engineers who created popular software called Mosaic. It offers more nonstandard features that make its enhanced pages much more attractive to the eye.

Microsoft Internet Explorer

Anything with Microsoft tag is bound to create hype. It is fast becoming popular. It is no doubt a viable competitor Netscape.

See here how the HTML file called Index looks like as a text document in WordPad and how it and how it looks in Netscape Communicator and Microsoft Internet Explorer.

Different versions of HTML

While the original HTML standard was developed by CERN, new versions are hashed out through a series of online meetings open to anyone on the Internet. Then they are analyzed, discussed, decided, and published by the W3 Consortium, led by the Laboratory for Computer Science at the Massachusetts Institute of Technology (MIT) and INRIA, a French technology group, in collaboration with CERN.

The W3C has been particularly successful with HTML 4, the current version, and the one that this book describes, which incorporates many of the previously proprietary extensions into a shared standard. In this book, you will learn both standard HTML as well as the extensions developed by Netscape and Microsoft that have made these browsers so popular.

DHTML

It is actually a combination of technologies to create dynamic content on Web pages. It draws on HTML 4, Cascading Style Sheets, and usually JavaScript. These elements are supposed to work together following the rules of the Document Object Model (DOM), to change the content of a page even after it has been loaded into the browser.

XML

HTML forms part of a large more powerful mark-up language called SGML. In an effort to harness the power of SGML without losing the simplicity of HTML, the W3C has developed Extensible Markup Language, or XML, that can be used to create your own mark-up languages, specifically suited to your topic.

Now we take a look at the various terms used while writing HTML.

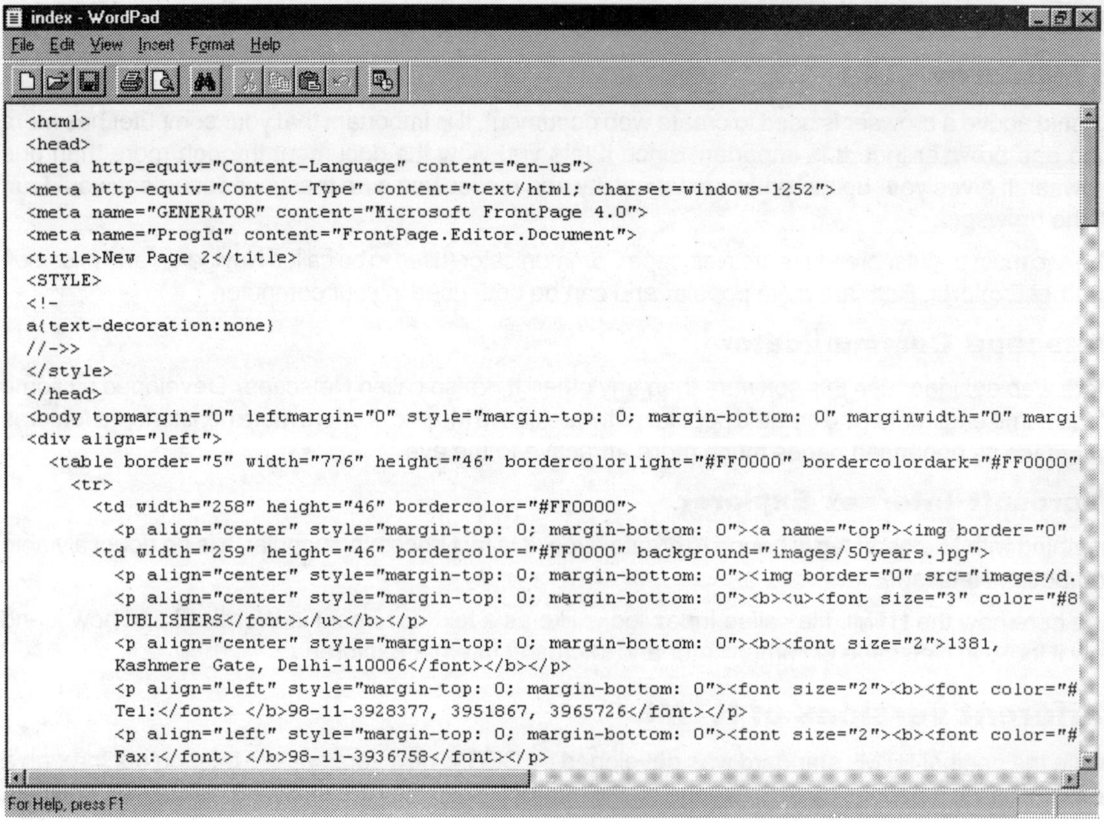

index - WordPad

File Edit View Insert Format Help

```
<html>
<head>
<meta http-equiv="Content-Language" content="en-us">
<meta http-equiv="Content-Type" content="text/html; charset=windows-1252">
<meta name="GENERATOR" content="Microsoft FrontPage 4.0">
<meta name="ProgId" content="FrontPage.Editor.Document">
<title>New Page 2</title>
<STYLE>
<!-
a{text-decoration:none}
//->>
</style>
</head>
<body topmargin="0" leftmargin="0" style="margin-top: 0; margin-bottom: 0" marginwidth="0" margi
<div align="left">
  <table border="5" width="776" height="46" bordercolorlight="#FF0000" bordercolordark="#FF0000"
    <tr>
      <td width="258" height="46" bordercolor="#FF0000">
        <p align="center" style="margin-top: 0; margin-bottom: 0"><a name="top"><img border="0"
      <td width="259" height="46" bordercolor="#FF0000" background="images/50years.jpg">
        <p align="center" style="margin-top: 0; margin-bottom: 0"><img border="0" src="images/d.
        <p align="center" style="margin-top: 0; margin-bottom: 0"><b><u><font size="3" color="#8
        PUBLISHERS</font></u></b></p>
        <p align="center" style="margin-top: 0; margin-bottom: 0"><b><font size="2">1381,
        Kashmere Gate, Delhi-110006</font></b></p>
        <p align="left" style="margin-top: 0; margin-bottom: 0"><font size="2"><b><font color="#
        Tel:</font> </b>98-11-3928377, 3951867, 3965726</font></p>
        <p align="left" style="margin-top: 0; margin-bottom: 0"><font size="2"><b><font color="#
        Fax:</font> </b>98-11-3936758</font></p>
```

For Help, press F1

New Page 2 - Netscape
File Edit View Go Communicator Help
Back Forward Reload Home Search Guide Print Security Stop
Bookmarks Location: file:///C|/My Documents/My Webs/devra/DEVRA/index.html
Search Instant Message Members WebMail Connections BizJournal SmartUpdate Mktplace RealPlayer
Devra Books
EDUCATIONAL PUBLISHERS
1381, Kashmere Gate, Delhi-110006
Tel: 98-11-3928377, 3951867, 3965726
Fax: 98-11-3936758
Mail: devra@vsnl.com
1950-2000 50 YEARS OF EDUCATIONAL EXELLENCE
Buy Books Place your order for our publications. Click Here
Guest Book Comments, Suggestions & Queries
Our Publications Catalogue of our various publications
New Arrivals
Reviews
Applet fphover running
Address C:\My Documents\My Webs\devra\DEVRA\index.html
Go
Devra Books
EDUCATIONAL PUBLISHERS
1381, Kashmere Gate, Delhi-110006
Tel: 98-11-3928377, 3951867, 3965726
Fax: 98-11-3936758
Mail: devra@vsnl.com
1950-2000 50 YEARS OF EDUCATIONAL EXELLENCE
Buy Books Place your order for our publications. Click Here
Guest Book Comments, Suggestions & Queries.
Our Publications Catalogue of our various publications
New Arrivals
Reviews
Done
My Computer

HTML tags

These are commands written between less than (<) and greater than (>) signs, also known as angle brackets, that indicate how the browser should display the text. There are opening and closing versions for many (but not all) tags, and the affected text is contained within the two tags. Both the opening and closing tags use the same command word but the closing tag carries an initial extra forward slash symbol /.

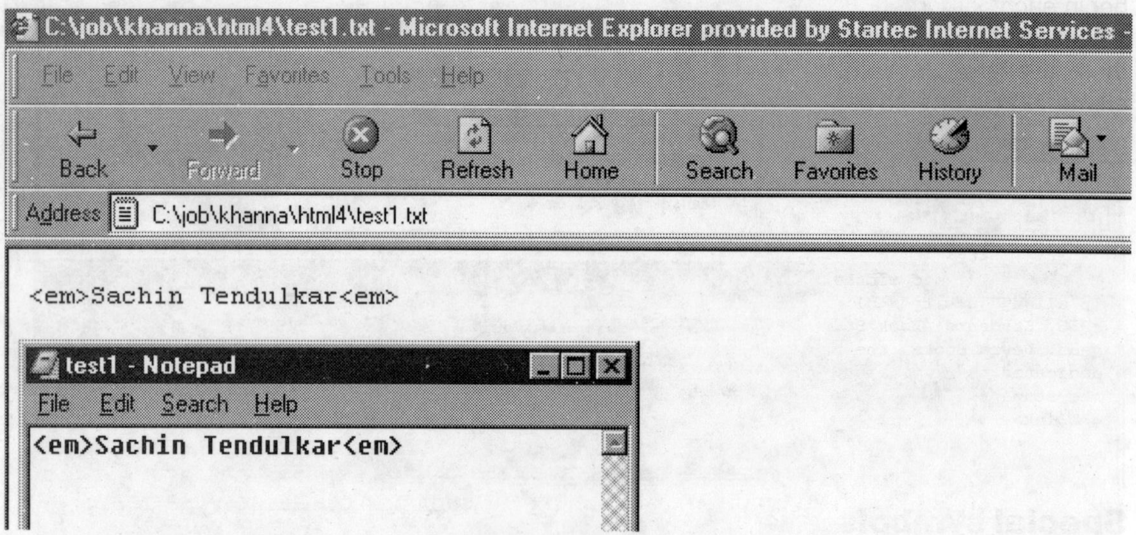

Attributes

Some tags have special attributes that offer a variety of options for the contained text. The attribute is entered between two symbols, less than and greater than. Often, you use a series of attributes in a single tag. Simply write one after the other—in any order—with a space between each one.
<TABLE BORDER>

Nesting tags

In some cases, you may want to modify your page contents with more than one tag. For example, you may want to add italic formatting to a word inside a header. There are two things to keep in mind here.

First, not all tags can contain all other kinds of tags. As a general rule, those tags that affect entire paragraphs can contain tags that affect individual words or letters (inline), but not vice versa. Second, order is everything. Whenever you use a closing tag it should correspond to the last unclosed opening tag. In other words, first A then B, then /B, and then /A.

<H1>Sachin<EM>Tendulkar</EM></H1>

Quotation marks

Mostly all values should be enclosed in straight quotation marks " " (NOT curly ones “ ”, note the difference). However, you can omit the quote marks if the value only contains letters (A-Z, a-z), digits (0-9), a hyphen (-), or period (.).

Spacing

The best part of spacing is that HTML browsers will ignore any extra spaces that exist between the tags in your HTML document. So you can make use of this and add spaces and returns to help view the elements in your HTML document more clearly while you are writing and designing your page. Also you would not be able to count returns or spaces to format your document. A return at the end of a paragraph in your HTML document will not appear in the browsed page, instead, use a P tag to begin each paragraph. You cannot repeat several P tags to add spaces between paragraphs. The extra tags are simply ignored. How do you control spacing then? Style sheets (you will read about them little later), let you specify with absolute precision just how much space should go between elements in your Web page. In addition, Netscape has developed some tags for the same purpose.

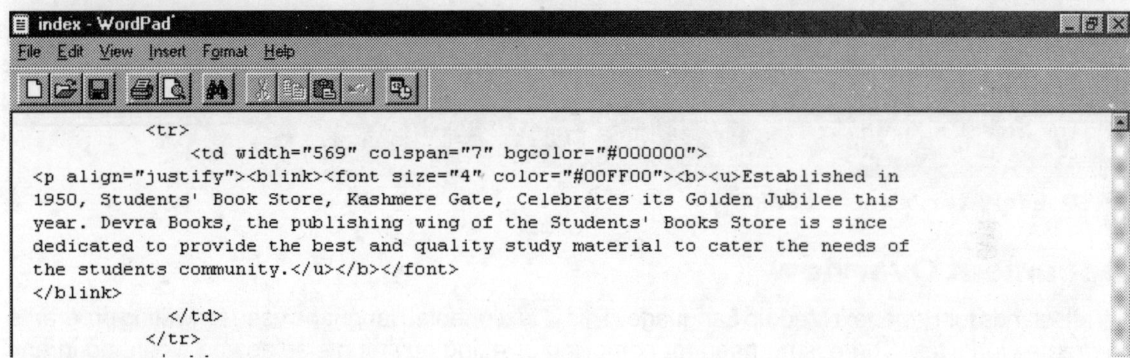


```
        <tr>
            <td width="569" colspan="7" bgcolor="#000000">
<p align="justify"><blink><font size="4" color="#00FF00"><b><u>Established in
1950, Students' Book Store, Kashmere Gate, Celebrates its Golden Jubilee this
year. Devra Books, the publishing wing of the Students' Books Store is since
dedicated to provide the best and quality study material to cater the needs of
the students community.</u></b></font>
</blink>
            </td>
        </tr>
```

Special symbols

Besides the usual ASCII set of characters, HTML supports other characters too. The standard ASCII set characters can be used as it is and will be displayed as it is in the browser.

There are four symbols that have special meanings in HTML documents. These are the greater than (>), less than (<), straight double quotation marks ("), and the ampersand (&). If you simply type them in your HTML document, the browser may attempt to interpret them. To show the symbols themselves, use a name or number code.

Look at the following example, where <BR> has resulted in giving a line break.

Tags with Automatic Line Breaks

Block-level tags generally include automatic, logical line breaks. For example, you don't need to use a new paragraph tag after a header, since a header automatically includes a line break. In fact, you only need to insert a new paragraph tag if you're using the tag to apply styles. Some common block-level tags are P, H1, BR, UL, and TABLE. Inline tags that affect only begin on a new line. Some common inline tags are EM, B, and IMG.

Values

Attributes can have values. In some cases, you must pick a value from small group if choices. For example, the CLEAR attribute for the BR tag can take values of left, right, or all. Other attributes are more strict about the type of values they accept. For example, the HSPACE attribute of the IMG tag will accept only integers as its value, and the SRC attribute of the IMG tag will only accept URLs for its value.

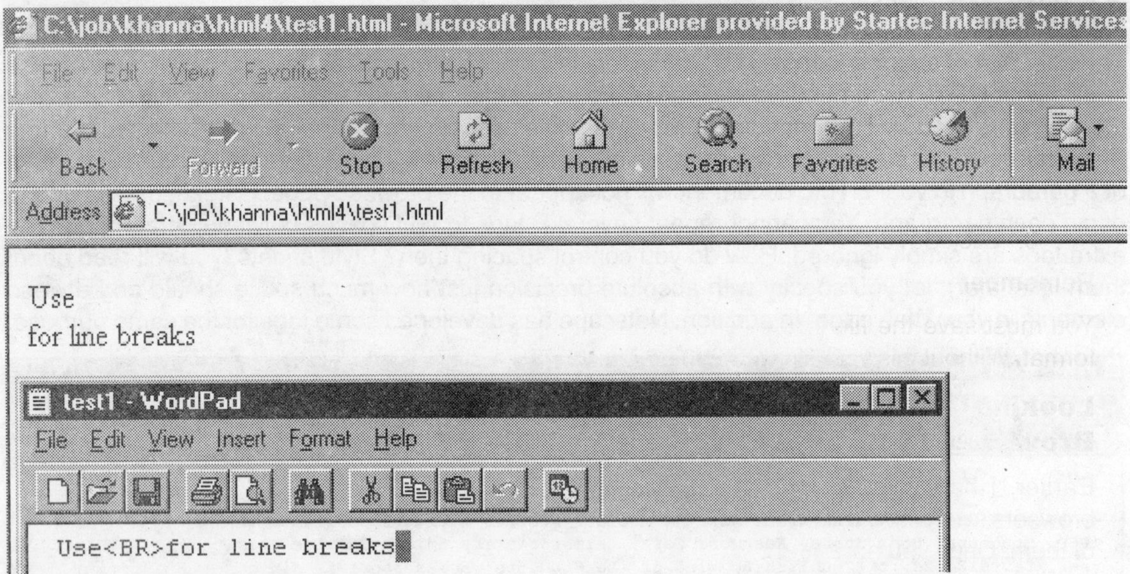

Document Overview

HTML stands for Hypertext Markup Language. This is text oriented language used for writing programs to create web sites. There is no need for compiling or doing anything else as you would do in any other programming language. However, there are few things which you have to take care of as you do in your text related documents. You can have Headings, Paragraphs, Bulleted text, etc. You can even have tables too. However, there is no fix page layout as such. HTML does not support it. You can have the text in Bold, Italics, etc. But, each one of the above set of text, has to be tagged accordingly, e.g., Heading, for the text to be taken as headings. In fact Markup means this only. But, remember the file has to be text and not a word processing document. A simple ASCII text file.

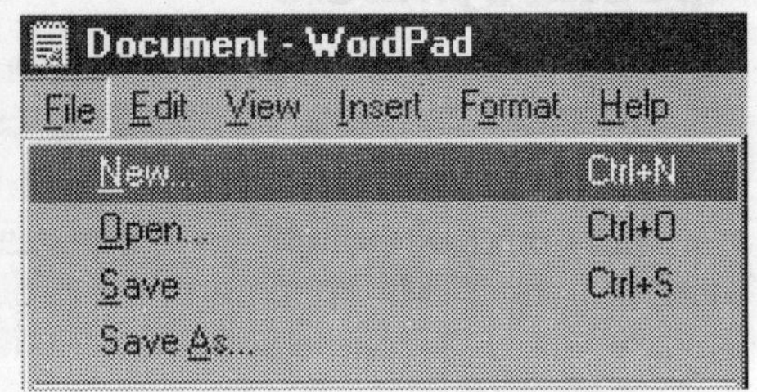

As said earlier, HTML is used to create web sites. Since we now know that the text files are created using HTML but how are these so-called text files going to be used. These text files are used with programs called HTML browsers, take these HTML files and create web sites.

Follow the following steps to create a new web page:

1. Open a text editor or word processor.
2. Choose file> New to create a new, blank document.
3. Type the following text to create the HTML content as shown next:

   Use
for line breaks
4. Choose File>Save As.
5. In the dialog box that appears, choose Text Document for the type.
6. Give the document the .htm or .html extension.
7. Choose the folder in which to save the Web page.

8. Click Save.

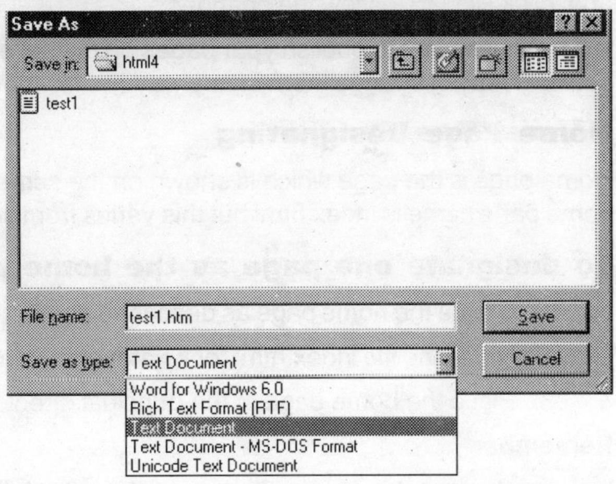

Remember

You must save the file in the Text Document format. Without this your file will have problem.

Looking at Your Page in a Browser

Earlier, I had talked about using both the browsers to see how the page looks like in both of them. Once you have created a page, you'll want to see how it looks like in a browser. In fact, since you do not know which browser the actual user will be using, it's a good idea to look at the page in several browsers.

Follow the following steps to look at your page in a browser:

1. Open your browser software (Netscape Communicator, Microsoft Internet Explorer, etc.)
2. Choose File > Open, Open File, or Open Page (just not Open Location), depending on the browser.
3. In the dialog box that appears find the Web page that you want as it will appear when you actually publish it on the server.

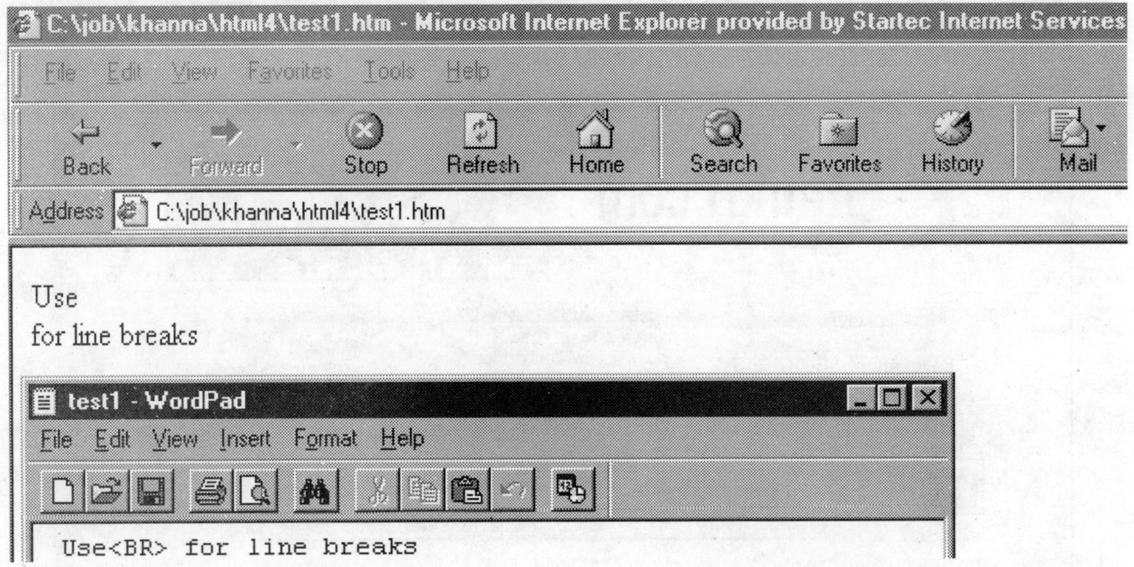

Remember

If your Web pages does not appear in the Open dialog box, make sure that you have saved it as Text Only and given it the .htm or html extension.

You don't have to close the document in the text editor before you view it with a browser. This makes editing much faster. You can switch to the editor, make your changes, and then back to the browser and click Reload to view the changes.

It is not necessary to publish your pages on the sever before you view them. Publishing them on the server does not change them in any way (it only moves them to a new location).

Home Page Designating

Home page is the page which is shown on the screen, when the site is opened. The most common home page name is index.html but this varies from server to server.

To designate one page as the home page:

1. Create the home page as described above.
2. Name the file index.html (or whatever the default home page name is for your server).
3. Place the home page in the principal directory that contains your Web site.

Remember

Your home page should have links to all the other information available at your Web site. Your home page should be a table of contents to your site, not just the first page.

If you want your users to be able to reach the other pages in your site, make your home page clean, neat, and fast.

See the home page of Naukri.com which is self explanatory.

Starting your Web page

When you want to see your page in the browser, the browser needs information right away about the kind of document it is, and how it should be displayed. You should begin every Web page with the !DOCTYPE and HTML tags.

Follow the following steps to start your Web page:

1. Type <!DOCTYPE HTML PUBLIC "-// W3C//DTD HTML at the top of your Web page, as shown on the next page.
2. Type the version of HTML you're using, in this case 4.0. If you are using lots of fonts and colours, use the word Transitional with 4.0. If the pages have lots of frames, use HTML 4.0 Frameset.
3. Type //EN"> <HTML>.
4. Now your text file is ready. Save it under a name.
5. Start your browser and open this particular file.
6. See that there is a blank web page since you have not given any substance to your page. This is what we will do in the next step.

Remember

While you can get away without either the !DOCTYPE or HTML tags, you should always include both to ensure that your page is correctly recognized, regardless of the browser used.

Create a template with the !DOCTYPE and HTML tags already typed in as a starting point for all your pages.

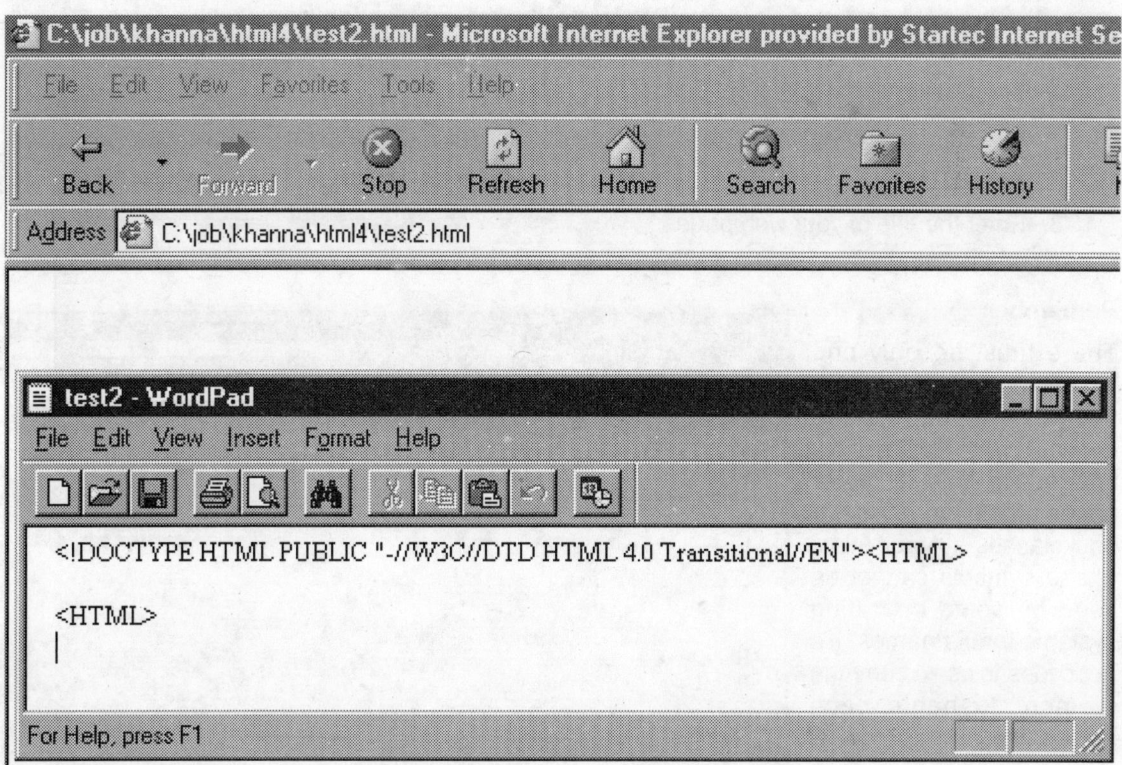

Header Elements

Nothing is complete without a head and a body. Look at yourself and you know what I am talking about. Similarly in HTML too we have Head and Body as the two main sections. The Head section provides information about the URL of your Web page as well as its relationship with the other pages at your site. The only element in the HEAD section that is visible to the user is the title of the Web page. The BODY of your HTML document contains the bulk of your Web page, including all the text, graphics, and formatting.

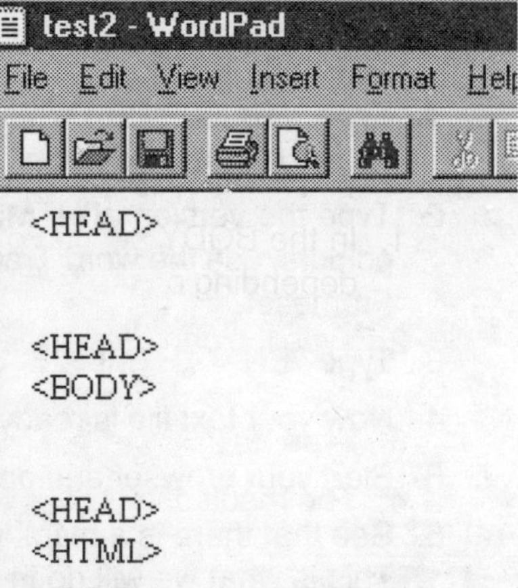

Follow the following steps to create the Head section:

1. Directly after the initial !DOCTYPE and HTML type <HEAD>.
2. Create the HEAD section, including the TITLE.
3. Type </HEAD>.

Follow the following steps to create the BODY:

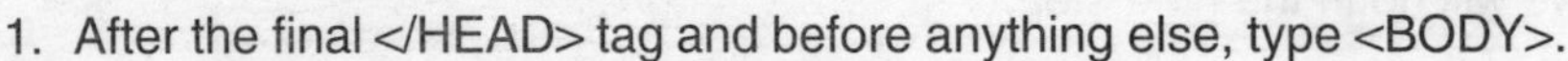

1. After the final </HEAD> tag and before anything else, type <BODY>.
2. Create the contents of your Web page.
3. Type </BODY>.

Creating a Title

Each page must have a title. A title should be short and descriptive. In some browsers, the title appears in the title bar of the window; in others, the title is centered at the top of the screen. The title is used in search indexes as well as in browsers' history lists and bookmarks.

Follow the following steps to create a title:

1. Place the cursor between the opening and closing HEAD tags, as shown on the next page.
2. Type <TITLE>.
3. Enter the title of your web page.
4. Type </TITLE>.

Remember

There must be only one TITLE tag in each HTML document. A title cannot contain any formatting, images, or links to other pages. Don't use colons or backslashes in your titles. These symbols cannot be used by some operating systems for file names. It's good idea to use a common element to begin each page's title.

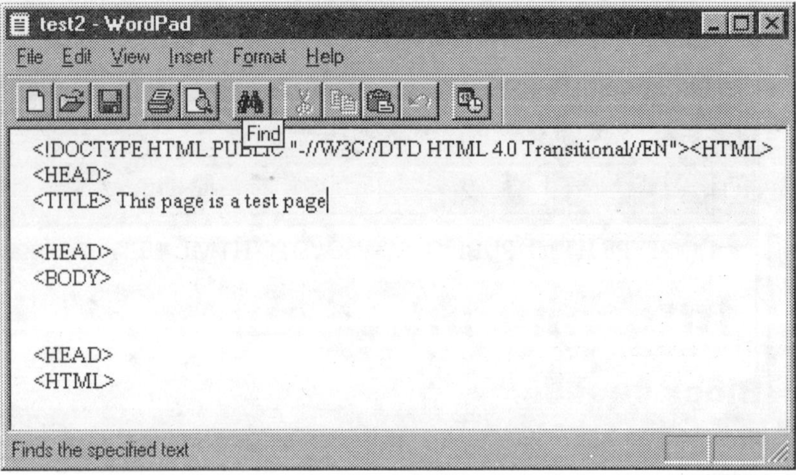

For example, you could begin each page with My Page: followed by the specific area described on that page.

Section headings

Various type of headings can be placed in the text. Maximum of 6 headings are allowed. But, you can use only 3 at a time. Each one of the heading is numbered from 1 onward. Obviously you cannot repeat the numbers.

Let us try to create a heading.

1. In the BODY section of your HTML document, type (Hn where n is a number from 1 to 6, depending on the level of header that you want to create). Here I have used it as H1.
2. Type the contents of the header, in this case I have typed it as "Information Technology Class X.
3. Type </Hn> where n is the same number used in step 1, in this case H1.
4. The result of this typing and result in browser are both shown on the next page.

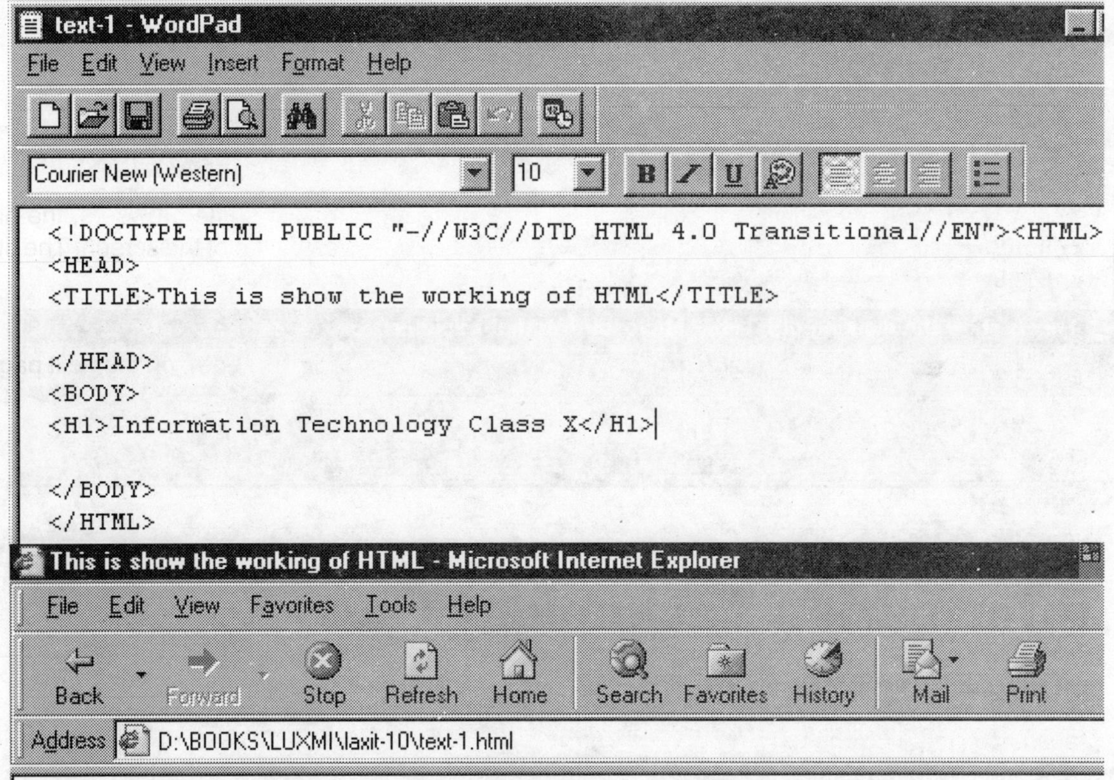

Block headings

For every new paragraph in HTML you must use the p tag, as was mentioned in the earlier paragraph. The following example shows how it is done.

1. After the heading line, press Enter to reach the next line in Word Pad.
2. Type <p> to start the new paragaph.
3. Type the text which you want to put in the paragraph. I have typed here KHANNA PUBLISHING PVT. LTD.
4. Close the paragraph by using the tag </p>.
5. The text in Word Pad and its effect in the browser are shown on the next page.

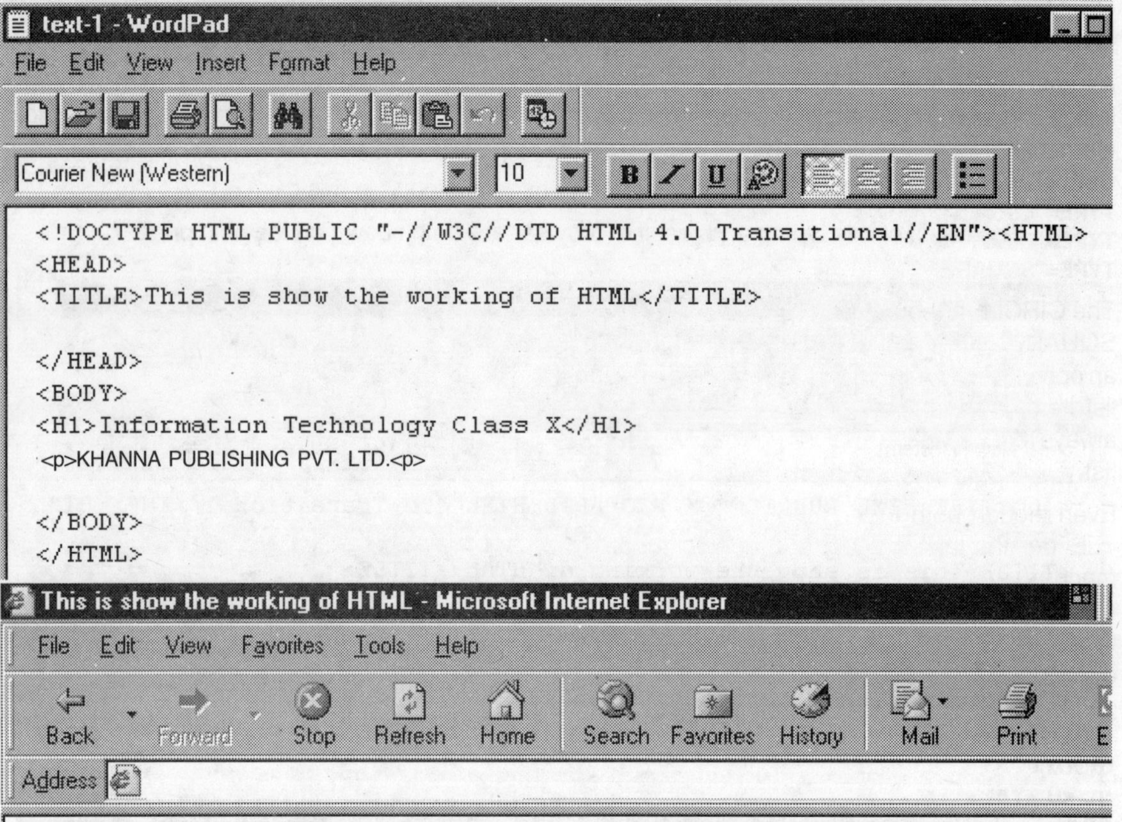

Lists

HTML support 3 type of lists: Unordered lists, Ordered lists and Definition lists. Ordered lists are numbered in some fashion, while Unordered lists are bulleted. Definition lists consist of a term followed by its definition.

Both ordered and unordered lists require start and end tags as well as the use of a special element to indicate where each list item begins (the <LI> tag). Unordered lists can be preceded by one of several bullet styles: a closed circle, an open circle, or a square.

The tags for an unordered list are <UL> and </UL>. Ordered lists can be preceded by Arabic numerals, uppercase or lowercase Roman numerals, or uppercase or lowercase alphanumeric characters. The tags for an ordered list are <OL> and </OL>. Definition lists require a start tag (<DL>) and end tag (</DL>) and two special elements: one for definition terms (the <DT> tag) and one for definitions (the <DD>) tag.

In addition to these three types of lists, HTML also allows two other types of lists that are much less commonly used: Directory lists (which use the <DIR> tag) and Menu lists (the <MENU> tag). However, these two types of lists are not very popular.

Unordered Lists

As said earlier, this list uses the <UL> and </UL> tags. The only element that can be contained inside the <UL> and </UL> tags is a list item, signified with the list item element. The list item element uses the <LI> tag (and optionally the </LI> tag) and contains the actual content of your lists. Both <UL> and <LI> have the same set of attributes:

TYPE="CIRCLE"
TYPE="DISC"
TYPE="SQUARE"

The CIRCLE attribute value is used for a hollow bullet, the DISC type creates a solid bullet, and the SQUARE value renderes a solid block. The default appearance for a list is with a disc. You can use an optional </LI> end tag at the end of each list item, however, the </UL> end tag is always required at the end of the unordered list.

Even though both the <UL> tag and the <LI> tag can take the TYPE attribute, it's much more common to use the attribute with the <UL> tag so that the entire list takes on the appearance you desire. For example, here is some HTML that generates two separate lists:

```
<TITLE>Menu Card</TITLE>
<BODY>
<p><Breakfast>
<UL>
<LI>Eggs
<LI>Milk
<LI>Apples
<LI>Coffee
</UL>
<p>Dinner
<UL TYPE="SQUARE">
<LI>Soup
<LI>Salad
<LI TYPE="CIRCLE"> Chicken
<LI TYPE="DISC">Rice
<LI TYPE="CIRCLE">Dessert
</UL>
```

See the result of this in the browser next.

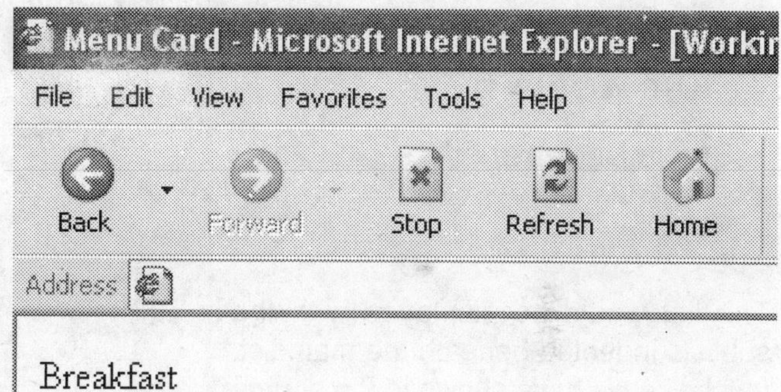

One important aspect of lists is that you can nest one list inside another to create a sublist. The default appearance of the sublists will vary from the main list, with the first sublist using circle bullets and the next nested list, using squares. For example:

```
<UL>
<LI>Microsoft
    <UL>
    <LI>Windows
    <LI>Office
        <UL>
        <LI>Microsoft Word
        <LI>Microsoft Excel
        </UL>
    <LI>Internet Explorer
    </UL>
<LI>Adobe
    <UL>
    <LI>PageMaker
        <UL>
        <LI>Version 7.0
        </UL>
    </UL>

<LI>CorelDRAW
    <UL>
    <LI>CorelDRAW 11
        <UL>
        <LI>CorelTRACE
        </UL>
    </UL>
</UL>
```

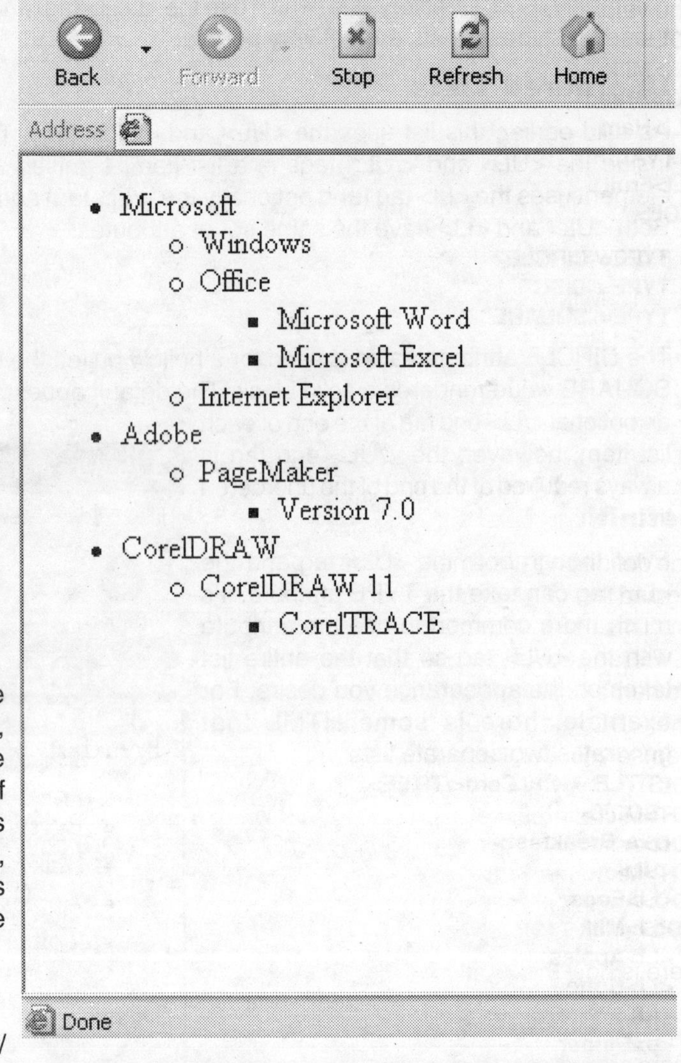

The list would be displayed with the sublists indented beneath the main list, much like we have shown in the source code for readability. There are a total of seven lists here. Each <UL> tag begins a new list. The main list (Microsoft, Adobe, and CorelDRAW) has six sublists - two per bulleted point. The above figure shows the same.

Creating Ordered Lists

The ordered list element's <OL> and </OL> tags are used to create ordered lists. Like unordered lists, ordered list must contain list item elements (with the <LI> tag) to contain your list's text. In fact, ordered lists are identical in behaviour to unordered lists, except they use numbers instead of bullets and you can use an attribute to start numbering at a number other than one.

Here are the attributes you can use with the <OL> tag:

TYPE="1" (Arabic numbers)
TYPE="a" (lowercase alphanumeric)
TYPE="A" (uppercase alphanumeric)

TYPE="i" (lowercase Roman numbers)
TYPE="I" (uppercase Roman numbers)
START="X" (Arabic numbers)

The Type attribute allows you to specify the numbering system you want to use. Arabic numbers are the default. In addition, when you are using ordered lists, the <LI> tag can use the VALUE attribute to make a particular list item have a certain number.

The value attribute is shown here:

```
<OL>
<LI>Milk
<LI>Bread
<LI>Ham
<LI VALUE="10">Coffee
<LI>Fruit
</OL>
```

In a browser the order of this list would appear as follows:

1. Milk
2. Bread
3. Ham
10. Dark Chocolate
11. Fruit

Definition Lists

The definition list element uses the <DL> start tag and the </DL> end tag to create a definition list. This list is rendered without bullets. The <DT> tag is used for definition terms (that is, the name or title of the item you are defining). The <DD> tag is used for the definitions themselves. For example,

```
<DL>
<DT>Microsoft Excel
<DD>A spreadsheet program
<DT>Microsoft Word
<DD>A word processing program
</DL>
```

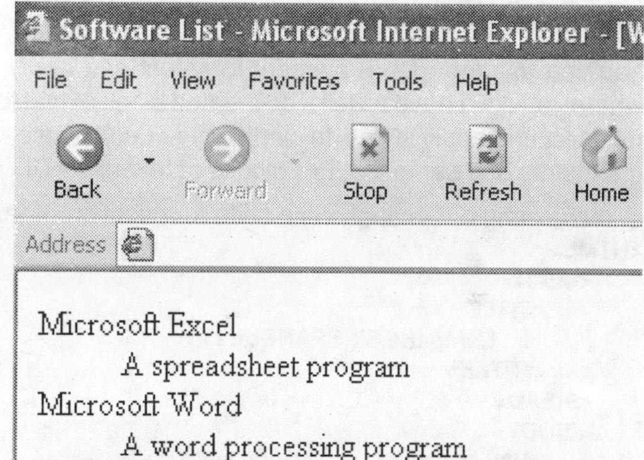

Here is how this would look in the browser.

Using Directory and Menu Lists

There are two other types of lists defined in HTML: Directory lists and Menus. However, these two types of lists are rarely used, and Navigator and Internet Explorer treat them identically to the way they treat unordered lists.

The directory list element is signified by the <DIR> and </DIR> tags. This element was intended to be used for directory lists of short

items (some sources recommend 20 or fewer characters so they can be listed in columns 24 characters wide). Here is a quick example of a directory list:

```
<DIR>
<LI>Item 1
<LI>Item 2
<LI>Item 3
</DIR>
```

Similarly the <MENU> and </MENU> tags make up the menu element, which is used for the menu lists. Menus can appear with different spacing results in different browsers, but Navigator and Internet Explorer don't display menu lists any differently than unordered lists. Here is a quick sample menu:

```
<MENU>
<LI>Milk
<LI>Butter
<LI>Bread
</MENU>
```

For both directory and menu lists, the only item that should be contained is a list-item element (the <LI> tag).

Inline elements

Inline elements can be explained in the following manner.

Formatting Inline Text

You can use the <DIV> element to select and apply styles to a block of text, but in those cases where you don't have a well – delineated block of text to work with, you can use the <SPAN> element to select and apply styles. to clarify – you usually use <SPAN> to apply styles inline, for example, in the middle of a sentence. For blocks of text use <DIV> for words or sentences use <SPAN>

In this example, I'm applying a style to a section of text using <SPAN>, turning that text red:

```
<HTML>
    <HEAD>
        <TITLE>
            Using the &lt;SPAN&gt; tag
        </TITLE>
    </HEAD>
    </BODY>
        <H2>
            Is my face <SPAN CLASS="redface" STYLE="color: red">red</SPAN>?
        </H2>
    </BODY>
</HTML
```

You can see the results of this HTML in browser. The word "red" appears in red and rest of text in black. Though it would look all black since the book is in black and white.

Creating Inline Layers

The <Ilayer> element is to the <LAYER> element as the <SPAN> tag is to the <DIV> element. It has much the same function, but rather than creating a stand alone HTML structure, the <ILAYER> is supposed to be used inline (in the middle of other text). The <ILAYER> element lets you create layers inline, as in this example, where I'm creating a new layer inline to move some text up a little higher on the page:

```
<HTML>
<HEAD>
<TITLE>Working with layer</TITLE>
</HEAD>
<BODY>
    <H1 ALIGN="CENTER">
        Working With Inline Layers
    </h1>
    <BR>
    <H1>
        This text includes some that
        <ILAYER NAME="LAYER1" LEFT = 5 TOP = 15>
        <I>pops up</I><ILAYER>
        a little.
    </H1>
</BODY>
</HTML>
```

You can see the results of this HTML in browser. I have positioned this layer using the Top and Left attributes to push some text up a little.

Creating Inline Or Floating Frames

Internet Explorer supports inline frames, also called floating frames. Inline frames can appear wherever you want them, including in the middle of the page. All you have to do is specify the height and width of the frame and the HTML page to load into it and you're set. Inline frames are supported with the <IFRAME> element, which in many ways is Internet Explorer's answer to Netscape Navigator's <LAYER> element. You can use inline frames to display a number of pages in one and if you turn the borders off the result can be much like using the <LAYER> element.

Here's an example showing how to create two inline frames:

```
<HTML>
    <HEAD>
        <TITLE>
            Inline Frames!
        </TITLE>
    </HEAD>
<BODY BGCOLOR="BLACK">
    <CENTER>
        <BR>
        <BR>
        <IFRAME WIDTH = 50% HEIGHT = 25% NAME = frame1
            SRC="frame1.htm">
        </IFRAME>
        <BR>
        <BR>
    </CENTER>
    </BODY>
</HTML>
```

You can see the results in the browser. Note that it's up to the browser to place these frames, much like other elements, such as images.

Creating Inline Video

You can also include videos in your Web pages. One way to do this in Internet Explorer is with the

<IMG> element, if you include the DYNSRC attribute. Here, I'm adding a Charlie Chaplin movie to a Web page:

```
<HTML>
    <HEAD>
        <TITLE>
            Video With the &lt;IMG&gt: Element
        </TITLE>
    </HEAD>
    <BODY>
        <CENTER>
            <H1>
                Video With the &lt; IMG&gt; Element
            </H1>
            <BR>
            <BR>
            <IMG DYNCSRC="chaplin.avi" ALIGN="Top">
            <BR>
            <BR>
        </CENTER>
    </BODY>
</HTML>
```

The results of this HTML can also be seen in the browser. As you can see, the short movie is playing.

Images

HTML supports all kind of graphic images. They can be logos or photographs. Graphics once inserted are displayed on the site page. Follow the following steps to insert an image on a page:

1. Place the cursor where you want the image to appear.
2. Type <IMG SRC=" image.ext" where image.ext indicates the location of the image file on the server.
3. In the example I have typed this after the <HEAD> line. Since I would be using GLOBE.JPG in the text, I have used its location.
4. If desired, type BRODER = n, where n is the thickness of the border in pixels. I have not used any.
5. Type the final >.
6. See the result, how it looks in the browser on the next page. Also note the text how it is typed in the word pad.

Remember

You can add graphics with extensions like, GIF, JPEG, PNG, or any other kind of images that the browser recognizes. Don't expect your visitors to he site to wait for more than 30 seconds to load and view your page.

So get the smaller images, create miniatures of large images and let visitors choose to view the larger images only if desired.

Specifying Size for Fast Viewing

As mentioned above your image should be small so that it can be loaded fast. If you specify the image's dimensions, the browser can fill in the text around the image as the image loads, so that

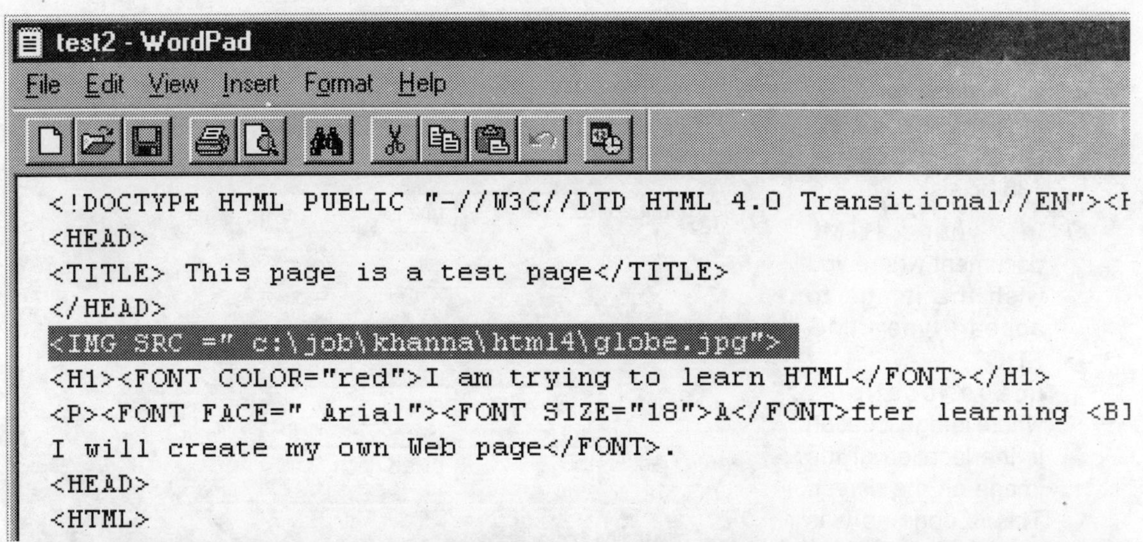

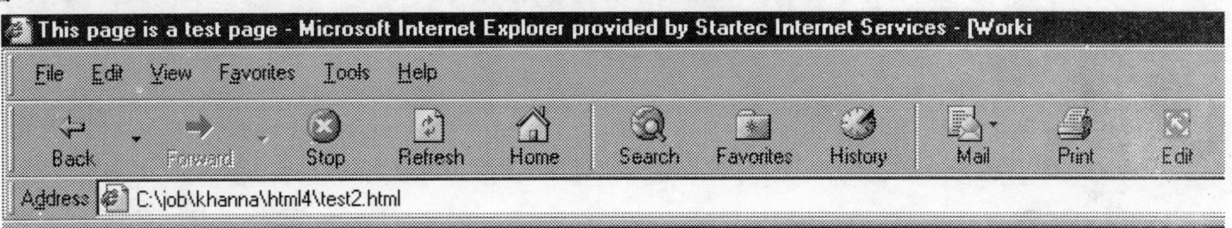

I am trying to learn HTML

After learning **HTML**, I will create my own Web page.

your visitors can read the page without having to wait for the images. Follow the following steps to specify the size of your image:

1. Open the image in Photoshop, or other image editing program.
2. Choose Image > Image Size.
3. Choose pixels for the unit of measure in both the Width and Height pop-up menus.

4. Write down the values shown in the Width and Height text boxes. In this case the width is 150 and height is 185 pixels.
5. In your HTML document where you wish the image to appear, type <IMG SRC = " image.location", where image.location is the location of the image on the server. This is done as was done in the previous case.

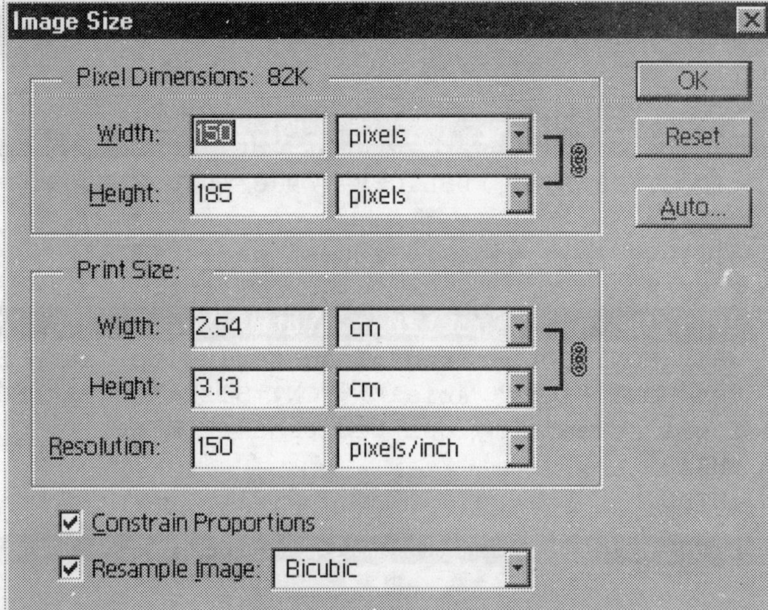

6. Type WIDTH=x HEIGHT=y, using the values you jotted down in step 4 to specify the values for x and y (the width and height of your image) in pixels. In this case width would be 150 and height to be 185.
7. Add other image attributes as described and then type the final >.
8. See how the figure looks like on the next page. I have deliberately used width as 100 and height as 145, since if I had used the same options, you would not have noticed the difference in the picture. Text is also given there.

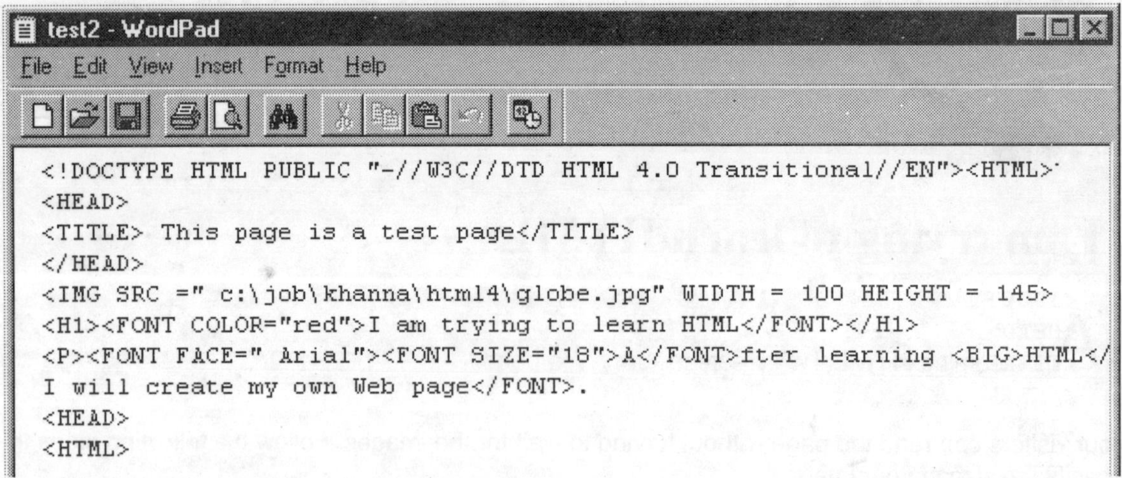

Wrapping Text around Images

It is not that you would always have the image covering the full width of the screen. You can use the ALIGN attribute (with the left and right variables only) to wrap text around an image.

1. Type <IMG SRC = "image.location" as was done earlier.

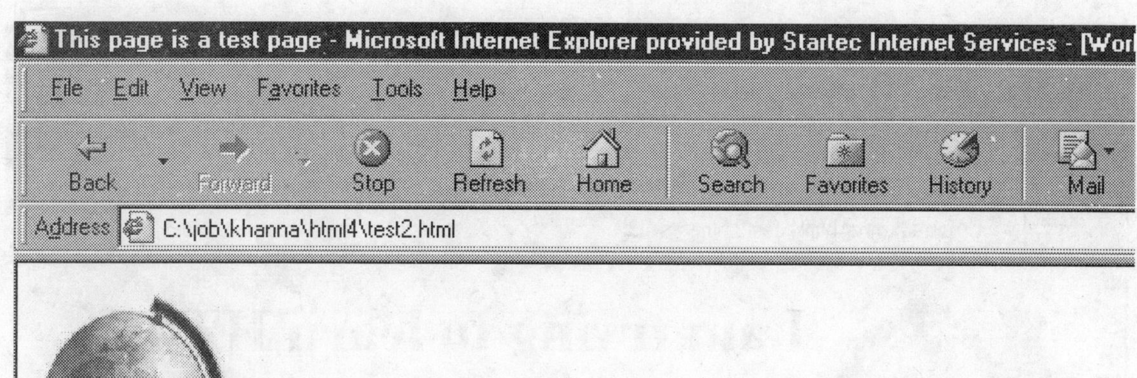

2. Either type ALIGN=left to align the image to the left of the screen while the text flows to the right or type ALIGN= right to align the image to the right edge of the screen while the text flows on the left side of the image.
3. Add other image attributes, as described in other parts of this chapter, if desired.
4. Type the final >.
5. See the result in the browser, as shown on the next page alongwith the HTML text .

In this case I have used the option of aligning the image to the left and let the text flow on the right.

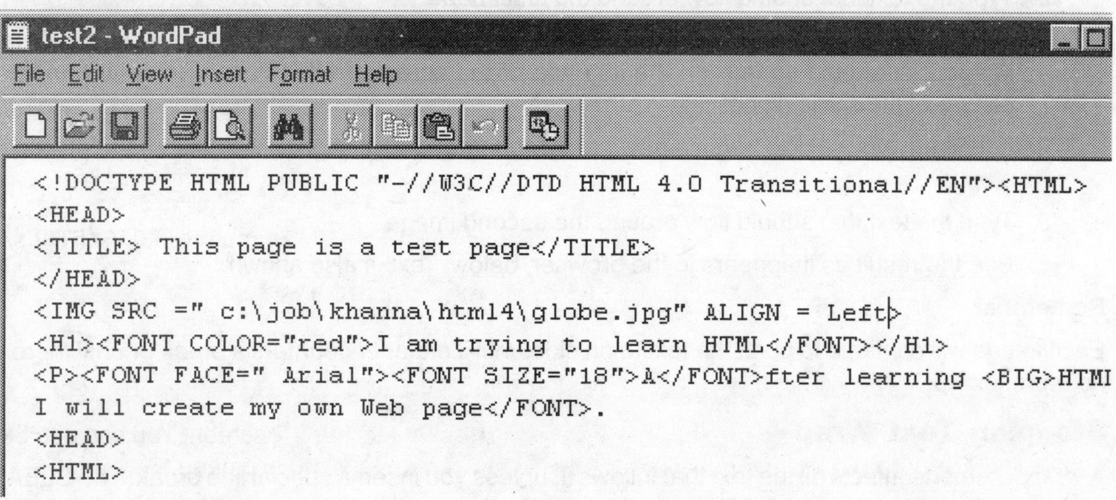


```
<!DOCTYPE HTML PUBLIC "-//W3C//DTD HTML 4.0 Transitional//EN"><HTML>
<HEAD>
<TITLE> This page is a test page</TITLE>
</HEAD>
<IMG SRC =" c:\job\khanna\html4\globe.jpg" ALIGN = Left>
<H1><FONT COLOR="red">I am trying to learn HTML</FONT></H1>
<P><FONT FACE=" Arial"><FONT SIZE="18">A</FONT>fter learning <BIG>HTMI
I will create my own Web page</FONT>.
<HEAD>
<HTML>
```

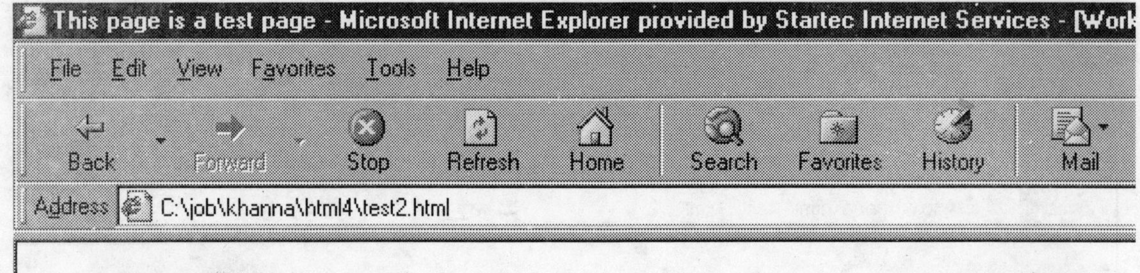

Remember

When you choose ALIGN=right, it's the image that goes to the right (while the text goes to the left). When you choose ALIGN=left, again the image will be on the left side with the text flowing around the right side.

Wrapping Text between Two Images

1. Type <IMG SRC=" right.image" ALIGN=right> where right.image indicates the location of the image that should appear on the right side of the screen. In this case I would be using the above Globe.JPG on the right.
2. Type the text that should flow around the first image.
3. Type <IMG SRC=" left.image". ALIGN=left> where left.image indicates the location of the image that should appear on the left side of the screen. In this case I am using another picture of the same globe.
4. If desired, type <P> to begin a new paragraph. that will be aligned with the image placed in step 3.
5. Type the text that should flow around the second image.
6. See the result as it appears in the browser, below. Text is also shown.

Remember

Each image will continue to push the text to on side until it either encounters a break or until there is no more text.

Stopping Text Wrap

A wrapped image affects all the text that follows it, unless you insert a special line break. The CLEAR attribute added to the regular BR tag indicates that the text should not begin until the specified margin is clear (that is, at the end of the image or images).

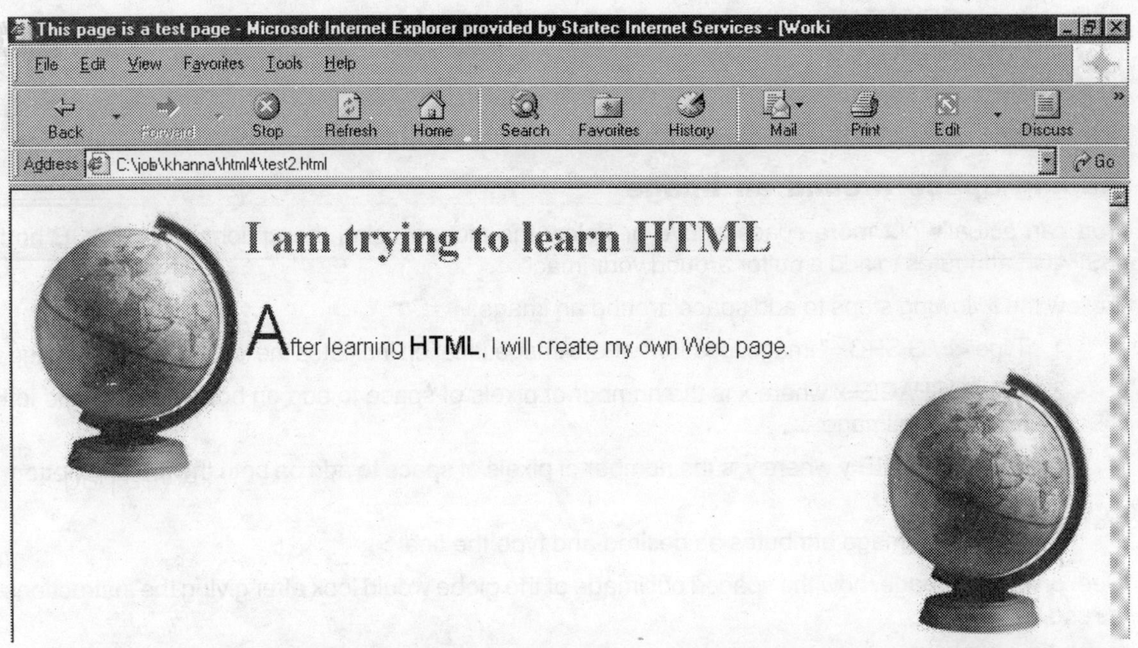

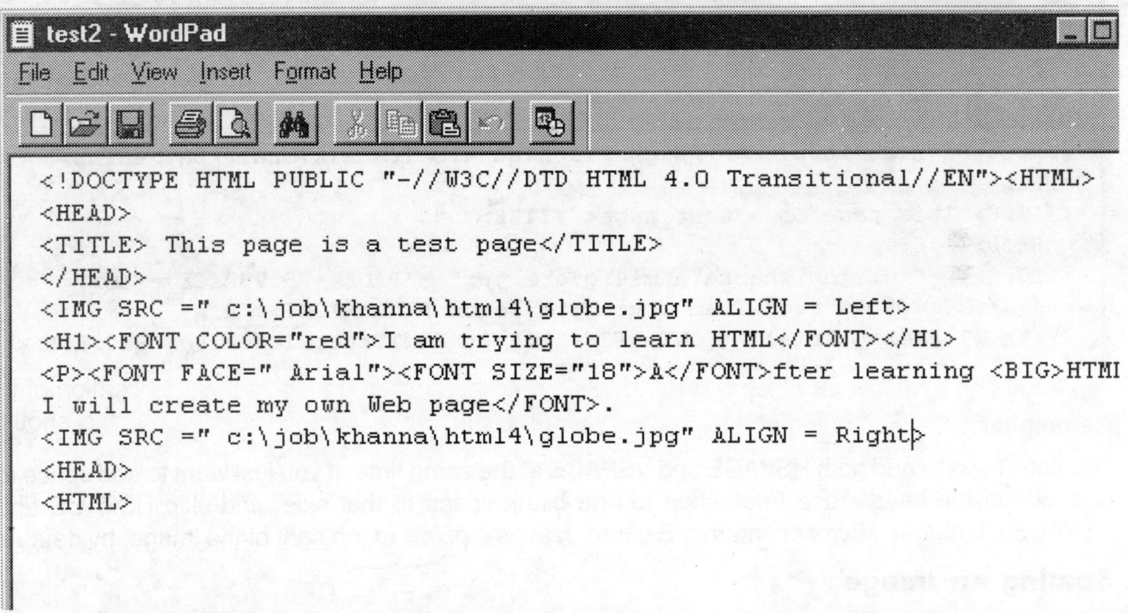

Follow the following steps to stop the text from wrapping:

1. Create your image and the text, as was done before.
2. Place the cursor where you want to stop wrapping text to the side of the image.
3. Either type <BR CLEAR= left> to stop flowing text until there are no more images aligned to the left margin.

   Or

type <BR CLEAR = right> to stop flowing text until there are no more images aligned to the right margin.

Or

type <BR CLEAR = all> to stop flowing text until there are no more images on either margin.

Adding Space around an Image

You can actually put more space above or below the picture using the options of VSPACE and HSPACE attributes to add a buffer around your image.

Follow the following steps to add space around an image:

1. Type <IMG SRC =" image.location" where image.location indicates the location of your image.
2. Type HSPACE=x where x is the number of pixels of space to add on both the right and left sides of the image.
3. Type VSPACE=y where y is the number of pixels of space to add on both the top and bottom of the image.
4. Add other image attributes as desired and type the final >.

See, on the next page, how the spaced out image of the globe would look after giving the instructions as shown below.

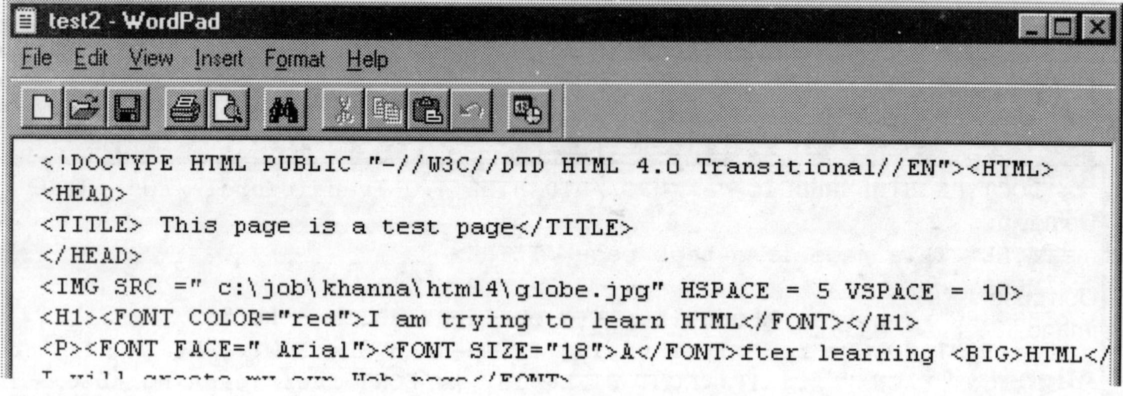

Remember

You don't have to add both HSPACE and VSPACE at the same time. If you just want to add space to one side of the image, use Photoshop to add blank space to that side, and skip HSPACE and VSPACE altogether. Microsoft Internet Explorer leaves 4 pixels to the right of the image, by default.

Scaling an image

Supposing you want to change the size of an image, you can do it by just specifying a new height and width in pixels. This way you can have large images on your page without long loading times.

Follow the following steps to scale an image:

1. Type <IMG SRC =" image.location where image.location is the location of the image.
2. Type WIDTH=x HEIGHT=y where x and y are the desired width and height, respectively, in pixels, of your image.
3. Add any other image attributes as discussed earlier and then type the final >.

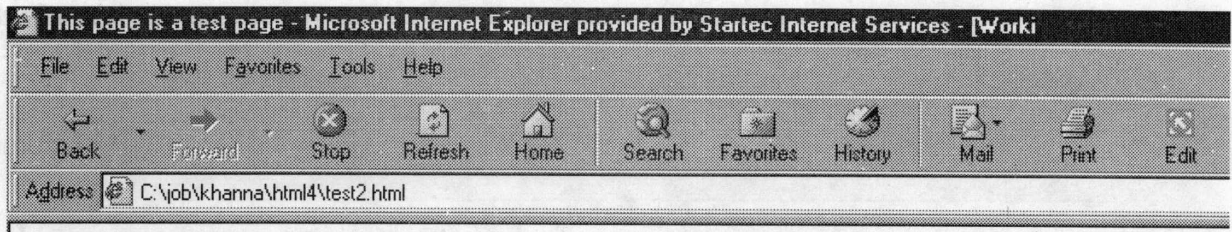

Remember

Do not increase the size of the image too much, it will take more time to load and will make the image messy.

Don't use the WIDTH and HEIGHT extensions to reduce the image size. Instead, created a smaller image. It will load faster and look better..

Aligning Images

We had earlier wrapped the image to the left and right of the text. Now we will make use of the ALIGN attributes for aligning images with text. You can align an image in various ways to a single line in a paragraph.

Follow the following steps to align an image with text:

1. Type <IMG SRC=" image.location" where image.location indicates the location of the image.
2. Type ALIGN= direction where direction is one of the attributes: texttop, top, middle, absmiddle, bottom, or absbottom.
3. Add other attributes as desired and then type the final >.
4. Type the text with which you wish to align the image.

In my example, I have taken the same globe and positioned it at various places with respect the text. See how it looks next. HTML text is shown on the next page.

Remember

You may not align an image and wrap text around it at the same time. Internet Explorer has trouble

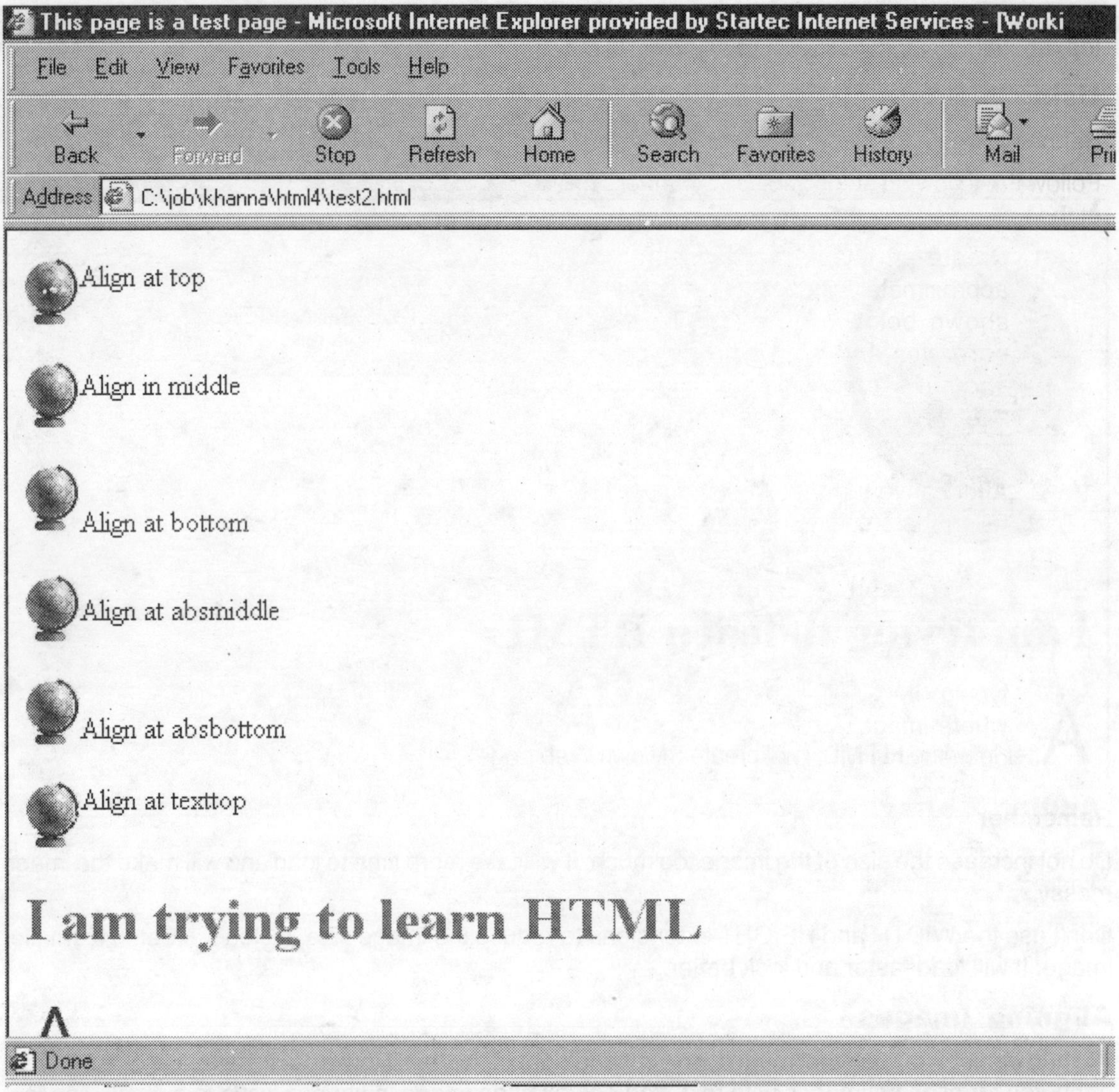

This page is a test page - Microsoft Internet Explorer provided by Startec Internet Services - [Worki

File Edit View Favorites Tools Help

Back Forward Stop Refresh Home Search Favorites History Mail Pri

Address C:\job\khanna\html4\test2.html

Align at top

Align in middle

Align at bottom

Align at absmiddle

Align at absbottom

Align at texttop

I am trying to learn HTML

A

Done

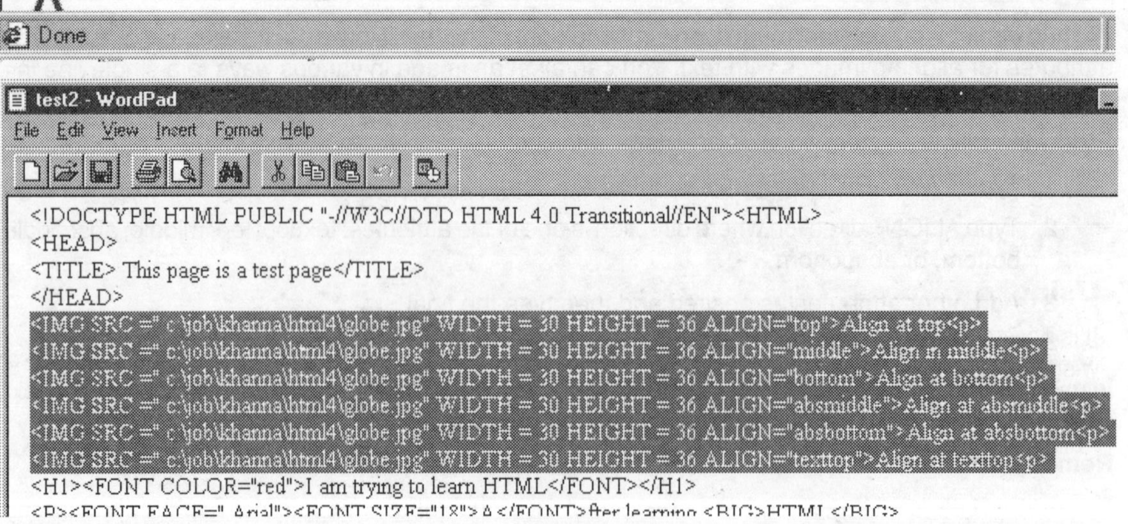

test2 - WordPad

File Edit View Insert Format Help

```
<!DOCTYPE HTML PUBLIC "-//W3C//DTD HTML 4.0 Transitional//EN"><HTML>
<HEAD>
<TITLE> This page is a test page</TITLE>
</HEAD>
<IMG SRC =" c:\job\khanna\html4\globe.jpg" WIDTH = 30 HEIGHT = 36 ALIGN="top">Align at top<p>
<IMG SRC =" c:\job\khanna\html4\globe.jpg" WIDTH = 30 HEIGHT = 36 ALIGN="middle">Align in middle<p>
<IMG SRC =" c:\job\khanna\html4\globe.jpg" WIDTH = 30 HEIGHT = 36 ALIGN="bottom">Align at bottom<p>
<IMG SRC =" c:\job\khanna\html4\globe.jpg" WIDTH = 30 HEIGHT = 36 ALIGN="absmiddle">Align at absmiddle<p>
<IMG SRC =" c:\job\khanna\html4\globe.jpg" WIDTH = 30 HEIGHT = 36 ALIGN="absbottom">Align at absbottom<p>
<IMG SRC =" c:\job\khanna\html4\globe.jpg" WIDTH = 30 HEIGHT = 36 ALIGN="texttop">Align at texttop<p>
<H1><FONT COLOR="red">I am trying to learn HTML</FONT></H1>
<P><FONT FACE=" Arial"><FONT SIZE="18">A</FONT>fter learning <BIG>HTML</BIG>
```

with aligning more than one image on a line. The results are erratic. In addition, it treats texttop as top, absmiddle as middle and absbottom as Bottom.

Using a Banner

Banner is what you place at the top of the web site. It is reflected on all the pages of your site.

Follow the following steps to place a banner at the top of each page:

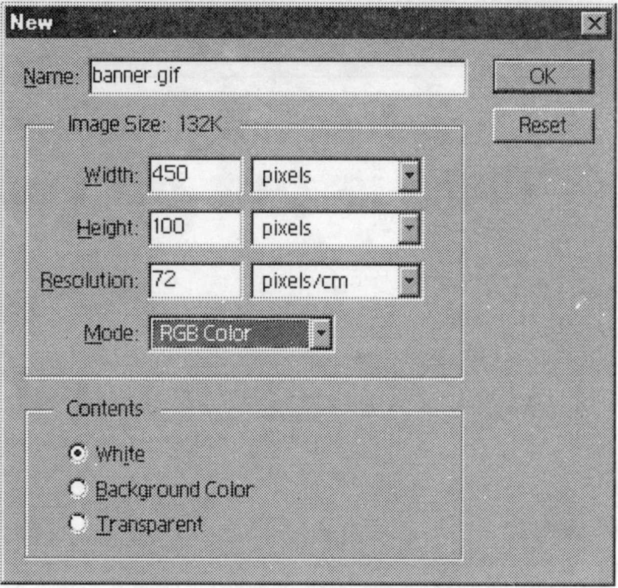

1. Create an image that measures approximately 450 x 100 pixels, as shown below. You can make it narrower and shorter, but you shouldn't make it much wider, otherwise it won't fit easily on most screens.
2. After converting it to indexed color, using the smallest bits/pixels ratio you can stand, save it as a GIF image.
3. Use this exact same image at the top of each of your Web pages, by typing <IMG SRC=" image.name">, where image.name is the location on the server of the banner.

Adding Horizontal Rules

This graphic element is completely supported by the majority of the browsers is the horizontal rule. There are several attributes you can use to jazz up horizontal rules. Follow the following steps to insert a horizontal rule:

1. Type <HR where you want the rule to appear. The text that follows will appear in a new paragraph below the new line.
2. If desired, type SIZE=n, where n is the rule's height in pixels.
3. If desired, type WIDTH=w, where w is the width of the rule in pixels, or as a percentage of the document's width.
4. If desired, type ALIGN=direction, where direction refers to the way a rule should be aligned on the page; either left, right or center.
5. If desired, type NOSHADE to create a solid bar, with no shading.
6. Type the final > to complete the horizontal rule definition.

Using Alternate Text

It is possible that your browser may not support images at all. Or even if it does it is so slow that you wish that the graphics was not there at all. For this you can create a text that will appear even if the image, for whatever reason, does not.

Follow the following steps to offer alternate text when images don't appears:

1. Place the cursor where you want the image (or alternate text) to appear.

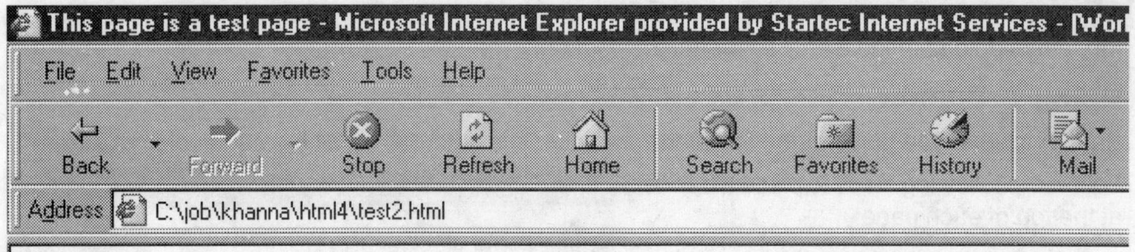

This is put on top of page

I am trying to learn HTML

After learning **HTML**, I will create my own Web page.

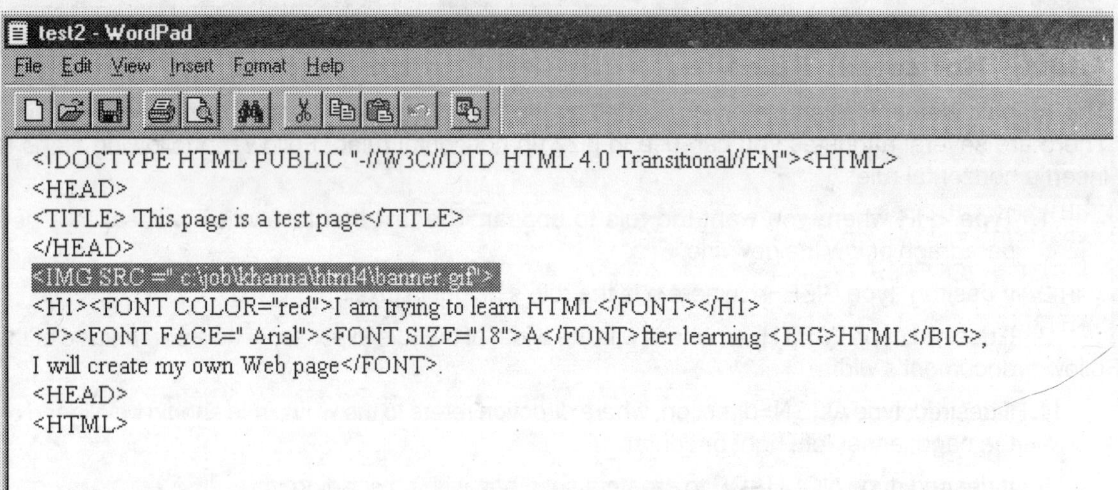


```
<!DOCTYPE HTML PUBLIC "-//W3C//DTD HTML 4.0 Transitional//EN"><HTML>
<HEAD>
<TITLE> This page is a test page</TITLE>
</HEAD>
<IMG SRC =" c:\job\khanna\html4\banner.gif">
<H1><FONT COLOR="red">I am trying to learn HTML</FONT></H1>
<P><FONT FACE=" Arial"><FONT SIZE="18">A</FONT>fter learning <BIG>HTML</BIG>,
I will create my own Web page</FONT>.
<HEAD>
<HTML>
```

2. Type <IMG SRC = "image.ext", where image.ext is the location of the image.
3. Type ALT =".
4. Type the text that should appear if, for some reason, the image itself does not.
5. Type ">.

Linking to External Images

Linking allows you to link instead of actually copying the image. This is done specially in the cases where the image is of a very large size. You can create an icon of the image and place it in your page.

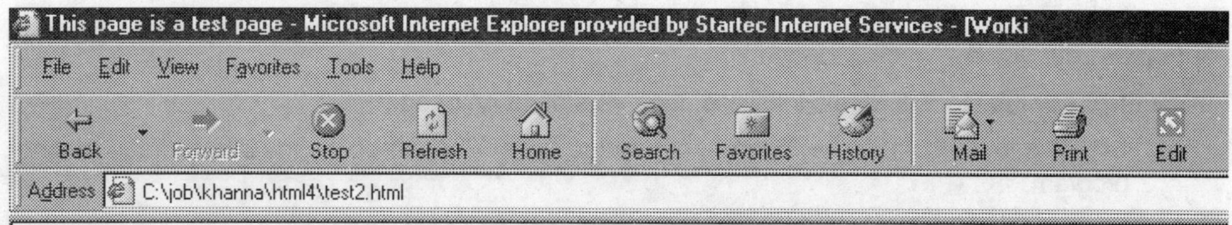

This is put on top of page

I am trying to learn HTML

After learning **HTML**, I will create my own Web page.

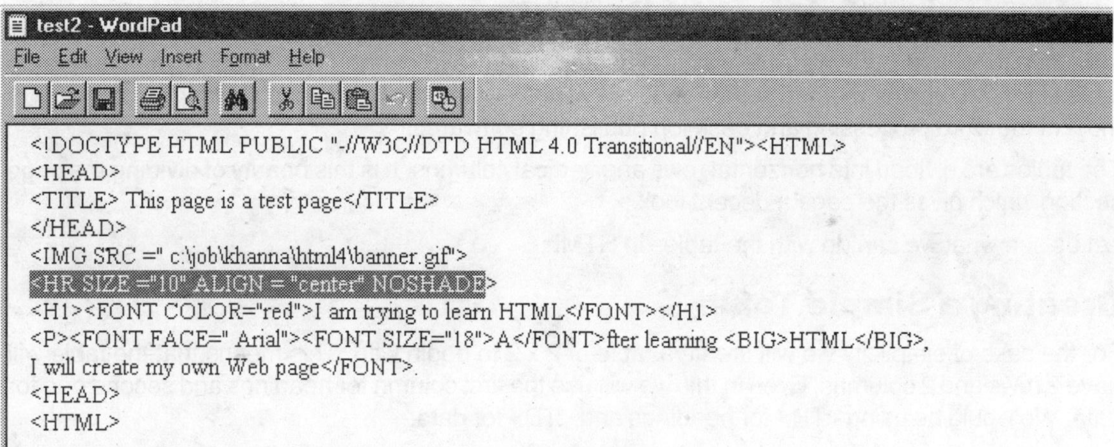

Follow the following steps to link a small icon to your larger image:

1. Place the cursor in your HTML text where you wish the icon to be placed.
2. Type <A HREF + image.location"> where image.location is the location of the full sized image on your screen.
3. Type <IMG SRC =" icon.location", where icon.location is the location of your icon.
4. If desired, type ALT =" alternate text" where alternate text is the text that should appear if, for some reason, the icon does not.
5. Type the final > of the icon definition.
6. Type </A> to complete the link to the full sized image.

Using Low Resolution Images

You can have reference to both high and low resolution versions of your image so that the low resolution image loads quickly and keeps the visitor's interest while the high resolution version impresses your visitors, once it loads in.

Follow the following steps to use a low resolution version of an image:

1. Create a low resolution version of your image using Photoshop or any other image editing program.
2. Place the cursor where you want the full resolution image to appear.
3. Type <IMG SRC = "image.gif", where image.gif is the location on the server of the high resolution image.
4. Type LOWSRC = "imagelow.gif", where imagelow.gif is the location on the server of the low resolution image.
5. Type HEIGHT=x WIDTH=y, where x and y are the height and width in pixels, respectively of the original image. If you do not specify these values, browsers use the size of the low resolution image for both images.
6. If desired, type ALT="substitute text", where substitute text is the text that will appear if the visitor can't view images with their browser.
7. Type the final >.

Working with Tables

As we all know Tables are a way of representing the data in the tabular form which not only is attractive to look at but it also easy to understand when read by the user. Though you can create tables in HTML but the method somehow is not very easy and the result much below the standard of most of the word processing and desk top publishing software.

The tables are divided into horizontal rows and vertical columns. It is this beauty of dividing the page section which gives the page a decent look.

Let us see what we can do with the tables in HTML.

Creating a Simple Table

For the sake of simplicity we will create a table of 2 x 2 to begin with. This means that the table will have 2 rows and 2 columns. Even in this we will use the first column for headings and second one for data. We would be using <TH> for headings and <TD> for data.

Follow the following steps to create a simple table:

1. Type <TABLE>.
2. Type <TR> to define the beginning of the first row.
3. Create a header cell in the first row of elements by typing the headings.
4. Type the contents of the first header.
5. Create a regular cell after the header.
6. Type the contents of the regular cell.
7. Repeat steps 2-6 for each row.
8. To finish the table, type </TABLE>.

It may not have been very clear unless you look at it practically. For this we will go back to our old example and create a table in it. Here I am using the marks received by the student in various subjects. Look at the text which has been typed in the HTML and its result in the browser.

Remember

If you want you can close each cell with </TH> or </TD>. Close each row with a </TR>. But, these tags are all optional.

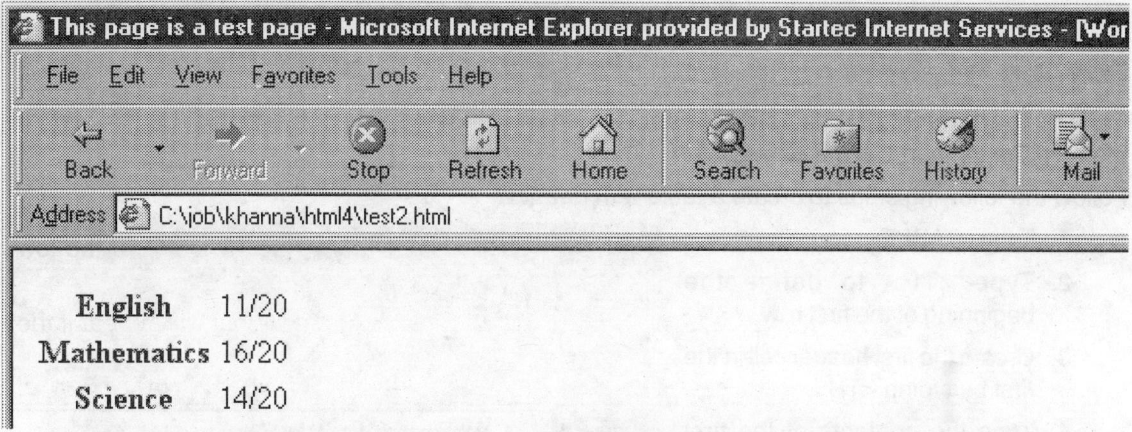

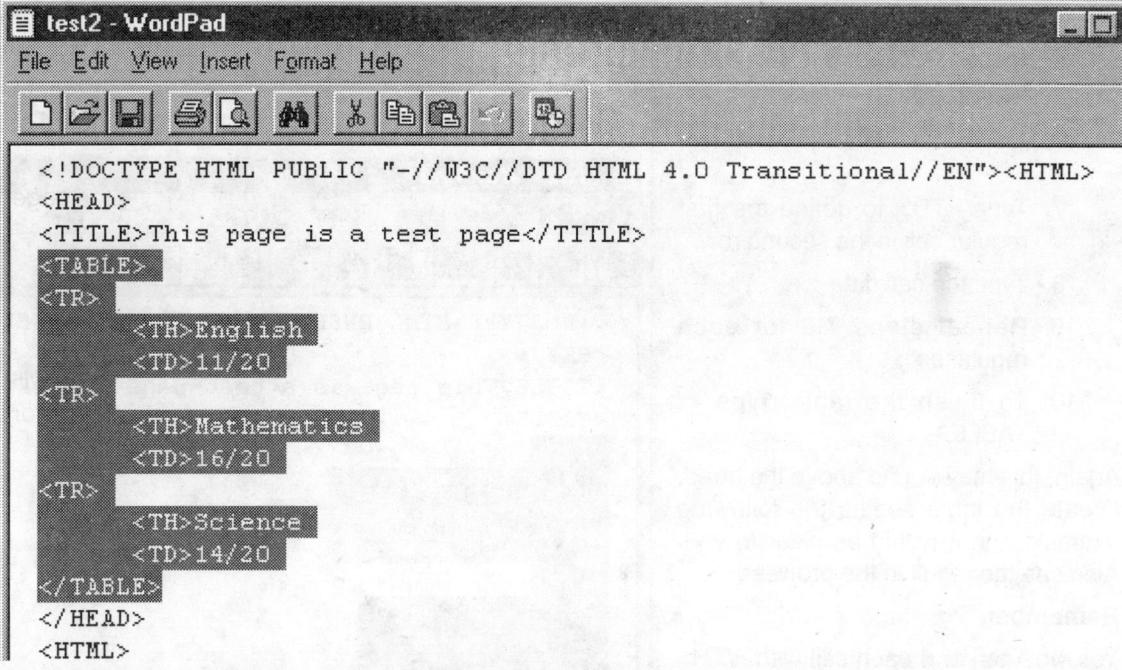

Changing a Table's Width and Height

The table created in the above example looks okay but is somewhat crumbersome. So let us increase its height and width to make it more readable. For this we will use the WIDTH and HEIGHT attributes to resize the whole table.

Follow the following steps to change a table's size:

Within the initial TABLE tag, type WIDTH=x HEIGHT =y, where *x* and y are either absolute values in pixels for the height and width of the table.

Remember

If the table is too small for the data, the browser will just ignore you. WIDTH is a standard of HTML 4, HEIGHT is not. However, the major browsers recognize both attributes. If the table is narrower than the window, you can center it by adding ALIGN=center to the TABLE tag.

Putting Headers

Earlier in our example we had put the headings on the left hand side of the marks. Let us now put them on the top and call them headers.

Follow the following steps to create a table with header:

1. Type <TABLE>.
2. Type <TR> to define the beginning of the first row.
3. Create the first header cell in the first by typing <TH>.
4. Type the contents of the first header cell.
5. Repeat steps 3-4 for each header cell.
6. Type <TR> to begin the second row.
7. Type <TD> to define the first regular cell in the second row.
8. Type the cell data.
9. Repeat steps 7-8 for each regular cell.
10. To finish the table, type </TABLE>.

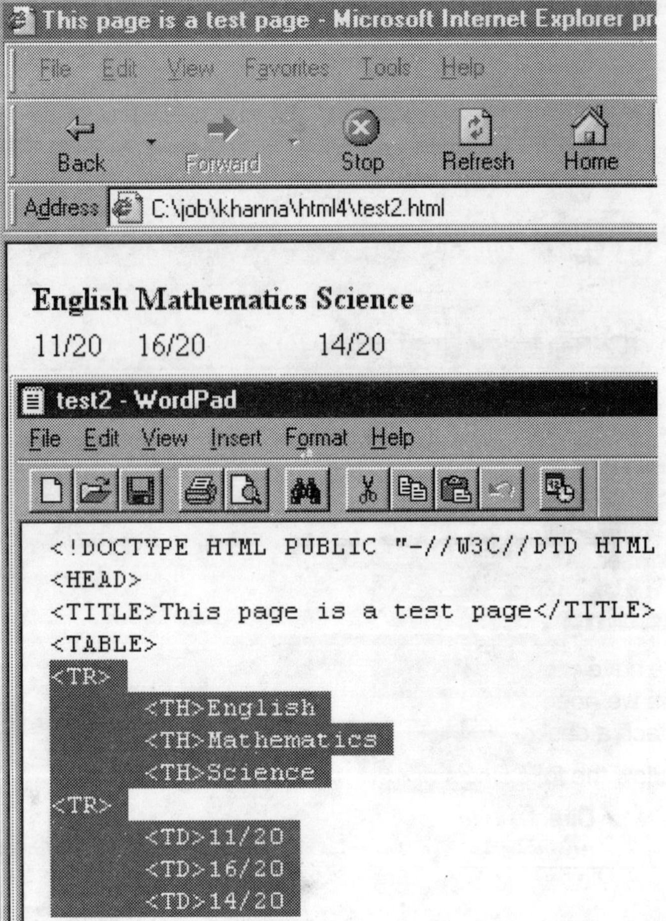

Again, this may sound above the head. Create the table seeing the following example and it would be clear to you. Also see the result in the browser.

Remember

Yes, you can end each cell with </TH> or </TD> and each row with </TR> but only if you want to, there is no compulsion.

Putting Headers on Top and Left

First we put the headings on the left and then we put them on the top. Let us now put them on both sides, i.e., both left and top.

Follow the following steps to create a table with header cells on top and down the left side:

1. Type <TABLE>.
2. Type <TR> to define the beginning of the first row.
3. Create the empty cell in the top left corner by typing <TD>
4. Create a header cell by typing **<TH>cell contents** where cell contents is the data of the cell.
5. Repeat step 4 for each header cell in the first row.
6. Type <TR> to begin the second row.
7. To define the first header on the left side, type **<TH>cell contents**, where cell contents is the data of the cell.
8. To create a regular cell after the header cell in the second row, type **<TD>cell contents**, where cell contents is the data that the cell should contain.
9. Repeat step 8 for each remaining regular cell in the row.
10. Repeat steps 6-9 for each remaining row.
11. To finish the table, type </TABLE>.

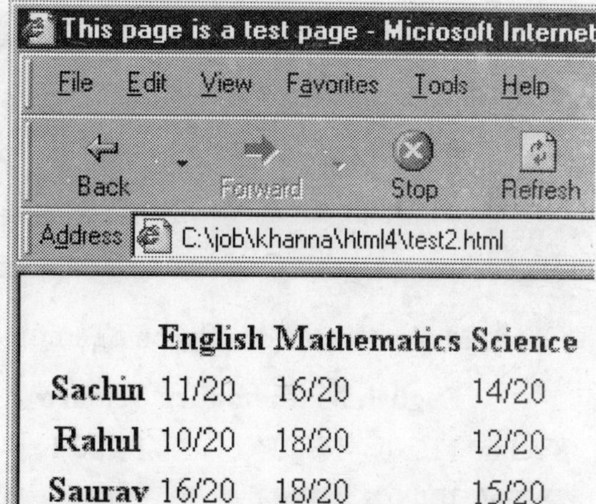

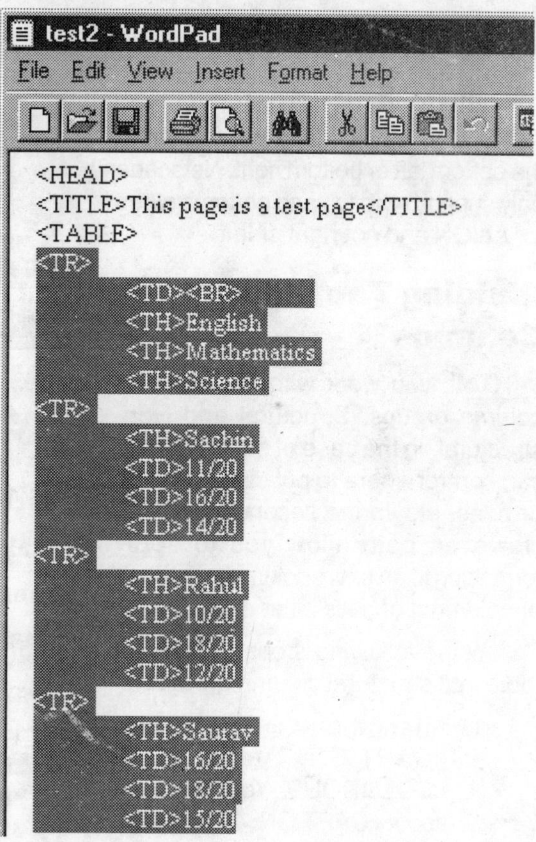

I know it is very difficult to understand this without the example. Follow the example given on the next page and then try to follow these steps to understand it better. Also don't forget to see the result in the browser.

Adding a caption

We have created a table but have not named it. For this we need a tag called CAPTION which lets us attach a descriptive title to our table.

Follow the following steps to create a caption

1. Directly after the initial <TABLE> tag, but not inside any row or cell tags, type <CAPTION.
2. If desired, type ALIGN=direction, where direction is either top, bottom, left, or right.
3. Type the final >.
4. Type the caption for the table which is taken as the heading of the table.
5. End with typing </CAPTION>.

Let us put one in our example and see the result shown below. The HTML text is shown on the next page.

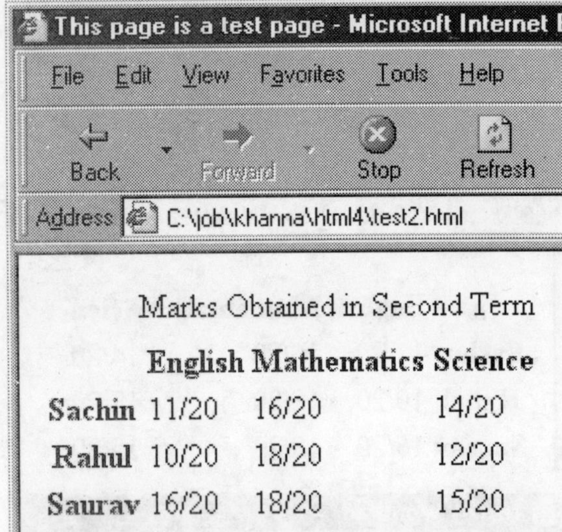

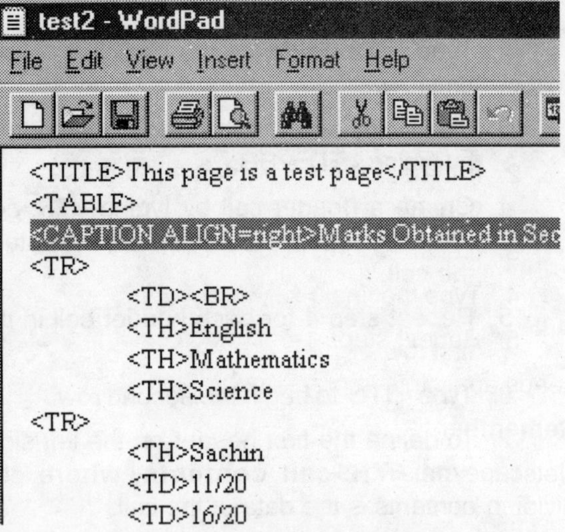

Remember

There is no way to place the caption at the bottom left or bottom right. Netscape only supports the top and bottom values for ALIGN and not right or left.

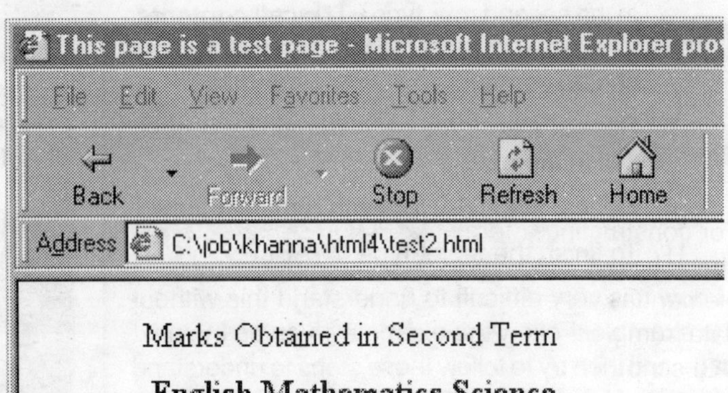

Dividing Table into Columns

In HTML you work with two kinds of column groups: Structural and Non-structural. In the case of structural you can control where to put dividing lines, or rules, etc. In the second you can't. However, both allow you to apply formatting to an entire column (or groups of columns) of cells all at once.

Follow the following steps to divide a table into structural column groups:

1. After the <TABLE> and <CAPTION> tags, type <COLGROUP. You will read about <CAPTION>, a little later.
2. If the column group has more than one column, type SPAN=n, where *n* is the number of columns in the group.
3. If desired, define the attributes for the column group.

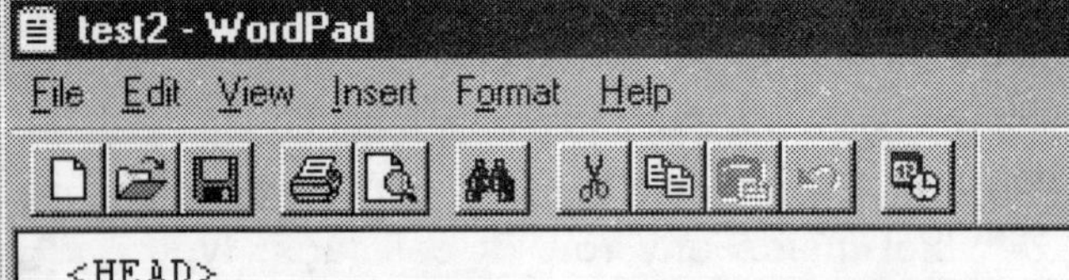

4. Type the final >.
5. Repeat steps 1-4 for each column group that you wish to define.

Follow the following steps to divide a table into non-structural column groups:

1. After the <TABLE> and <CAPTION> tags, type <COL.
2. If the column group has more than one column, type SPAN=n, where *n* is the number of columns in the group.
3. If desired, define the attributes for the column group.
4. Type the final >.
5. Repeat steps 1-4 for each column group that you wish to define.

Let us apply this to our example and see the result.

Remember

Netscape does not support any one of these. Use COLGROUP when you want to determine where dividing lines (rules) should go. Use COL for everything but deciding where dividing lines go. If the column group contains only one column, you don't need to use the SPAN attribute. Its default is 1.

Working with Forms

Forms are used to get information from the visitors of your site. There are two basic parts of a form: the structure or shell, that consists of fields, labels, and buttons that the visitor sees on a page and hopefully fills out, and the processing script that takes that information and converts it into a format that you can read or tally.

For constructing a form's shell you can create text boxes, special password boxes, radio buttons, check boxes, drop-down menus, larger text areas, and even clickable images. You will give each element a name that will serve as a label to identify the data once it is processed. Processing the data from a form is a bit more complicated. Before I go any further let me tell you something about the CGI script first.

CGI Script

CGI, which stands for Common Gateway Interface, is simply a standardized way for sending information between the server and the script. So, to resume, a CGI script is a program (usually written in a programming language called Perl) that communicates with the server in a standard (CGI-like) way. Each element on your form will have a name and a value associated with it. The name identifies the data that is being sent. It might be something like visitor-name. The value is the data; and can either come from you, the Web page designer, or from the visitor who types it in a field.

When a visitor clicks the submit button or an active image, the name-value pair of each form element is sent to the server. The CGI script takes all the name-value pairs and separates them out into something a real human can read and understand.

There are two ways to send information to the server: GET and POST.

The GET method appends the name-value pairs to the end of the URL. This information is then passed to the QUERY_STRING variable. The CGI script must then breaks down the QUERY_STRING environment variable to analyze the incoming data. The principal disadvantage of the GET method is that the amount of data you can retrieve is limited. Its chief advantage is that you can create a link that accesses a CGI script.

The POST method sends a data file with the name-value pairs to the server's standard input, together with the Content Type and the Content Length in bytes. The CGI script then takes the data from standard input and analyzes it. The principal advantage of this method is that it has no size limit.

If you want to make a link to a CGI script, use GET. If you're worried that the data will be truncated, use POST. The most important thing is that the CGI script that you use knows how to parse information gathered with the method that you choose.

Security

You must know that GGI scripts can leave your server wide open to invaders. That's one reason many ISPs do not allow their users to use CGI scripts. If your ISP does allow you to run CGI-scripts, you should still read up on security issues. If your ISP okays your use of CGI scripts, your next step is to get your hands on one. If you are a programmer, you can write your own. If not, you can find scripts on the Web and adapt them for your own use.

Perl is the most common language used for CGI scripts, partly because it's easily ported from one platform to another, partly because it's great for massaging data into understandable information. You can use other programming languages, like C++, Visual Basic, or even AppleScript to create CGI scripts. One extra nice thing about Perl programmers is that they like to share. You can find tons of ready-to-use CGI scripts written in Perl all over the Web.

Getting a Script

Whether you've written your own script or are using someone else's, you'll have to do a couple of things to get it ready for use with your form.

If you're using a Perl script that you've downloaded from someone else, you'll have to open it up and see what variables and path names it uses. You will have to change these to reflect your particular situation.

The next step is to copy the script to the server, generally with an FTP program like Fetch or WS-FTP. Some servers require that all CGI scripts be located in the cgi-bin directory. Others provide a personal cgi-bin directory for each user. Still others let you store CGI scripts wherever you want, as long as you add a particular extension to them for identification. You'll have to ask your ISP. If your page is on a Unix server, you will have to use chmod to make the CGI script accessible and executable.

So you've got your CGI script on your server, ready to go. The only thing left is to add it to your form, so lets do it.

Creating a form

Form has 3 important parts:

FORM tag, which includes the URL of the CGI script that will process the form;

Form elements, like fields and menus;

Submit button, which sends the data to the CGI script on the server.

Follow the following steps to create a form:

1. Type <FORM.
2. Type METHOD=method, where method is either GET or POST.
3. Type ACTION="script.url"> where script.url is the location on the server of the CGI script that will run when the form is submitted.

4. Create the form's contents.
5. Type </FORM> to complete the form.

Let us use this information to create a form and fill the information as shown in the HTML text shown below and see it in the browser.

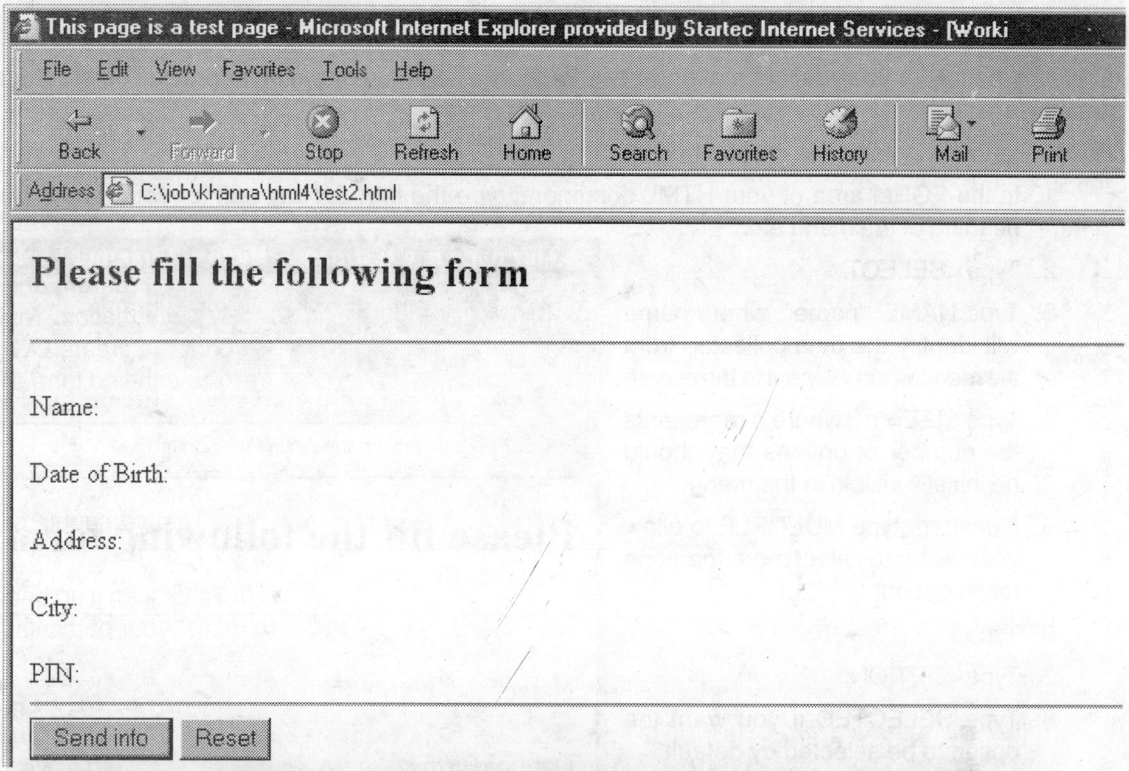

This page is a test page - Microsoft Internet Explorer provided by Startec Internet Services - [Worki
File Edit View Favorites Tools Help
Back Forward Stop Refresh Home Search Favorites History Mail Print
Address C:\job\khanna\html4\test2.html
Please fill the following form
Name:
Date of Birth:
Address:
City:
PIN:
Send info
Reset

test2 - WordPad
File Edit View Insert Format Help

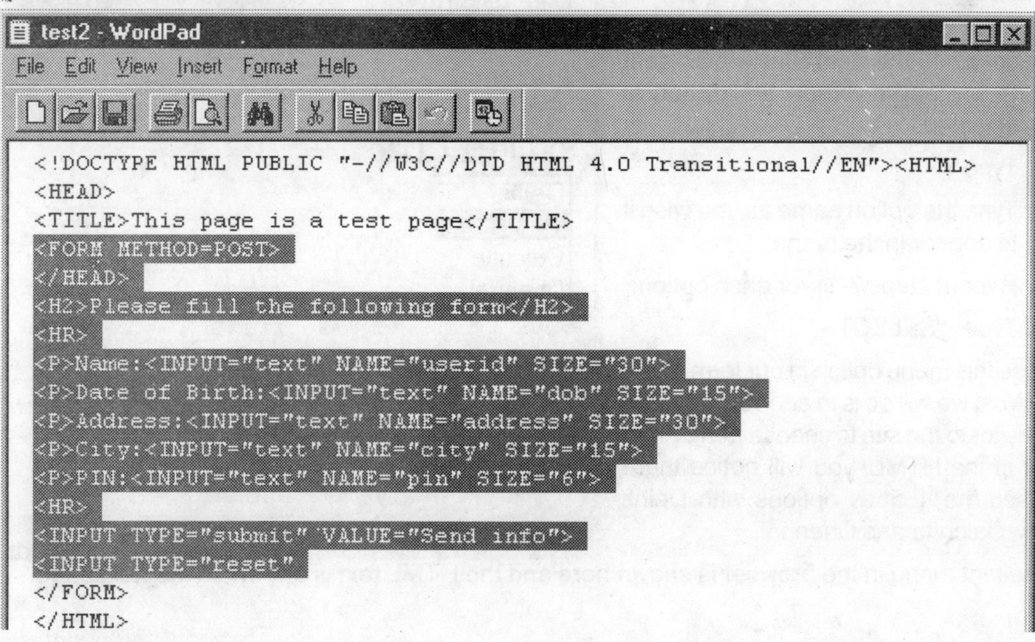

```
<!DOCTYPE HTML PUBLIC "-//W3C//DTD HTML 4.0 Transitional//EN"><HTML>
<HEAD>
<TITLE>This page is a test page</TITLE>
<FORM METHOD=POST>
</HEAD>
<H2>Please fill the following form</H2>
<HR>
<P>Name:<INPUT="text" NAME="userid" SIZE="30">
<P>Date of Birth:<INPUT="text" NAME="dob" SIZE="15">
<P>Address:<INPUT="text" NAME="address" SIZE="30">
<P>City:<INPUT="text" NAME="city" SIZE="15">
<P>PIN:<INPUT="text" NAME="pin" SIZE="6">
<HR>
<INPUT TYPE="submit" VALUE="Send info">
<INPUT TYPE="reset"
</FORM>
</HTML>
```

Remember

In order for your visitor to send you the information on the form, you'll need either a submit button (if your form contains fields, buttons, and other elements that your visitors will fill in) or an active image. You can use tables to layout your form elements more precisely.

Creating menus

Creating menus for your visitors makes it easy for them to enter information or provide criteria for a search.

Follow the following steps to create menus:

1. In the FORM area of your HTML document, type the introductory text for your menu, like heading of the menu etc.
2. Type <SELECT.
3. Type NAME="name", where name will identify the data collected from the menu when it is sent to the server.
4. Type SIZE="n", where n represents the number of options that should be initially visible in the menu.
5. If desired, type MULTIPLE to allow your visitor to select more than one menu option.
6. Type >
7. Type <OPTION.
8. Type SELECTED if you want the option to be selected by default.
9. Type VALUE="value", where value specifies the data that will be sent to the server when the option is selected.
10. Type >.
11. Type the option name as you wish it to appear in the menu.
12. Repeat steps 7-11 for each option.
13. Type </SELECT>

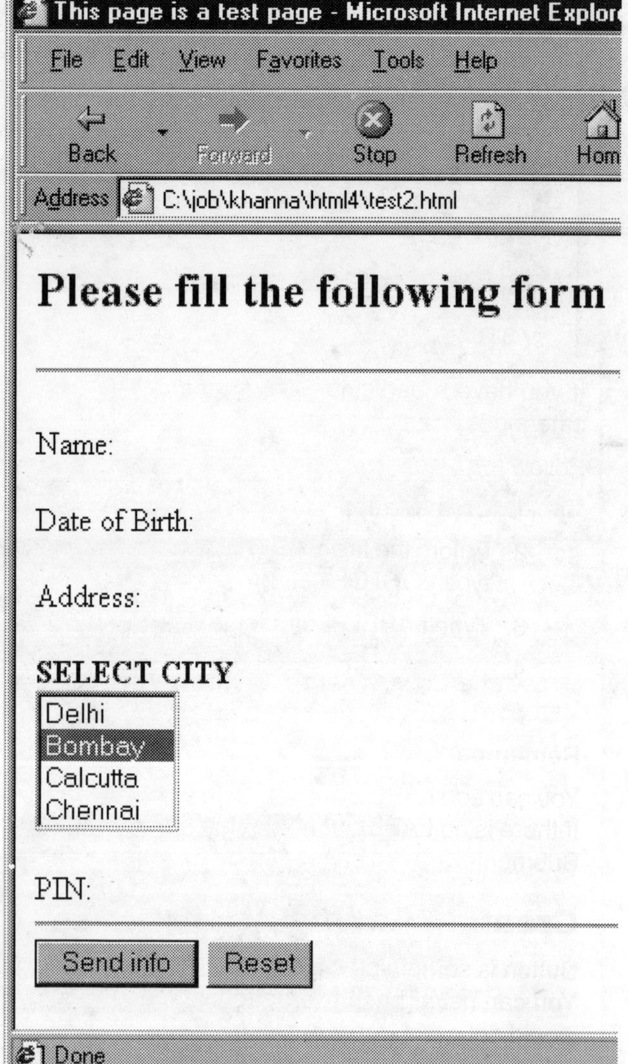

Let us use this menu option in our form made earlier. What we will do is to add various cities for the visitor to the site to choose from. From the text of the HTML, you will notice that I have used the 4 cities options with Delhi, Bombay, Calcutta and Chennai.

The resultant menu in the browser is shown here and the HTML text is shown on the next page.

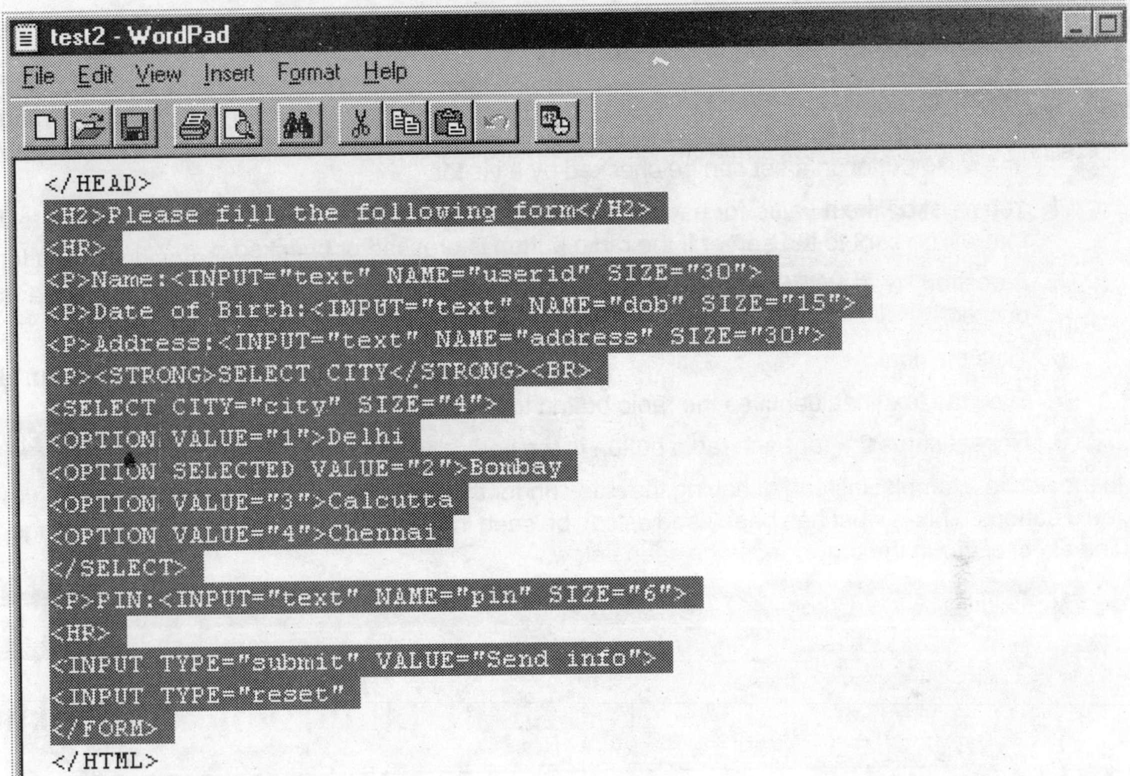


```
</HEAD>
<H2>Please fill the following form</H2>
<HR>
<P>Name:<INPUT="text" NAME="userid" SIZE="30">
<P>Date of Birth:<INPUT="text" NAME="dob" SIZE="15">
<P>Address:<INPUT="text" NAME="address" SIZE="30">
<P><STRONG>SELECT CITY</STRONG><BR>
<SELECT CITY="city" SIZE="4">
<OPTION VALUE="1">Delhi
<OPTION SELECTED VALUE="2">Bombay
<OPTION VALUE="3">Calcutta
<OPTION VALUE="4">Chennai
</SELECT>
<P>PIN:<INPUT="text" NAME="pin" SIZE="6">
<HR>
<INPUT TYPE="submit" VALUE="Send info">
<INPUT TYPE="reset"
</FORM>
</HTML>
```

If you have a particularly large menu with many options, you may want to group the options into categories and place them in submenus.

Follow the following steps to create a menu with submenus:

1. Create a menu as described above.
2. Before the first <OPTION> tag in the first group that you wish to place together in a submenu, type <OPTGROUP.
3. Type LABEL="submenutitle">, where submenutitle is the header for the submenu.
4. After the last <OPTION> tag in the group, type </OPTGROUP>.
5. Repeat steps 2-4 for each submenu.

Remember

You can add LABEL="option_name" to the OPTION tag to specify what the menu option should say. If there is no LABEL attribute, the browser automatically uses the text that follows the OPTION tag. Submenus are not supported by Explorer or Netscape.

Creating Radio Buttons

Button is something which you press to get result as you do on most of the push button machines. You can never push two buttons at once. Radio buttons on forms work the same way, except you would not get a Pepsi or Coke.

Follow the following steps to create a radio button:

1. In the FORM area, type the introductory text for your radio buttons, which may be like **Select one of the following options**.
2. Type <INPUT TYPE="radio".
3. Type NAME="radioset", where radioset identifies each radio button in a particular set. Only one radio button in a set can be checked by a visitor.
4. You must define a value for each radio button. Type VALUE="data", where data is the text that will be sent to the server if the radio button is pressed or checked.
5. If desired, type PRESSED to make the radio button active by default when the page is opened.
6. Type the final >.
7. Type the text that identifies the radio button to the visitor.
8. Repeat steps 2-7 far each radio button in the set.

In the above example, instead of having the cities under the menu option, we can have them on the radio buttons. This is what has been used as can be seen in the HTML text shown on the next page. The effect of this in the browser can be seen below.

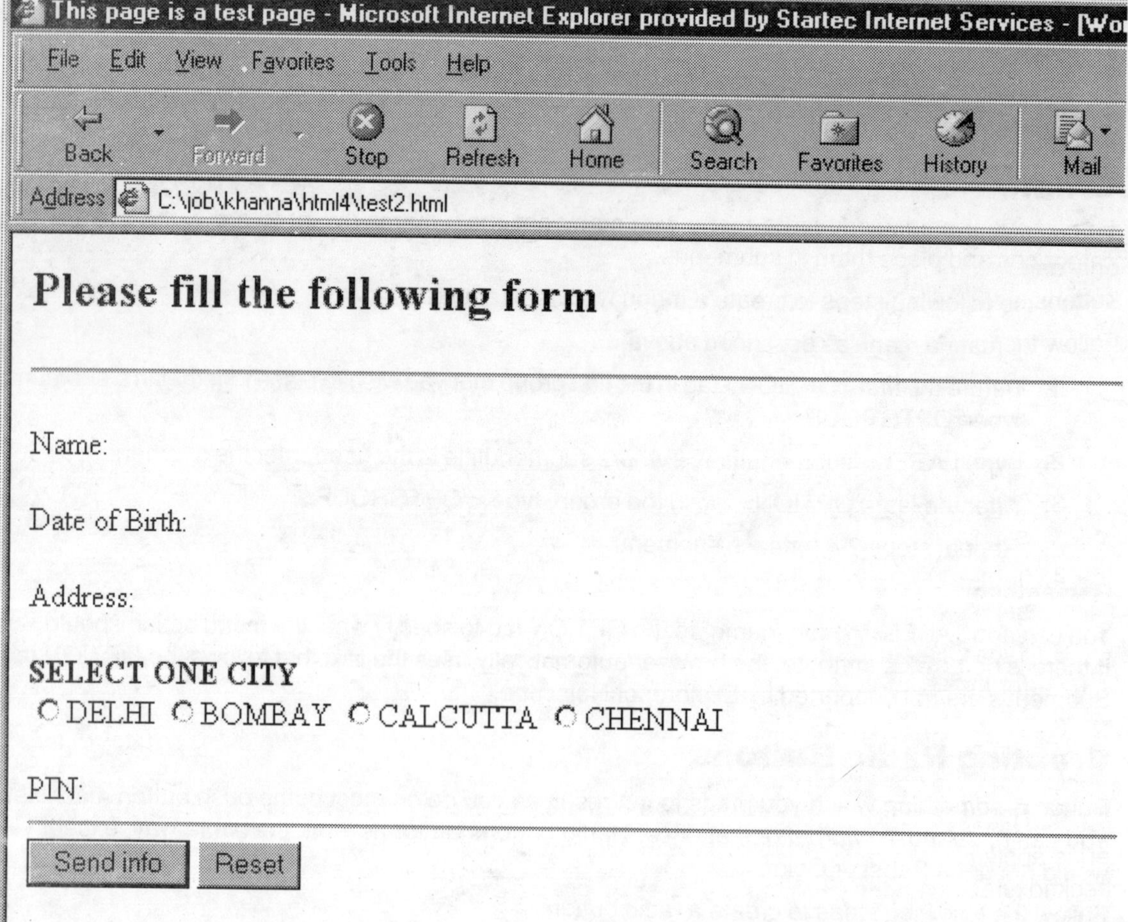

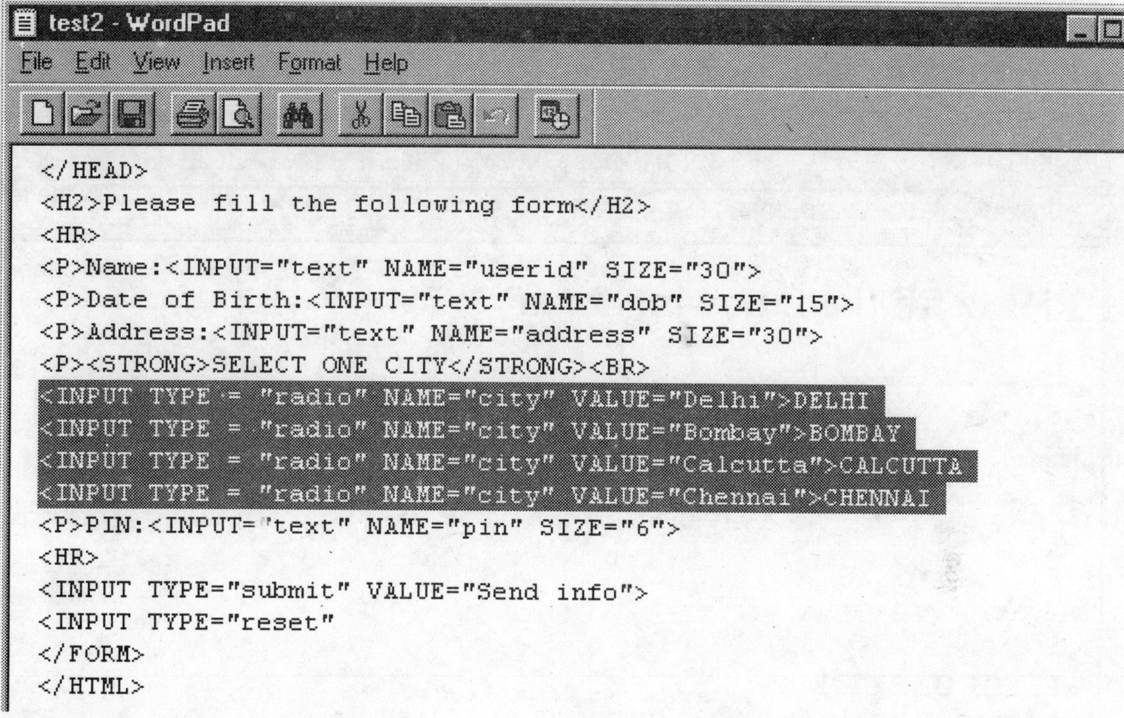


```
</HEAD>
<H2>Please fill the following form</H2>
<HR>
<P>Name:<INPUT="text" NAME="userid" SIZE="30">
<P>Date of Birth:<INPUT="text" NAME="dob" SIZE="15">
<P>Address:<INPUT="text" NAME="address" SIZE="30">
<P><STRONG>SELECT ONE CITY</STRONG><BR>
<INPUT TYPE = "radio" NAME="city" VALUE="Delhi">DELHI
<INPUT TYPE = "radio" NAME="city" VALUE="Bombay">BOMBAY
<INPUT TYPE = "radio" NAME="city" VALUE="Calcutta">CALCUTTA
<INPUT TYPE = "radio" NAME="city" VALUE="Chennai">CHENNAI
<P>PIN:<INPUT="text" NAME="pin" SIZE="6">
<HR>
<INPUT TYPE="submit" VALUE="Send info">
<INPUT TYPE="reset"
</FORM>
</HTML>
```

Creating Check Boxes

The main difference between a check box and a radio button is that while radio buttons can accept only one answer per set, a visitor can check as many check boxes in a set as they like. Like radio buttons, check boxes are linked by the value of the NAME attribute.

Follow the following steps to create check boxes:

1. In the FORM area of your HTML document, type the introductory text like **Select one or more of the following** for your check boxes.
2. Type <INPUT TYPE="checkbox". where checkbox is a single word.
3. Type NAME="title", where title is the word that identifies all of the check boxes in a particular set.
4. Type VALUE= "value" to define a value for each check box. The value will be sent to the server if the check box is checked.
5. Type CHECKED to make the check box checked by default when the page is opened. You (or the visitor) may check as many check boxes as desired.
6. Type > to complete the check box.
7. Type the text that identifies the check box to the user.
8. Repeat steps 2-7 for each check box in the set.

Let us apply this too on our example and see the result. Instead of radio in this case we will have checkbox, rest everything will remain the same. See the result on the next page and the HTML text also.

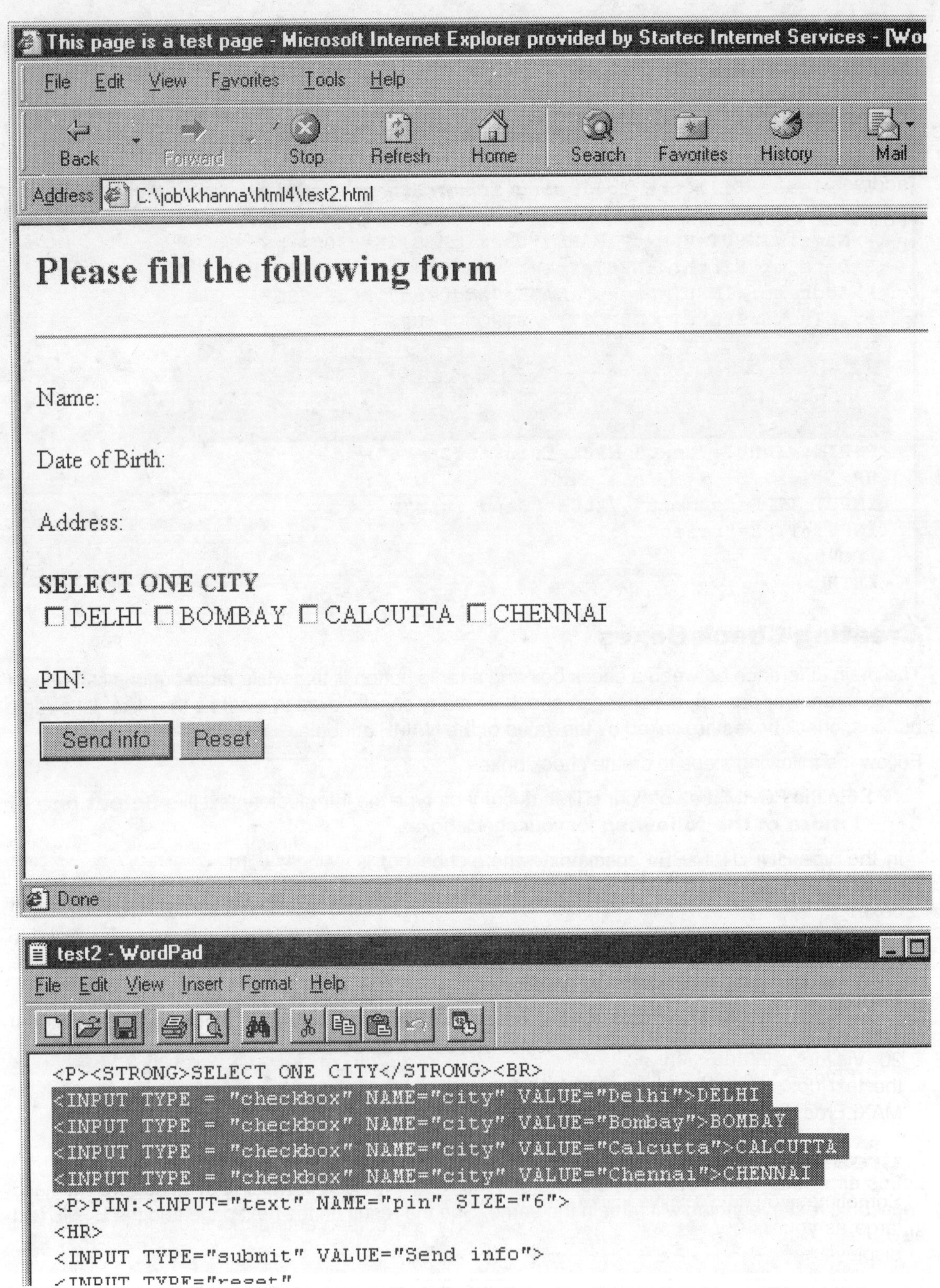
This page is a test page - Microsoft Internet Explorer provided by Startec Internet Services - [Wo
File Edit View Favorites Tools Help
Back Forward Stop Refresh Home Search Favorites History Mail
Address C:\job\khanna\html4\test2.html
Please fill the following form
Name:
Date of Birth:
Address:
SELECT ONE CITY
DELHI BOMBAY CALCUTTA CHENNAI
PIN:
Send info
Reset
Done
test2 - WordPad
File Edit View Insert Format Help
<P><STRONG>SELECT ONE CITY</STRONG>
<INPUT TYPE = "checkbox" NAME="city" VALUE="Delhi">DELHI
<INPUT TYPE = "checkbox" NAME="city" VALUE="Bombay">BOMBAY
<INPUT TYPE = "checkbox" NAME="city" VALUE="Calcutta">CALCUTTA
<INPUT TYPE = "checkbox" NAME="city" VALUE="Chennai">CHENNAI
<P>PIN:<INPUT="text" NAME="pin" SIZE="6">
<HR>
<INPUT TYPE="submit" VALUE="Send info">

Remember

You can create a separate keyboard shortcut for each check box in a set.

Creating Text Boxes

Text boxes are like forms only where you provide a box for the visitor to fill in the data like name, address, etc.

Follow the following steps to create a text box:

1. Inside the FORM area of your HTML document, type the title of the text box.
2. Type <INPUT TYPE="text".
3. Type NAME="title", where title is the word that identifies the data entered in the text box when it is collected by the server.
4. If desired, define the size of the box on your form by typing SIZE="n", replacing n with the desired width of the box, measured in characters.
5. If you want you can even set the maximum number of characters that can be entered in the box by typing MAXLENGTH="n", where n is the desired maximum length in characters.
6. Finish the text box by typing a final >.

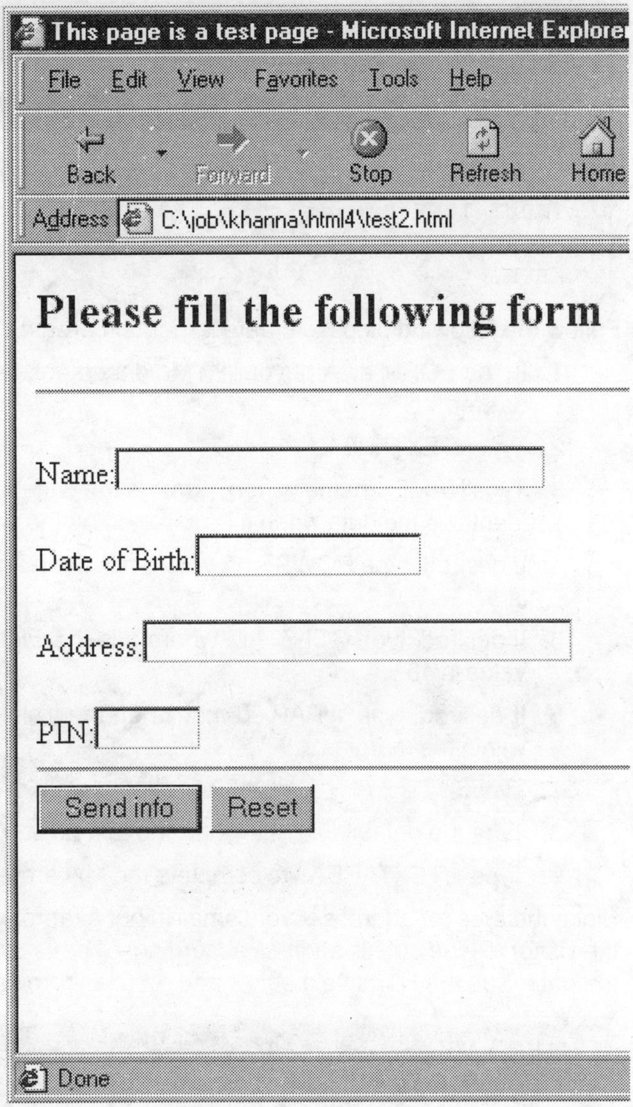

In the same example, now replace each option with INPUT TYPE to have the text boxes to be filled in by the visitor to the site. See the required HTML text below and its effect in the browser on the next page.

Remember

The default value for the SIZE attribute is 20. Visitors can add more text than fits in the text box, up to the value defined for MAXLENGTH.

Creating Larger Text Areas

Sometimes you have to give visitor a lot of room for typing in the information. Text areas may be as large as your page, and will expand as needed if the person enters more text than can fit in the display area.

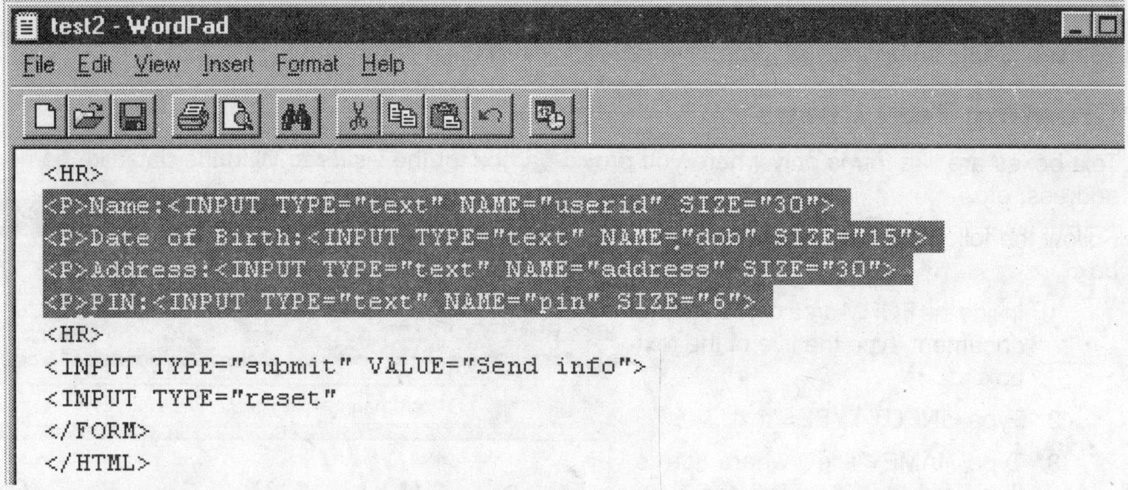

test2 - WordPad

File Edit View Insert Format Help

```
<HR>
<P>Name:<INPUT TYPE="text" NAME="userid" SIZE="30">
<P>Date of Birth:<INPUT TYPE="text" NAME="dob" SIZE="15">
<P>Address:<INPUT TYPE="text" NAME="address" SIZE="30">
<P>PIN:<INPUT TYPE="text" NAME="pin" SIZE="6">
<HR>
<INPUT TYPE="submit" VALUE="Send info">
<INPUT TYPE="reset"
</FORM>
</HTML>
```

Follow the following steps to create larger text areas:

1. In the FORM area of your HTML document, type the introductory text for your text area, if desired.
2. Type <TEXTAREA.
3. Type NAME="name" where name is the variable name you give to your text area. The name identifies the data when it is collected by the server.
4. If desired, type ROWS="n", where n is the height of the text area in rows. The default value is 4.
5. If desired, type COLS="n", where n is the width of the text area in characters. The default value is 40.
6. If desired, type WRAP so that when the visitor types the lines are automatically wrapped within the margins.
7. Type >.
8. Type the default text, if any, for the text area. You may not add any HTML coding here.
9. Type </TEXTAREA> to complete the text area.

Since the area for address is very small in our example, so we will create a larger text area block for the visitor to type out his address accordingly. I have used the area of 4 rows and 60 cols for filling in the data. See the HTML text below and resultant browser result on the next page.

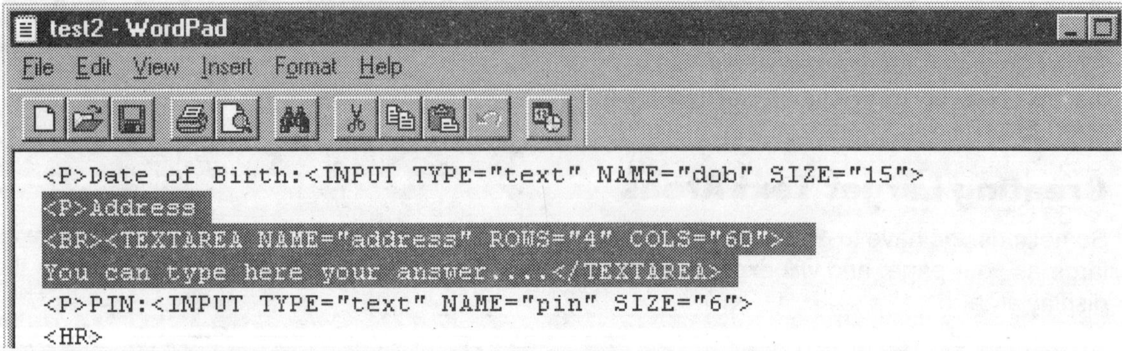

test2 - WordPad

File Edit View Insert Format Help

```
<P>Date of Birth:<INPUT TYPE="text" NAME="dob" SIZE="15">
<P>Address
<BR><TEXTAREA NAME="address" ROWS="4" COLS="60">
You can type here your answer....</TEXTAREA>
<P>PIN:<INPUT TYPE="text" NAME="pin" SIZE="6">
<HR>
```

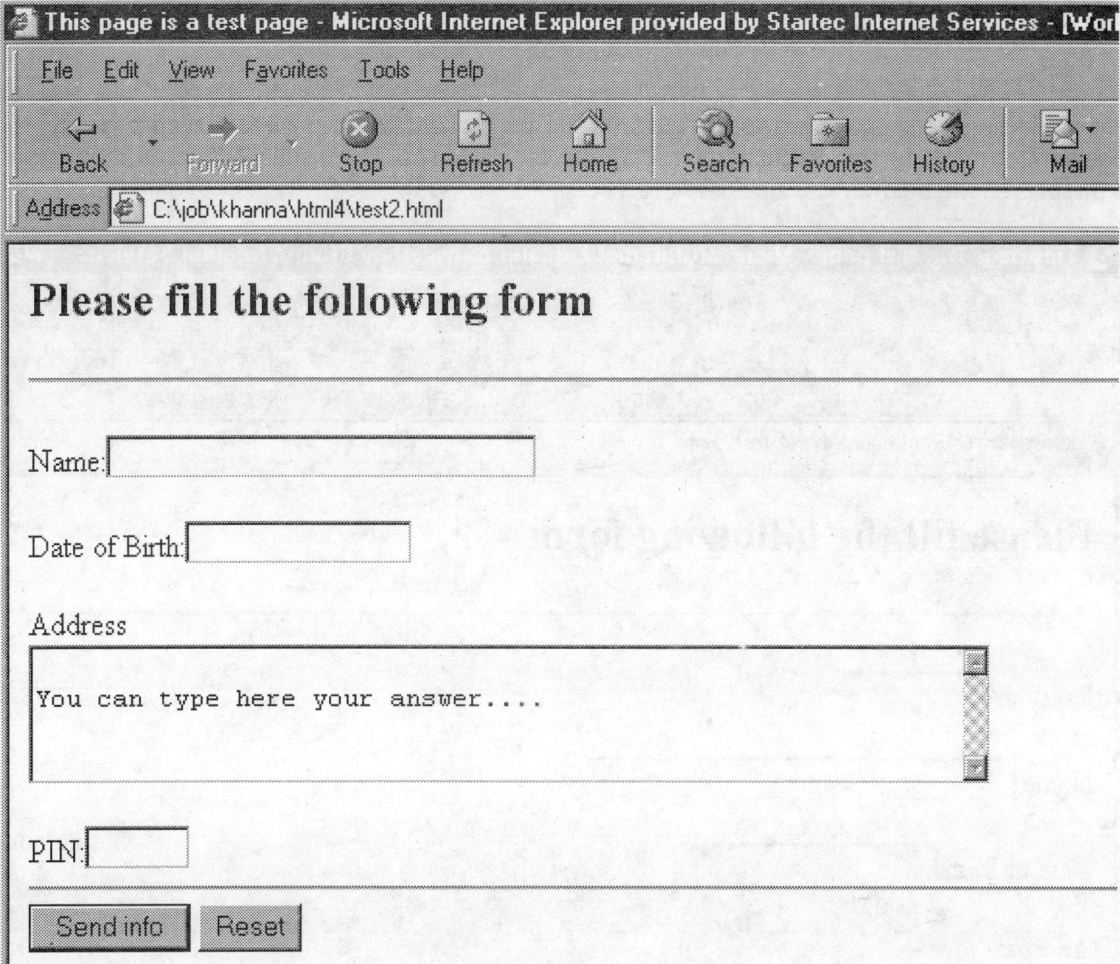

Remember

Visitors can enter up to 32,700 characters in a text area. Scroll bars will appear when necessary. Explorer automatically wraps text even without the nonstandard WRAP.

Creating Password Boxes

A password box is similar to a text box, but when the visitor types in it, the letters are hidden by bullets or asterisks.

Follow the following steps to create password entry boxes:

1. Inside the FORM area of your HTML document, type the title of the password box, for example, **Enter password**.
2. Type <INPUT TYPE="password".
3. Give the password box a name by typing NAME="title", where title is the word that identifies the data entered in the password box when it is collected by the server.
4. If desired, define the size of the box on your form by typing SIZE="n", replacing n with the desired width of the box, measured in characters.

5. If desired, define the maximum number of characters that can be entered in the box by typing MAXLENGTH="n", replacing n with the desired value.
6. Finish the text box by typing a final >.

We would also like to put a password in our form. So just add password in the form before the name. See how the HTML text would change, as shown on the next page. The resultant effect as seen in the browser is shown below.

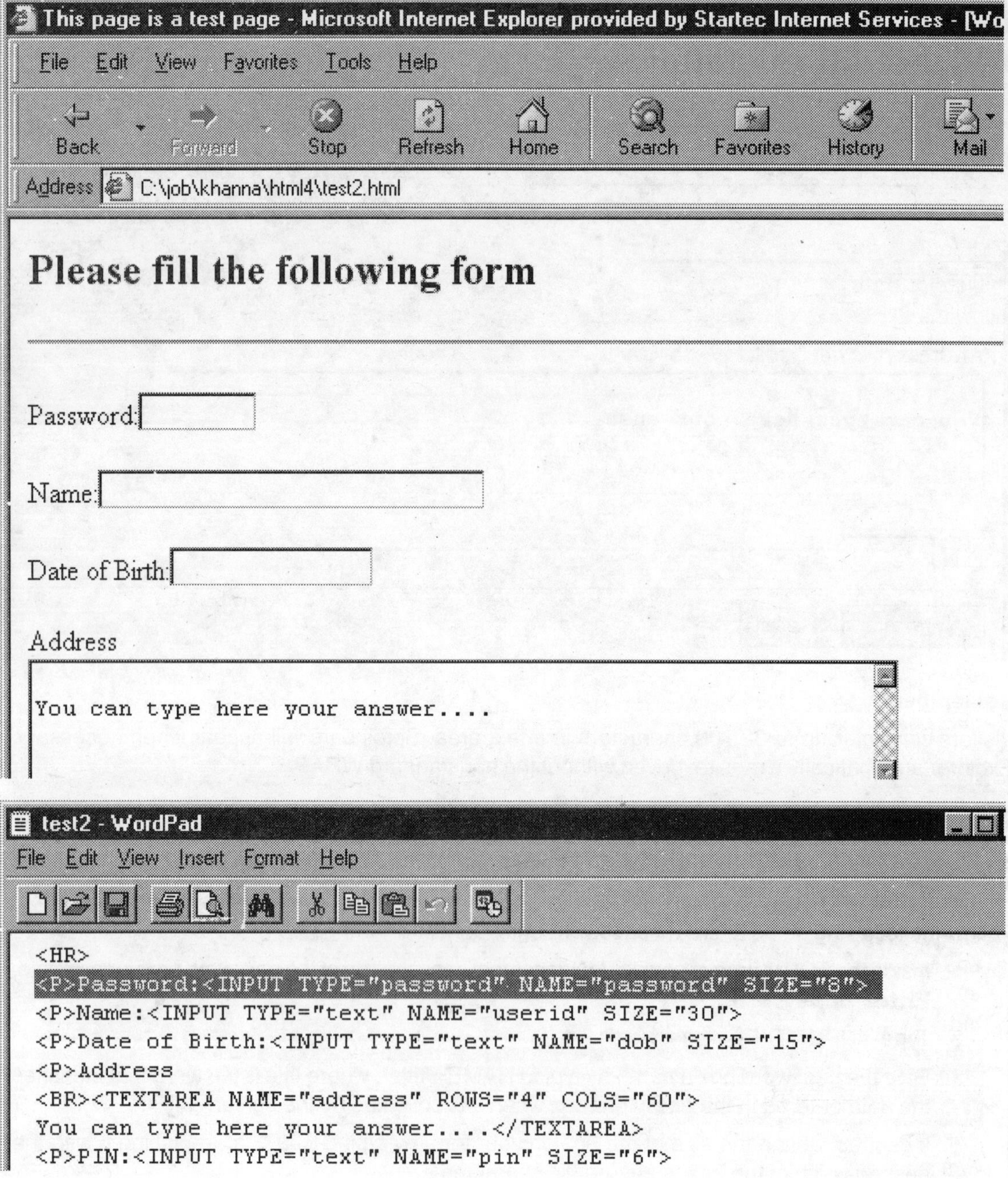

Remember

Since the data is not encrypted when the information is sent to the server, moderately experienced crackers can discover the password without much trouble.

Submit Button

This button is used to send the information to the server by the visitor. You can create a submit button simply without an image or style information, or you can use HTML tags to make a button so gorgeous that your visitors won't be able to resist pressing it.

Follow the following steps to create a submit button:

1. Type <INPUT TYPE="submit".
2. If desired, type VALUE="submit message" where submit message is the text that will appear in the button.
3. Type the final >.

HTML adds several new tags that let you create prettier submit buttons You can add an image, change the font, or even change the background colour. That'll get them to submit that form!

Follow the following steps to create a submit button with an image

1. Type <BUTTON NAME="submit" VALUE="submit" TYPE="submit".
2. If desired, type STYLE="font: 24pt Arial Black; background: yellow" to change the appearance of the button text.
3. Type >
4. Type the text, if any, that should appear on the left side of the image in the button.
5. Type <IMG SRC="image.gif" where image.gif is the name of the image that will appear on the button.
6. If desired, add any other image attributes.
7. Type > to complete the image.
8. Type the text, if any, that should appear on the right side of the image in the button
9. Type </BUTTON>

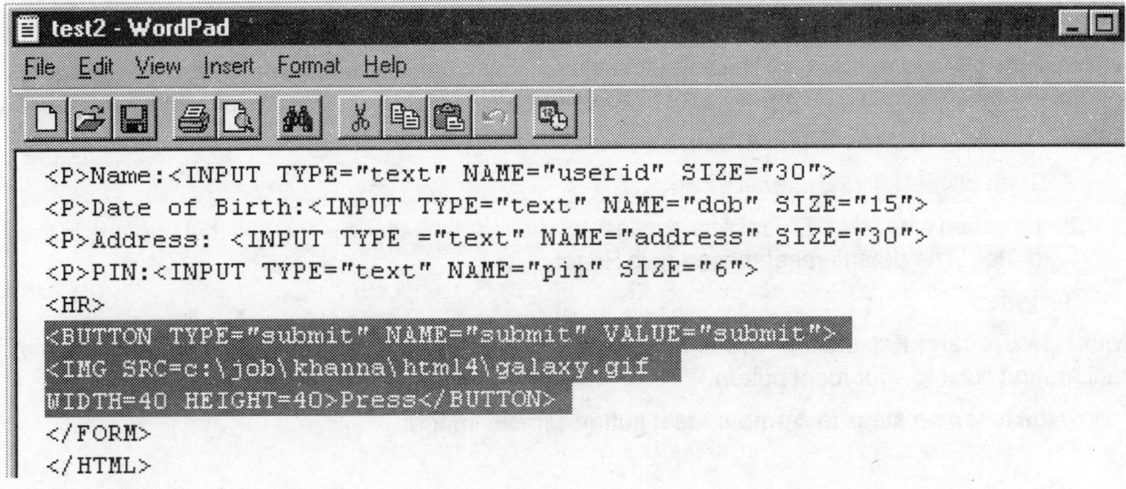

Let us put a button in our form too. I have used here the galaxy.gif for the image of the button and have asked visitor to press it to submit. See the text of the HTML and the its resultant effect in the browser below.

Remember

You can use the BUTTON tag to create a submit button without an image. Only Microsoft Internet Explorer supports the BUTTON tag.

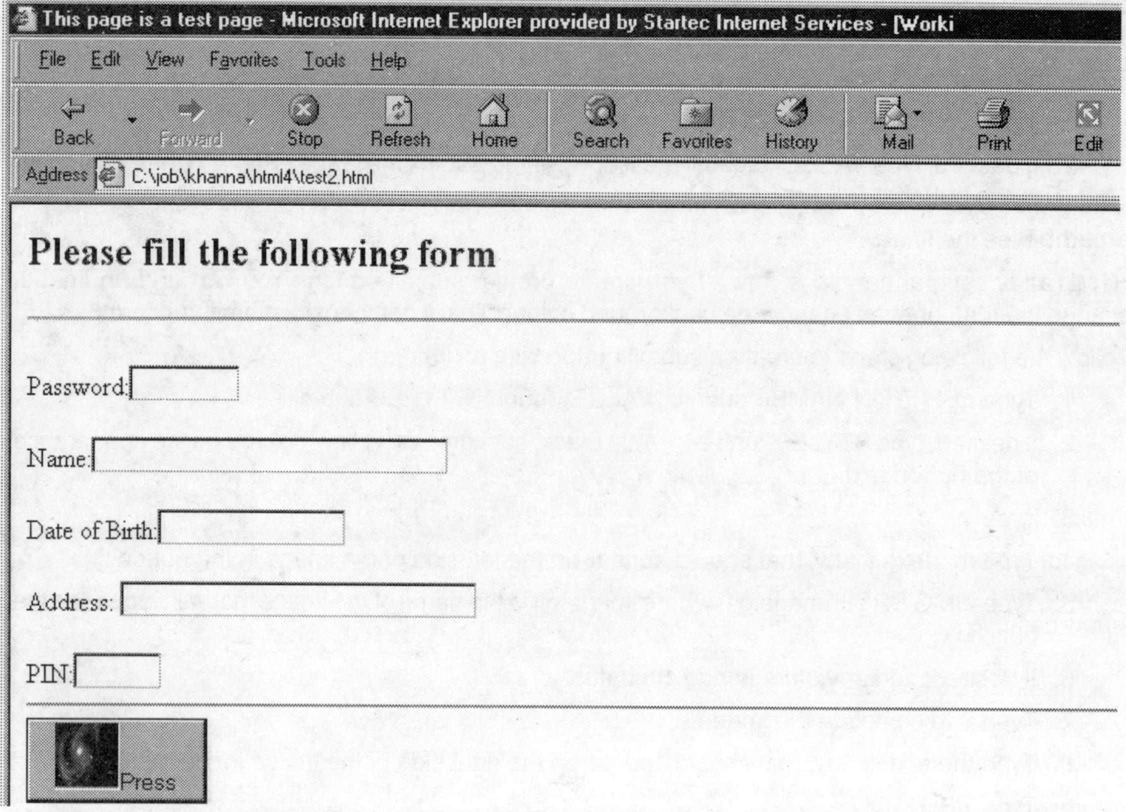

Resetting the Form

While filling up the form, it is very likely that the visitor may make a mistake which can of spelling, or second thought. You can give your visitors a reset button so that they can start over with a fresh form.

Follow the following steps to add a reset button:

1. Type <INPUT TYPE="reset".
2. If desired type VALUE= "reset message" where reset message is the text that appears in the button. The default reset message is Reset.
3. Type >.

With HTML 4 (and Explorer for the time being), you can add images, font choices, and even a background color to your reset button.

Follow the following steps to create a reset button with an image:

1. Type <BUTTON NAME="reset" VALUE="reset: TYPE="reset".
2. If desired, type STYLE="font: 24pt ArialBlack; background:yellow" to change the appearance of the button text.
3. Type >.
4. Type the text, if any, that should appear on the left side of the image in the button.
5. Type <IMG SRC="image.gif", where image.gifis the name of the image that will appear on the button.
6. If desired, add any other image attributes.
7. Type > to complete the image.
8. Type the text, if any, that should appear on the right side of the image in the button.
9. Type </BUTTON>.

Remember

You can also use the BUTTON tag to create a reset button without an image.

Working with Frames

Everybody wants to give more and more information on a web page. Well, the size of the page is fixed. So what you do. You divide the page into various frames and show the various topics on the same page at the same time. By dividing a page, called a frameset, into frames, you can allow the visitor to see more than one page at a time, without completely cluttering up their screen. Each frame contains its own Web page, and theoretically could be viewed independently in a separate window.

This is the beauty of frames. You can have a stationary banner frame across the top of the window that includes your company name and logo. Meanwhile, a dynamic frame on the left side of the window can include a table of contents.

Finally, the main frame of the window will be devoted to the contents frame, whose data changes each time your visitor clicks on a new topic in the table of contents.

Creating Frameset

A frameset is a door within window. Each door shows different information. You decide how many doors your window will have, what size each door will be, how its borders should look and if it should have scroll bars or not.

Follow the following steps to create a simple frameset:

1. Type <FRAMESET after the </HEAD> tag on the frameset page.
2. Type ROWS= "a where a is the height of the first row. The value may either be a percentage, an exact number of pixels, or completely variable with an asterisk: * depending on the size of the other rows.
3. Type, b where b is the height of the second row, again expressed as a %, an absolute value in pixels, or a variable with an asterisk *.
4. Repeat step 3 for each additional row.
5. Type "> to complete the row definition.
6. Type <FRAME to assign a URL and other attributes to the top row/pane.

7. Type NAME="name" where name is a word that identifies this particular frame's use, like banner; index, or contents.
8. Type SRC="content.html"> where content.html is the URL for the page that will be initially displayed in this frame when the visitor first navigates to this frameset.
9. Repeat steps 6-8 for each row you defined in steps 2-4.
10. Type </FRAMESET> to complete the frameset and the construction of your "window".
11. Create the pages that will be displayed initially in the frames, that is, those referenced by the SRC tag in step 8. This is the "landscape" behind the window.

In the example shown here, I have divided the page into 2 rows of 50 each. The above one has the information from the file called TEST1.HTML and the second one has the picture directly from the source. See both the HTML text and the its result in the browser on the next page.

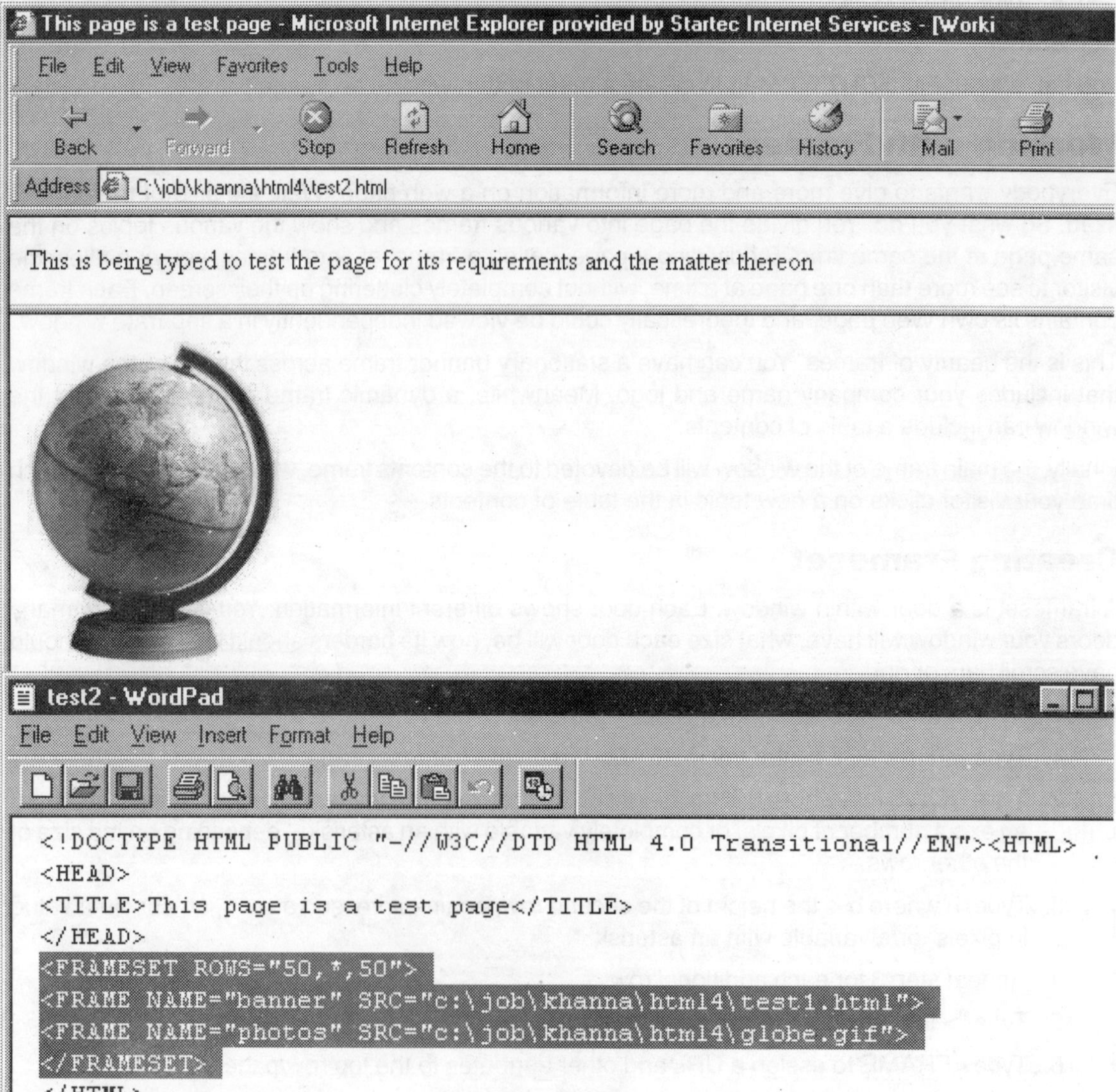

Remember

Asterisk (*) is used to allocate to a frame whatever leftover space there is available in the window. If the first two frames occupy 40 and 60 pixels respectively, and the window size is 250 pixels, the frame with the asterisk will occupy 150 pixels. You can use more than one asterisk at a time. The remaining space will be divided equally among the frames marked with an asterisk. The BODY tag is not used at all in frameset pages.

Creating Columns

You can divide the page into columns too instead of rows.

Follow the following steps to create frames in columns:

1. Type <FRAMESET after the </HEAD> tag in the frameset page.
2. Type COLS="a,b"> where a and b represent the width of the corresponding columns as percentage, number of pixels or variable.
3. Type <FRAME to define the leftmost frame/column.
4. Type NAME="name" where name is a word that identifies this particular frame's use, like banner, index, or contents.
5. Type SRC="content.html"> where content.html is the URL of the page that you want to be displayed in this frame when the visitor initially navigates to this frameset.
6. Repeat steps 3-5 for each frame/ column.
7. Type </FRAMESET> to close the frame.
8. See the result in the browser.

With everything else remaining the same, I have divided the page into 2 columns of 100 each. See the result in the browser here and the HTML text on the next page.

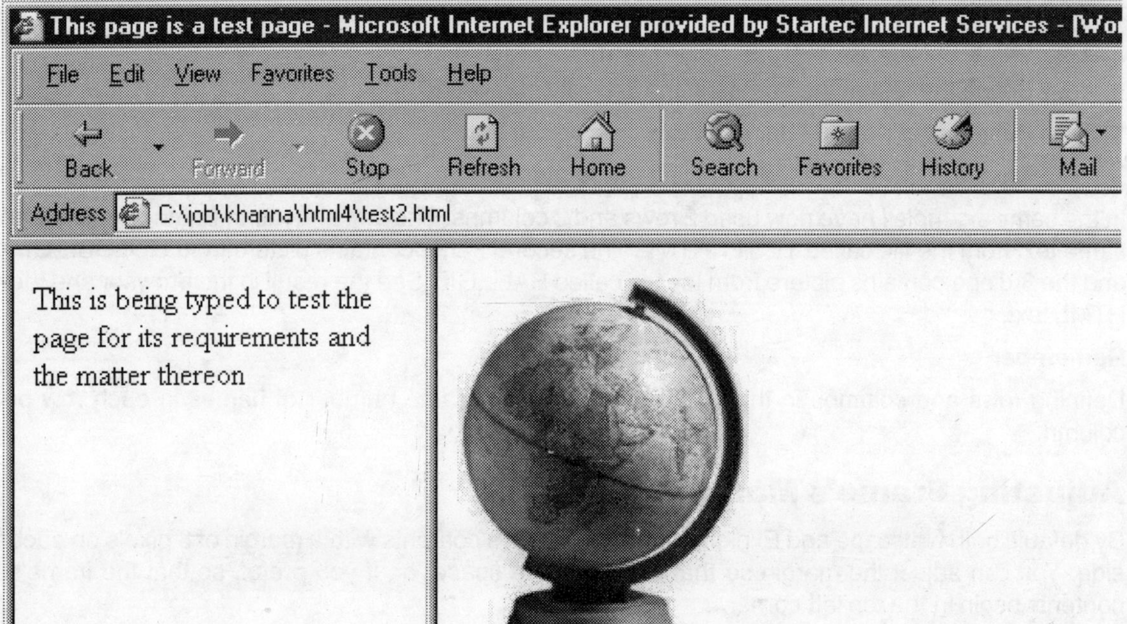

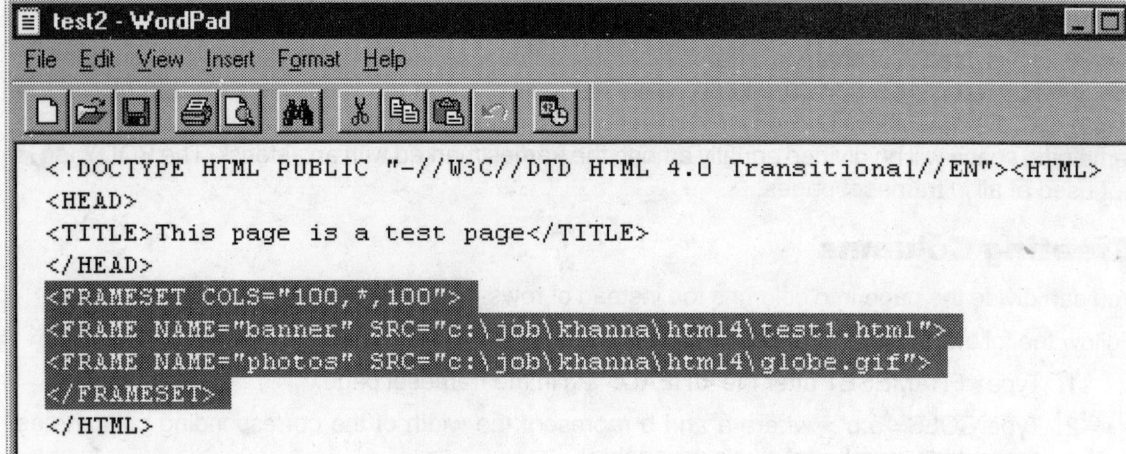

Creating Rows and Columns

So far I have talked about Rows and Columns. There maybe a time when you have to divide the page into both rows and columns. You can create both rows and columns in the same frameset to accommodate different kinds of information.

Follow the following steps to create a frameset with both rows and columns:

1. Type <FRAMESET to begin.
2. Type ROWS ="a, b" where a and b (and any others) represent the height of the corresponding rows.
3. Type COLS="x, y" where x and y (and any others) represent the width of the corresponding columns.
4. Type >.
5. Define the first frame in the first row by typing <FRAME NAME="name" SRC="initialurl.html">.
6. Define the rest of the frames in the first row from left to right.
7. Repeat steps 4 and 5 for each row, from top to bottom.
8. Type </FRAMESET > to complete the frameset.

In the same example I have now used 2 rows and 2 columns. The first and fourth frame contains the same text from the file called TEST1.HTML. The second frame contains picture from GOLOBE.GIF and the 3rd one contains picture from the file called BALL.GIF. See the result in the browser and the HTML text.

Remember

Defining rows and columns in this way limits you to the same number of frames in each row or column.

Adjusting Frame's Margins

By default, both Netscape and Explorer display a frame's contents with a margin of 8 pixels on each side. You can adjust the margin so that there is more space, or, if you prefer, so that the frame's contents begin in the top left corner.

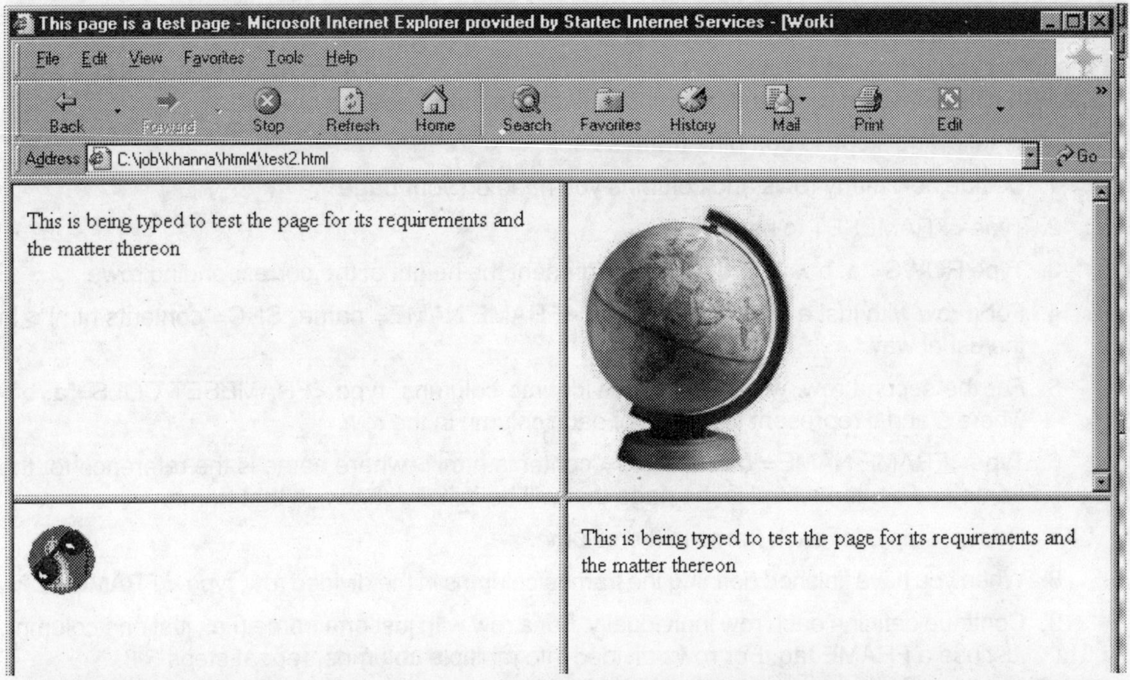

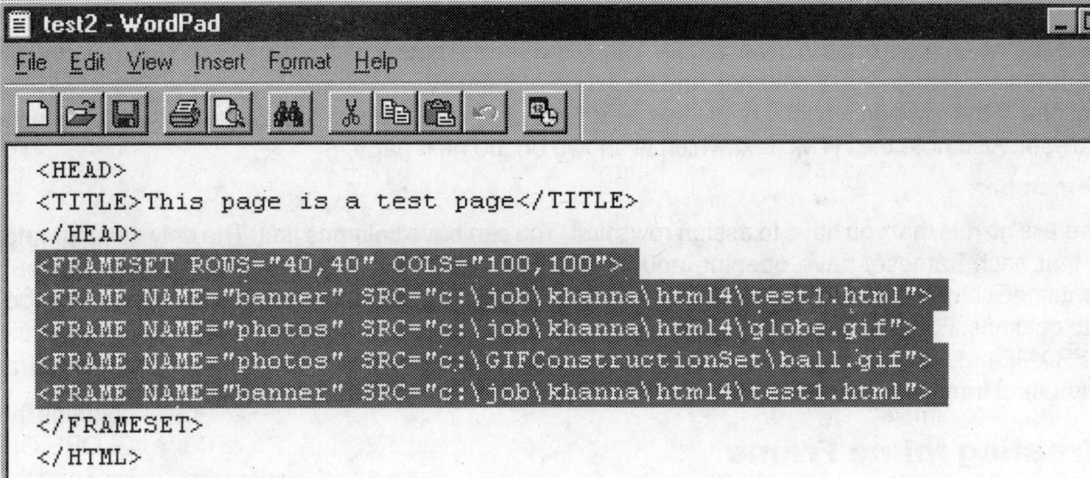

Follow the following steps to adjust a frame's margins:

1. In the desired FRAME tag, before the final > , type MARGINWIDTH=w where w is the desired amount of space, in pixels, between the left and right edges of the frame and the frame's contents.
2. Type MARGINHEIGHT =h where h is the desired amount of space, in pixels, between the top and bottom edges of the frame and the frame's contents.

Remember

The margin is transparent and thus always appears to be the same color as the background of the page displayed in the frame.

Combining Framesets

It is not that you are going to have an equal number of row and columns. Mostly you will see a web page with different set of rows and columns. Let us see how these are created.

Follow the following steps to combine framesets:

1. Decide how many rows and columns you need on your page.
2. Type <FRAMESET to begin.
3. Type ROWS="a, b"> where a and b represent the height of the corresponding rows.
4. For a row with just a single frame, type <FRAME NAME="name" SRC="contents.html"> in the usual way.
5. For the second row, which we will divide into columns, type <FRAMESET COLS="a, b"> where a and b represent the width of each column in the row.
6. Type <FRAME NAME="name" SRC="contents.html"> where name is the reference for the frame and contents.html is the page that will be initially shown in that frame.
7. Repeat step 6 for each frame in the column.
8. When you have finished defining the frames/columns in the divided row, type </FRAMESET>.
9. Continue defining each row individually. For a row with just one frame (i.e., just one column), just use a FRAME tag. For rows divided into multiple columns, repeat steps 5-8.
10. Type </FRAMESET> to complete the outer frameset.

Now see what I have done in my example. I have divided the page into 2 rows first and then divided the intermediate section into 2 columns and each one of them has a picture, i.e., a globe and a ball. The last one of the row again contains the text which is there in the first row too. See the result here in the browser. Also note the HTML text which is shown on the next page.

Remember

There is no rule that you have to assign rows first. You can have columns first. The only thing to watch is that each frameset have opening and closing <FRAMESET> tags, and that each frameset is contained completely with no overlapping. It is important to stress that not every row need be divided into columns. For rows with a single frame, i.e., that span the entire window from left to right, just use a FRAME tag. For rows divided into columns, use an inner frameset. If you want to create the same number of frames in each row and each column, you don't need to combine multiple framesets.

Creating Inline Frame

This is used to create an inline frame which is nothing but a frame which allows you to mix text, graphics, and a frame all on one page.

Follow the following steps to create an inline frame:

1. In the container page, type <IFRAME, SRC="frame.url", where frame.url is the page that should be initially displayed in the inline frame.
2. Type NAME="name", where name is a word that identifies this inline frame's use.
3. Type WIDTH=x HEIGHT=y where x and y represent the width and height respectively of the inline frame as a percentage or as an absolute value in pixels.
4. If desired, type ALIGN=LEFT or ALIGN=RIGHT to wrap the text that comes after the frame around the frame.

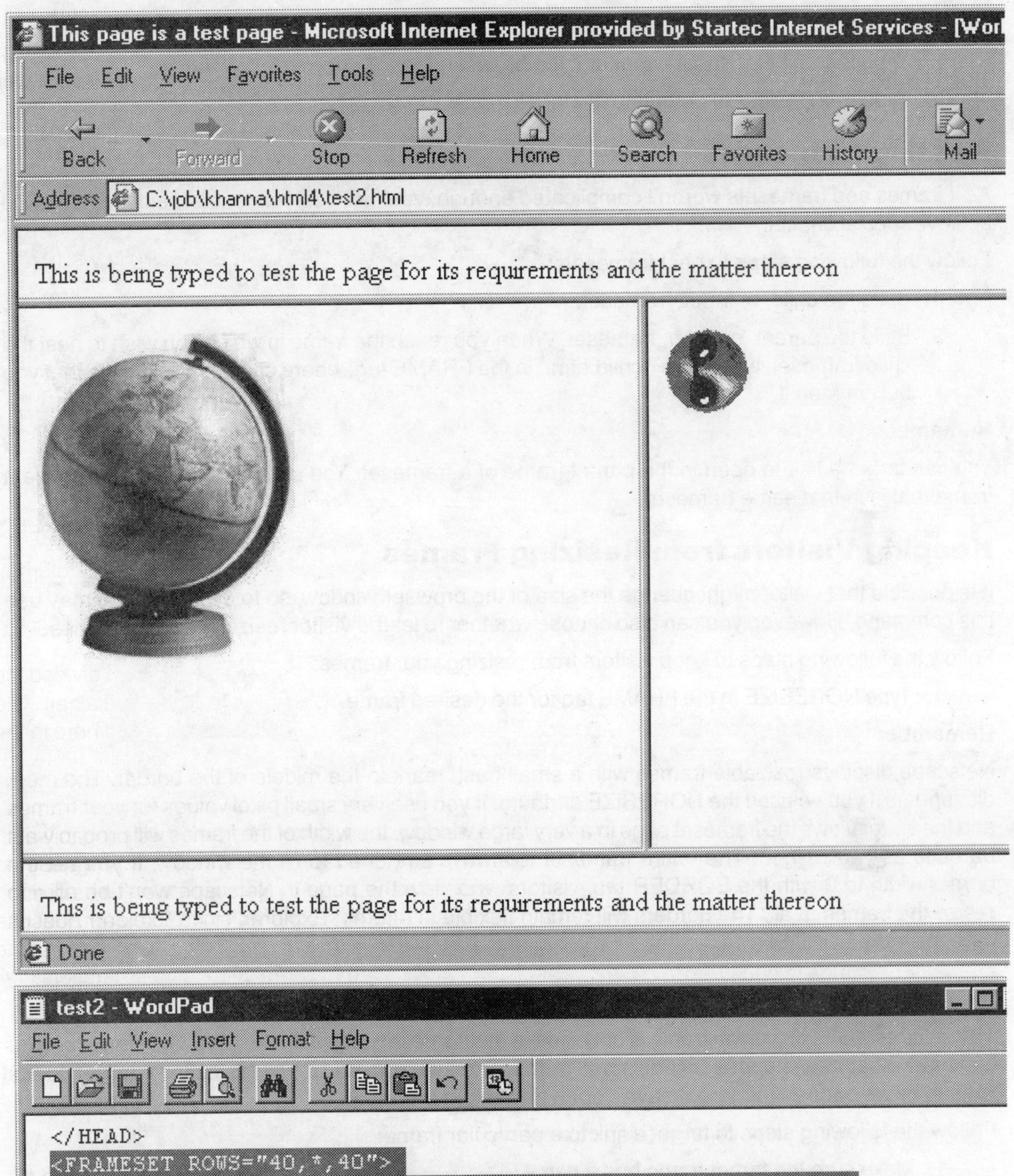
This page is a test page - Microsoft Internet Explorer provided by Startec Internet Services - [Wor
File Edit View Favorites Tools Help
Back Forward Stop Refresh Home Search Favorites History Mail
Address C:\job\khanna\html4\test2.html
This is being typed to test the page for its requirements and the matter thereon
This is being typed to test the page for its requirements and the matter thereon
Done
test2 - WordPad
File Edit View Insert Format Help
</HEAD>
<FRAMESET ROWS="40,*,40">
<FRAME NAME="banner" SRC="c:\job\khanna\html4\test1.html">
<FRAMESET COLS="100,*,100">
<FRAME NAME="photos" SRC="c:\job\khanna\html4\globe.gif">
<FRAME NAME="photos" SRC="c:\GIFConstructionSet\ball.gif">
</FRAMESET>
<FRAME NAME="banner" SRC="c:\job\khanna\html4\test1.html">
</FRAMESET>

5. Type >.
6. Type the text that should appear if the browser doesn't support inline frames.
7. Type </IFRAME> to complete the inline frame.

Nesting framesets

As if frames and framesets weren't complicated enough, you can nest framesets inside of frames to achieve special effects.

Follow the following steps to nest framesets:

1. Build the child, or inner, frameset.
2. Build the parent, or outer, frameset. When you reach the frame in which you wish to nest the child frameset, type SRC="child.html" in the FRAME tag, where child.html is the file that you built in step 1.

Remember

You can target a link to open in the parent frame of a frameset. You can't nest a frameset inside a frame that is in that same frameset.

Keeping Visitors from Resizing Frames

It is possible that visitor might change the size of the browser window, so to avoid this you may use this command. However, you can also choose whether to let the visitor resize individual frames.

Follow the following steps to keep visitors from resizing your frames:

1. Type NORESIZE in the FRAME tag for the desired frame.

Remember

Netscape displays resizable frames with a small hash mark in the middle of the border. The mark disappears if you've used the NORESIZE attribute. If you use very small pixel values for your frames and the visitor views the frameset page in a very large window, the width of the frames will probably not be quite as you wished. The entire frameset is always stretched to fill the window. If you set the border width to 0 with the BORDER tag, visitors who view the page in Netscape won't be able to resize the frames at all. The borders will remain flexible in Internet Explorer, since Explorer doesn't recognize the BORDER tag.

Targeting Links to Particular Frames

The initial contents of a given frame is specified in the frameset page with the SRC tag. However, you can have other pages appear in that same frame. The trick is to add a pointer, called a target, to the links to those pages. The target says, "open this link in frame xyz".

Follow the following steps to target a link to a particular frame:

1. Make sure the target frame has a name.
2. On the page where the link should appear, type <A HREF="page.html" where page.html is the file that should be displayed in the target frame.
3. Type TARGET="name" where name is the reference given to the target frame within the FRAME tag.
4. Add any other attributes as desired to the link and then type the final >.

Changing the Default Target

Coming back to links if a frame contains a link, the link will open up in that same frame, by default, unless you change the target earlier. You can change the default target for all of the links on a page by using the BASE tag,

Follow the following steps to change the default target:

1. In the HEAD section of the page that contains the links, type <BASE,
2. Type TARGET="name", where name is the word that identifies the frame or window in which you want the links to appear, by default,
3. Type > to complete the BASE tag,

Remember

You can override the target specified in the BASE tag by choosing a target in the link itself.

Revision

1. HTML is the language of Internet.
2. HTML tags are written within the less than (<) and greater than (>) signs.
3. Mostly all values should be enclosed in straight quotation marks.
4. There is no restriction on the number of spaces given by you.
5. HTML supports most of the special characters.
6. Any word processor can be used for creating a text file for HTML.
7. You can see your text in a browser.
8. Your text should have the html or htm extension.
9. The first line of the text defines the HTML.
10. Main body of the text is divided between <BODY> and </BODY>.
11. Headings in the text are divided between <HEAD> and </HEAD>.
12. You can have as many as 6 headings in the text.
13. You can have graphics too in your HTML files.
14. The tag for using graphics is IMG SRC.
15. You can change the weight and height of the image in HTML.
16. You can align your text with respect to the image.
17. You can add lines of various widths in your text.
18. You can wrap the text in between two images.
19. HTML support 3 type of lists: Unordered lists, Ordered lists and Definition lists.
20. Unordered lists can be preceded by one of several bullet styles: a closed circle, an open circle, or a square.
21. Ordered lists can be preceded by Arabic numerals, uppercase or lowercase Roman numerals, or uppercase or lowercase alphanumeric characters.
22. Definition lists require a start tag (<DL>) and end tage (</DL>) and two special elements: one for definition terms (the <DT> tag) and one for definitions (the <DD>) tag.
23. There are two other types of lists defined in HTML: Directory lists and Menus.
24. To create a table you use the tag <TABLE>.
25. You use COLGROUP to divide table in columns.
26. Horizontal rows are set using the tag <TR>.

27. Table headers are created using <TH>.
28. You can add a border too to the table.
29. You can add a background image to the table.
30. Using ALIGN you can align the text in the cells.
31. A form is used to get the feedback from the visitors to the site.
32. You can create a form using the tag <FORM>.
33. You can give various options using the tag <OPTION>.
34. Values can also be given to the Options.
35. Check boxes allow the visitor to click in them.
36. Text boxes are used to allow visitor to write text.
37. Large text can be left using the tag TEXTAREA.
38. Submit button can also be set by using the INPUT tag.
39. You can divide your web page into various frames so that you can add more information using these frames.
40. A frameset is a door within window. Each door shows different information.
41. You can divide the page into columns instead of rows.
42. You can create both rows and columns in the same frameset to accommodate different kinds of information.
43. It is not that you are going to have an equal number of row and columns. Mostly you will see a web page with different set of rows and columns.

Multiple Choice Questions

1. Full form of HTML is:
   (a) Hyper Text Markup Language (b) Hyper Text Monitor Language
   (c) Hyper Type Markup Language
2. HTML tags must be written within :
   (a) ( ) (b) [ ]
   (c) <>
3. Values in HTML are provided within :
   (a) Question mark (b) Quotation marks
   (c) Left and right brackets
4. Your text file should have the extension :
   (a) html (b) htm
   (c) any of the 2 above
5. HTML definition is given in the text in :
   (a) Last line (b) Not given
   (c) First line
6. What character is used to end the tag :
   (a) / (b) ?
   (c) !
7. Tag for using graphic is IMG followed by :
   (a) SRC (b) RCS
   (c) CRS

8. For giving a line in the page, we which tag:
   (a) Line (b) HR
   (c) SR
9. For giving space around the image we use:
   (a) VSPACE (b) HSPACE
   (c) Both
10. For creating a table, you would use the tag:
   (a) <TABLE> (b) <CREATE>
   (c) <TAB>
11. Table can be divided into cols and :
   (a) semi colons (b) rows
   (c) cells
12. Horizontal rows are created by using :
   (a) <TH> (b) <TR>
   (c) <CR>
13. You can put header to a table using the tag:
   (a) <TH> (b) <TR>
   (c) <CR>
14. A form is used for :
   (a) filling data (b) getting information
   (c) displaying data
15. Tag used for creating menu is :
   (a) OPTION (b) TYPE
   (c) MENU
16. Radio buttons are so called because :
   (a) They look like radio buttons (b) They were created by radio
   (c) You have to switch on radio for using them
17. Form is used for :
   (a) Giving information to the visitors (b) Getting information from the visitors
   (c) To display graphics
18. TEXTAREA is used to :
   (a) Providing text to the visitors (b) Providing information to the visitors
   (c) Provide space for visitor to fill up
19. You can divide the page into rows and :
   (a) Columns (b) Tables
   (c) Frames

Answers

1. a	2. c	3. b	4. c	5. c
6. a	7. a	8. b	9. c	10. a
11. b	12. b	13. a	14. b	15. c
16. a	17. b	18. b	19. a	

Subjective Questions

1. What is HTML used for?
2. How are HTML tags written?
3. How are the values given in HTML?
4. How is the text typed for HTML?
5. How would you start and end the body of the text?
6. How many headings you can have in a page?
7. What all goes in the first line of HTML text? Explain them.
8. Describe the process of inserting graphics.
9. How would you create a banner?
10. How would you add a horizontal line?
11. How would you wrap text around the image?
12. How would you set margins for the page?
13. Describe how would you create tables in HTML?
14. How would you divide the table in various cols?
15. How would you put background colour to the table?
16. What do you understand by Unordered list?
17. How would you create a Ordered list?
18. What do you understand by Directory and Menu lists?
19. Describe the process of dividing the table into columns?
20. How would you give headers to the table columns?
21. What is a Form?
22. How is a Form created?
23. How would you create menus?
24. What are radio buttons?
25. What do you understand by Frameset?
26. How would you combine various framesets?
27. How would you give space between framesets?
28. How would you nest various framesets?

3 Cascade Style Sheets (CSS)

Introduction to CSS

CSS is sometimes referred to as a style sheet language because the most visible part of CSS is the language in which one expresses style sheets. However, a major part of CSS is the mechanism that interprets style sheets and resolves conflicts between rules. All browsers that support CSS are required to use the dame mechanism as this. The mechanism has two main parts: cascading and inheritance.

The inheritance mechanism is used only when no rules in any of the style sheets in the cascade try to set a certain property value. The property value will then be inherited from the parent element. If the property does not inherit, the initial value will be used instead.

Cascading refers to the cascade of style sheets from different sources that may influence the presentation of a document. Style sheets come from the browser and the designer, and may come from the user as well. The cascading mechanism is designed to resolve conflicts between these style sheets. Compared to other style sheet proposals, CSS is quite traditional in the stylistic properties it supports. However, as far as we know, cascading is unique to CSS. To get down to the working of CSS, it is very important to have the working knowledge of HTML.

Creating Style Sheets

There are two ways to create CSSs. You can either use a normal text editor and write the style sheets or you can use a dedicated tool – for example a Web page design application – which supports CSS. The dedicated tools allow you to create style sheets without learning the syntax of the CSS language. For example,

```
H1 { color: blue }
```

This is just a simple CSS rule that contains one rule. A rule is a statement about one stylistic aspect of one or more elements. A style sheet is a set of one or more rules that apply to an HTML documents. The rule above sets the color of all first-level headings (H1).

A rule consists of two parts:

- Selector – the part before the left curly brace
- Declaration – the part within the curly braces

The selector is the link between the HTML document and the style. It specifies what elements are affected by the declaration. The declaration is that part of the rule that sets forth what the effect will be. In the example above, the selector is H1 and the declaration is **color:blue**. Hence, all H1 elements will be affected by the declaration, that is, they will be turned blue. The color property just aspects the foreground text color there are other properties for background, border, etc. The above selector is based on the type of the element: it selects all elements of type "H1." This kind of selector is called type selector. Any HTML element type can be used as a type selector.

A declaration has two parts separated by a colon:

- Property – that part before the colon

- Value – that part after the color.

The property is a quality or characteristic that something possesses. In the previous example, it is color. The value is a precise specification of the property. In the example, it is "blue", but it could just as easily be green, red, yellow or some other color.

Grouping Selectors and Rules

Short style sheets load faster than longer ones. CSS therefore includes several mechanisms to shorten style sheets by way of grouping selectors and declarations.

For example, consider these three rules:

```
H1 { font-weight: bold }
H2 { font-weight: bold }
H3 { font-weight: bold}
```

All three rules have exactly the same declaration – they set the font to be bold. Since all three declarations are identical, we can group the selectors into a comma-separated list and only list the declaration once, like this:

```
H1, H2, H3 { font-style: bold }
```

This rule will produce the same result as the first three.

A selector may have more than one declaration. For example, we could write a style sheet with these two rules:

```
H1 { color: blue}
H2 { text-align: center }
```

In this case, we set all H1s to be blue and to be centered on the canvas. But we can achieve the same effect faster by grouping the declarations that relate to the same selector into a semicolon-separated list, like this:

```
H1 {
      color: blue;
      text-align: center;
}
```

All declarations must be contained within the pair of curly braces. A semicolon separates the declarations and may – but doesn't have to – also appear at the end of the last declaration. Also, to make your code easier to read, we suggest you place each declaration on its own line, as we did here.

Now you have the basics of how to create CSS rules and style sheets. However, you are yet to put it in a HTML document.

Putting Style Sheets to the document

For any style sheet to affect the HTML document, it must be linked to the document. That is, the style sheet and the HTML document must be combined so that they can work together to present the document. This can be done in any of four ways:

1. Apply the basic, document-wide style sheet for the document by using the STYLE element.
2. Apply a style sheet to an individual element using the STYLE attribute.
3. Link an external style sheet to the document using the LINK element.
4. Import a style sheet using the CSS @import notation.

In the next section, we discuss the first method: using the STYLE element.

Linking by using the STYLE element

You can link the style sheet and the HTML document together by putting the style sheet inside a STYLE element at the top of your document. The STYLE element was introduced in HTML specifically to allow style sheets to be inserted inside HTML documents. Here's a style sheet (shown in italics) linked to a sample document by using the STYLE element.

```
<HTML>
    <TITLE>My home page</TITLE>
    <STYLE>
        H1, H2 { color: blue }
    </STYLE>
    <BODY>
        <H1>My home page </H1>
        <P>I have been writing books on computers and have written books for class:
    <UL>
        <LI>Class IX
        <LI>Class X
        <LI>Class XI
        <LI>Class XII
    </UL>
    <H2>General Books</H2>
    <P>A lot of general computer books.
    </BODY>
</HTML>
```

See the result as shown next. Both the headings appear in blue and the rest of the text in black. Notice that the STYLE element is placed after the TITLE element and before the BODY element. The title of a document does not show up on the canvas, so it is not affected by CSS styles.

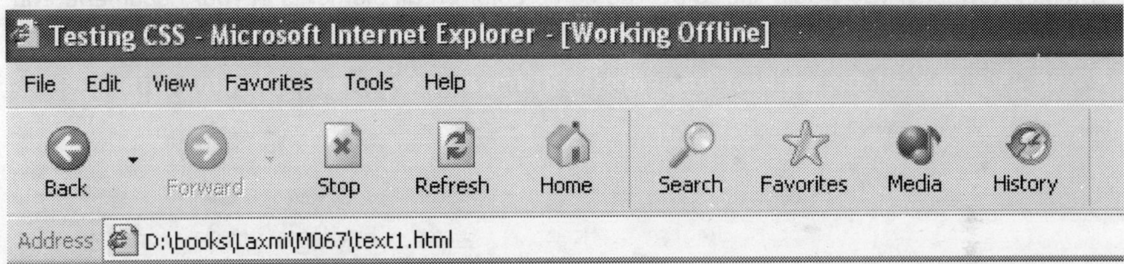

Tree Structures and Inheritance

There are many reason for having tree-structured documents in HTML. For style sheets, there is one very good reason; inheritance. Just as children inherit from their parents, so do HTML elements. Instead of inheriting genes and money, HTML elements inherit stylistic properties.

Let's start by taking a look at the sample document:

```
<HTML>
    <TITLE>My home page</TITLE>
    <BODY>
        <H1>My home page </H1>
        <P>I have been writing books on computers and have written books for class:
    <UL>
        <LI>Class IX
        <LI>Class X
        <LI>Class XI
        <LI>Class XII
    </UL>
    <H2>General Books</H2>
    <P>A lot of general computer books.
    </BODY>
</HTML>
```

Through inheritance, CSS property values set on one element will be transferred down the tree to its descendants. For example, our examples have up to now set the color to be blue for H1 and H2 elements. Now, say, you would like to set the same color on all elements in your document. You could do this by listing all element types in the selector:

```
<STYLE TYPE="text/css">
    H1, H2, P, LI { color: blue}
</STYLE>
```

However most HTML documents are more complex than our sample document and your style sheet would soon get long. There is a better – and shorter – way. Instead of setting the style on each element type, we set it on their common ancestor, the BODY element:

```
<STYLE TYPE="text/css">
    BODY { color: blue }
</STYLE>
```

Since other elements inherit properties form the BODY element, they will all inherit the color blue.

As you have seen above, inheritance is a transport vehicle that will distribute stylistic properties to descendants of an element. Since the BODY element is a common ancestor for all visible elements, BODY is a convenient selector when you want to set stylistic rules for the entire document.

Overriding Inheritance

In the previous example, all elements were given the same color through inheritance. Sometimes, however, children don't look like their parents. Not surprisingly, CSS also accounts for this. Say you would like for H1 elements to be blue while the rest should be green. This is easily expressed in CSS:

```
<STYLE TYPE="text/css">
    BODY { color: blue}
    H1 { color: red}
</STYLE>
```

Since H1 is a child element of BODY (and thereby inherits from BODY) the two rules in the above style sheets are conflicting. The first one sets the color of the BODY element – and thereby also the color of H1 through inheritance – while the second one sets the color specifically on the H1 element. Which rule will win? Let's find out:

The reason why the second rule wins is that it is more specific than the first. The first rule is very general – it affects all elements on the canvas. The second rule only affects H1 elements in the document and is therefore more specific. If CSS had been a programming language, the order in which the rules were specified would determine which of them would win. CSS is not a programming language and in the above example, the order is irrelevant. The result is exactly the same if we use this style sheet.

```
<STYLE TYPE="text/css">
    H1 { color: navy}
    BODY { color: green }
</STYLE>
```

CSS has been designed to resolve conflicts between style sheet rules like the one above.

Properties that don't Inherit

As a general rule, properties in CSS inherit from parent to child elements as described in the previous examples. Some properties, however, don't inherit and there is always a good reason why. We will use the background property as an example of a property that doesn't inherit. Let's say you want to set a background image for a page. This is a common effect on the Web. In CSS, you can write:

```
<HTML>
    <TITLE>My home page</TITLE>
    <STYLE TYPE="text/css">
        <BODY {
            background: url(texture.gif) white;
            color: black;
        }
```

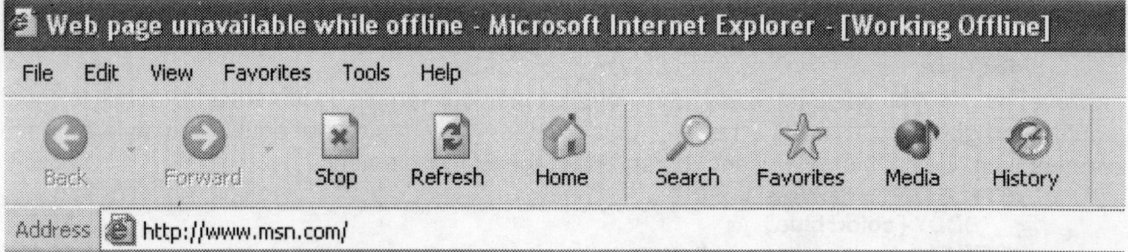


```
</STYLE>
    <BODY>
        <H1>My home page </H1>
        <P>I have been writing books on computers.
    </BODY>
</HTML>
```

The background property has a URL ("texture.gif") that points to a background image as value. There are a few noteworthy things in the above example:

- The background image covers the surface like a wallpaper – also the backgrounds of the H1 and P element have been covered. This is not due to inheritance, but to the fact that unless otherwise set, all backgrounds are transparent. So, since we haven't set the backgrounds of the H1 or P element to something else, the parent element, BODY, will shine through.
- In addition to the URL of the image, a color (white) has also been specified as the background. In case the image can't be found, you will see the color instead.
- The color of the BODY element has been set to black. To ensure contrast between the text and the background, it is a good habit to always set a color when the Background property is set.

So, exactly why doesn't the background property inherit? Visually, the effect of transparency is similar to inheritance: it looks like all elements have the same backgrounds. There are two reason: first, transparent backgrounds are faster to display (there is nothing to display!) than other backgrounds. Second, since background images are aligned relative to the element they belong to you would otherwise not always end up with a smooth background surface.

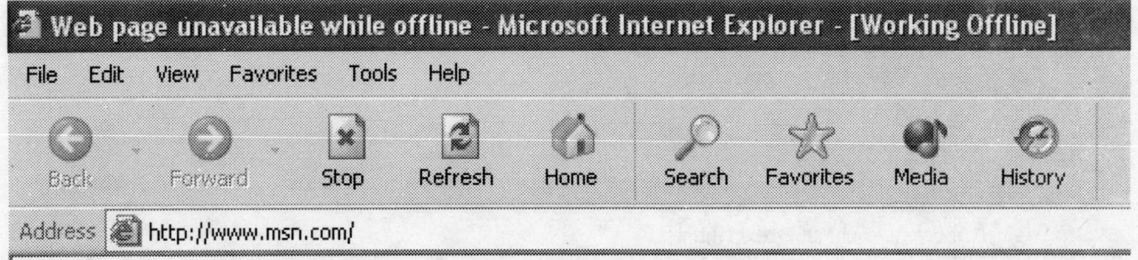

Common Tasks with CSS

Setting colors and backgrounds are among the most common tasks performed by CSS. Other common tasks include setting fonts and white space around elements. This section gives you a guided tour of the most commonly used properties in CSS.

Common tasks: fonts

Let's start with fonts. If you have used desktop publishing applications in the past, you should be able to read this little style sheet:

```
H1 { font: 36pt serif }
```

The rule above sets the font for H1 element. The first pat of the value – 36pt – sets the font size to be 36 points. The second part of the value – serif – tells the browser to use a font with serifs. The more decorated serif fonts suit My home page well since the modern sans-serif fonts (fonts without serifs) weren't used in his time. The result is shown above.

The font property is a shorthand property by setting several other properties at once. By using it, you can shorten your style sheets and set values on all properties it replaces. If you choose to use the expanded version, you would have to set all of these so replace the example above:

```
H1 {
    font-size: 36pt;
    font-family: serif;
    font-style: normal;
    font-weight: normal;
    font-variant: normal;
    line-height: normal;
}
```

Sometimes you only want to set one of these. For example, you may want to slant the text in some elements. Here is an example:

```
UL { font-style: italic }
```

The font-style property will not change the font size or the font family, it will only slant the existing font. When set on the UL element, the LI elements inside will become slanted, since font-style is inherited.

Similarly, the font-weight property is used to change the weight – thickness – of the letters. You can further emphasizes the list items by setting their ancestor to be old:

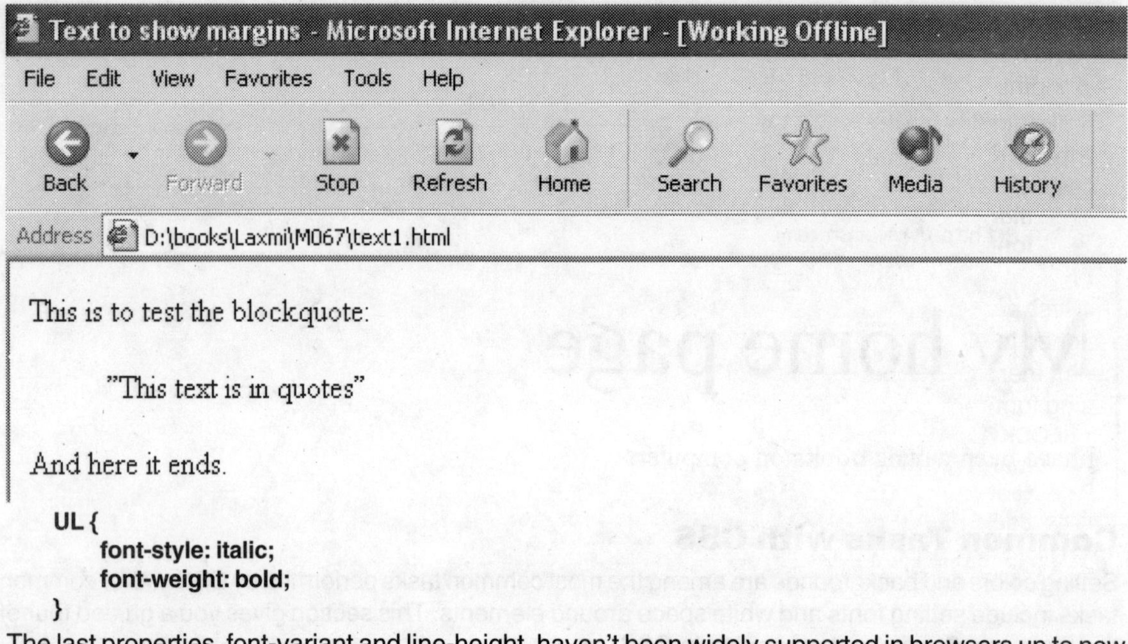


```
UL {
    font-style: italic;
    font-weight: bold;
}
```

The last properties, font-variant and line-height, haven't been widely supported in browsers up to now and are therefore not as commonly used yet.

Common tasks: Margins

Setting space around elements is a basic tool in typography. The headline above this paragraph has space above it and (slightly less) space below it. By default, your browser knows quite a bit about how to display the different kinds of elements in HTML for example, it knows that lists and BLOCKQUOTE elements should be indented to set them apart from the rest of the text. As a designer, you can build on these settings while at the same time provide your own refinements. Let's use the BLOCKQUOTE element as an example. Here's a test document.

```
<HTML>
    <TITLE>Text to show margins</TITLE>
    <BODY>
        <P>This is to test the blockquote:
        <BLOCKQUOTE>"This text is in quotes"</BLOCKQUOTE>
        <P>And here it ends.
    </BODY>
</HTML>
```

The screen-shot below is how a typical HTML browser would display the document.

As you can see, the browser has added space on all sides of the quoted text. In CSS, this space is called "margins" and all elements have margins on all four sides. The properties are called: margin-top, margin-right, margin bottom and margin-left. You can change how the BLOCKQUOTE element is displayed by writing a little style sheet:

```
BLOCKQUOTE {
    margin-top: 1em;
    margin-right: 0em;
    margin-bottom: 1em;
    margin-left: 0em;
    font-style: italic;
}
```

The em in the typographical terminology means the width of the character which has the maximum width and it is usually the width of character W. So, the above example will result in the vertical margins begins as high as the font size (1em) of the BLOCKQUOTE and horizontal margins having zero width. To make sure the quoted text can still be distinguished, it has been given in italic slant.

Just like font is a shorthand property to set several font-related properties at once, margin is a shorthand property which sets all margin properties. The above example can therefore be written:

```
BLOCKQUOTE {
    margin: 1em 0em 1em 0em;
    font-style: italic;
}
```

The first part of the value — 1em — is assigned to martin-top. From there its clockwise: 0em is assigned to margin-right, 1em is assigned to margin-bottom and 0em is assigned to margin-left. With the left margin set to zero, the quoted text needs more styling to set it apart from the rest of the text. Setting font-style to italic helps and adding a background color further amplifies the quote:

```
BLOCKQUOTE {
    margin: 1em 0em 1em 0em;
    font-style: italic;
    background: #EDB;
}
```

As expected, the background color behind the quote has changed. Unlike previous examples, the color was specified in red/green/blue (RGB) components.

One stylistic problem in the example above is that the background color or barely covers the quoted text. The space around the quote – the margin area – does not use the element's background color. CSS has another kind of space called padding which uses the background color of the element. In other respects the padding properties are like the margin properties: they add space around an element. Let's add some padding to the quote:

```
BLOCKQUOTE {
    margin: 1em 0em 1em 0em;
    font-style: italic;
    background: #EDB;
    padding: 0.5;
}
```

The result of setting the padding is added space between the text and the rectangle that surrounds it: Notice that the padding property was only given one value (0.5em). just like the margin property, padding could have taken 4 values which would have been assigned to the top, right, bottom and left padding respectively. However, when the same value is to be set on all sides, listing it once will suffice. This is true both for padding and margin.

Common tasks: links

To make it easier for users to browse in hypertext documents, the links should have a style that distinguishes them from normal text. HTML browsers have often underlined hyperlink text. Also, various color schemes have been used to indicate if the user has previously visited the link or not.

Since hyperlinks are such a fundamental part of the Web. CSS has special support for styling them. Here's a simple example:

```
A:link { text-decoration: underline }
```

The above example specifies that unvisited links should be underlined. The links are underlined, as we have specified, but they are also blue which we have not. When authors do not specify all possible styles, browsers use default styles to fill in the gaps. The interaction between author styles, browser default styles and user styles (the user's own preferences) is another example of CSS's conflict resolution rules. It is called the cascade (the "C" of CSS).

The selector (A:link) deserves special mentioning. You probably recognize "A" as being an HTML element, but the past part is new, ":link" is one of several so-called pseudo-classes in CSS. Pseudo-classes are used to give style to elements based on information outside of the document itself. For example, the author of the document can't know if a certain link will be visited or not.

A:visited { text-decoration:none }

This rule gives style to visited links just like A:link style to univisited links. Here is a slightly more complex example:

A:Link, A:visited { text-decoration: none }
A:hover { background: cyan }

The last rule introduces a new pseudo-class :hover. Assuming the user is moving a pointing device (like a mouse) the specified style will be applied to the element when the user moves the pointer over ("hovers" over) the link. A common effect is to change the background color. Here is what it looks like:

The :hover pseudo-class has an interesting history. It was introduced in CSS2 after the hover effect became popular among JavaScript programmers. The JavaScript solution requires complicated code compared to the CSS pseudo-class and this is an example of CSS picking up effects that have become popular among Web designers.

Colours

Printing colour is expensive. Hence, in books colour has always been used sparingly. On the Web, however, the use of colour is virtually free. Most Web users have colour monitors, so displaying colour costs nothing, although printing it is still costly. So there is no reason for Web designers not to incorporate colour into their designs.

Of course, colour can be overdone. As with all aspects of print and Web, design – whether fonts, space or images – colour should be used to achieve a purpose with the design. Colour throng in at random doesn't work well. Remember, too much variety obscures, rather than clarifies. Some combinations of colours are very hard to read, such as red type on a blue background. A background that differs from the foreground text only in colour and not in brightness also strains the eye. For some people dark letters on a light background are easier to read, while for others, the opposite is true. People may also have associations with certain colours that may either help or hinder their understanding the text. For example, red seems to be almost universal in marking something that is important. But the interpretations of other colours often depend on culture and even on the user's personal experiences. Colours and how we perceive them are both very technical and complex subjects, far beyond the scope of this book. We encourage interested readers to further explore the subject of colour, particularly to learn how to combine colours effectively to achieve the desired effects.

The range of colours that can be reproduced by a computer monitor is called its gamut. The gamut depends not only on the colour and brightness that a computer monitor can produce but also on the brightness of the light that reflects off the screen. The effective gamut is reduced as the light in the room gets brighter, the largest gamut is available in a darkened room. Note that colour printers work very differently from computers and their gamuts differ from that of computer monitors. Colours often appear one way on the monitor's screen and another when printed by a colour printed or by a traditional printing process.

Let us now see how CSS is used to specify the colour of text and borders. This is done with the colour property. Also, we describe how to set backgrounds, either to a certain colour or to an image. This is done with the various background properties. Before I describe the properties, let us look at various kind of colours supported by CSS.

Naming Colours

There are many ways in which colour can be specified. The English language – and all other languages as well – has words for the most common colours: for example "red", "blue" and "brown". If you tell someone that your house is red, your listener will get a general idea about the colour of your house, but there are many different shades of red which will all be described as "red".

Another way to describe a certain colour is to use numbers. Paint manufacturers often provide sets of numbered colour swatches so that you can ask for paint that is "blue number 216" for example. On computer screens the most common way to specify a colour is to give an RGB value which specifies the mixture of red, green and blue. However, this, too, is an imprecise way of selecting colours because the result depends on the type of screen you have and how much light is in the room in which the monitor is located.

A third way to specify a colour is to refer to the colour of something else. For example, you may want the colour of your curtains to be the same as the colour of your sofa. A more relevant example for a style sheet is to set the text colour of a document to be the same as a system colour, for example the foreground colour of the user's window system. CSS allows you to specify colours using all these methods. A small number of colour names and system colours are defined and other colours can be specified as RGB values.

Colour Names

CSS predefines 16 colour names; ***aqua*** (a light greenish blue, sometimes called cyan), ***black***, ***blue***, ***fuchsia*** (light purple/pink) ***gray***, ***green***, ***lime*** (light green) ***maroon*** (dark red), ***navy*** (dark blue), ***olive***, ***purple***, ***red***, ***silver*** (light gray) ***teal*** (blue-green) ***yellow***, ***white***.

Here is a simple example of a style sheet which sets the colour and background using the colour names:

```
BODY {
        colour: black;
        background: yellow;
}
```

The predefined colour names are easy to use for those who write style sheets in a simple text editor. More advanced tools allow designers to graphically select colours which are then turned into RGB colours.

RGB colours

Computer monitors commonly use the RGB – red, green, blue – colour model to display colour. In addition to RGB, there were are a few other colour models which could have been used. One of them is the HSB (hue-saturation-brightness) model, which is often used by artists since it is similar to how artists mix colours. Another is CMYK (cyan-magenta-yellow-black) a colour model commonly used by professional printers. We chose the RGB model because it represents the how colour is displayed on a colour video monitor.

Red, green and blue light is mixed in specified properties to represent colours on the monitor. In the RGB colour model, each of the three colours is represented by a value between 0 and 100%, where 100% represents the maximum brightness of a colour. The values are arranged as a triplet, where the first number represents red, the second green and the last blue. Black is represented as the triplet (0, 0, 0) – zero amounts of all three colours – and white by (100%, 100% 100%) – 1000% of all three colours. Every triplet with equal amounts of each colour is a shade of gray. For example (90%, 90%, 90%) is a light gray, while (40%, 40%, 40%) is a dark gray. When the values are not equal, it is often hard to predict what the colour looks like. That (100%, 0, 0) is red is not too hard to see. But that a colour with the values (65%, 16%, 16%) is a shade of brown is not so obvious.

The RGB values of CSS1 are called "sRGB," for "standard-RGB", sRGB is a colour space which ensures that all colours specified in CSS are exactly defined. This means that the computer which displays the document will know exactly what colours are specified in the style sheet and all Web devices should be able to display the exact colour. As noted above, however, limitations in computer hardware make this a hard promise to fulfill.

To specify an RGB colour in CSS, three values – a triplet – must be provided. That can be done by using any of three methods. The first two we've already discussed. Which you use is a matter of taste, since they all produce the same result.

1. Percentages: for example, a colour specification like rgb (100% 35,5%, 10%) specifies a maximum amount (100% of red light, 35.5% of green light and 10% of the blue light. The result is a deep orange red.
2. Numbers in the range of 0 to 255: thus rgb (255, 91, 26) should be the same colour as rgb (100%, 35.5%, 10%) in the percentage example.
3. Hexadecimal numbers: for example, #FF5B1A which will produce the same shade of red as in (1) and (2).

In either of the first or second methods, if you enter a value that is outside the acceptable range, for example 125% in method one or 300 in method two, the value will be "clipped". That is, the errant value will be reduced to the maximum value allowed by the Web device. Since different devices have different ranges, you should avoid values outside the acceptable range. The third method for specifying a colour needs a bit more explanation. You use the same numbers as with the second method, 0 to 255, but you write them as a single hexadecimal number preceded by a hash mark (#); for example, #FF5B1A, which produces the same colour red as rgb (100%, 35.5%, 10%) and rgb (255, 91, 26). This notation is not very intuitive, but it included since it is also used in HTML. The hexadecimal notation can be written in two variations that use either three of six hexadecimal digits. The three-digit form defines the same colour as the six-digit form does but with all digits doubled; that is #A84 is the same colour as #AS8844.

Properties

We have already seen some examples of the colour property and the background properties in use. Now let us understand the various properties in details.

Colour Property

The colour property sets the colour of the text of an element. It is the foreground colour. It can also set the colour of text decorations, "Fonts" such as underline, as well as borders that have been created with the border properties. The property takes one value – a colour – which is specified by one of the methods discussed in the previous section. The following are examples of rules to specify a colour for text or a text decoration:

```
EM { colour: red }
P { colour: rgb(255, 0, 0) }/* red */
H1 { colour: #f00 }/* also red */
H1 { colour: #ff000 }/* same as above */
```

A colour is inherited by child elements. The default colour is set in the user's default style sheet, if one exists. Otherwise, the default colour depends on the browser.

Setting the Colour of a border

Setting the colour of a border can be done in two ways. One is to set it directly by including the colour via the border properties, as done here:

```
P { border: medium double red }
```

Another way is to set the colour of the text that is the content of the element, as done here:

```
P {
    border: thin dotted;
    colour: blue;
}
```

In this case, no colour has been set on the border, so it will assume the colour of the text, which is blue.

If we set a colour on the border itself, as done here:

```
P {
    border: thin dotted red;
    colour: blue;
}
```

Colour of Hyperlinks

Until CSS, the most common use of colour in Web browsers (other than in images) was to draw attention to hyperlinks. Some browsers use two colours: one for links to documents that the user has traversed before and another for links that he hasn't tried yet. Others use a third colour for the "active" link; that is, for the short duration while the user keeps the mouse button pressed over the text. In CSS rules for these can be written through a pseudo-class on an A element, as follows:

```
A;link { colour:blue }/* unvisited links */
A:visited { colour: red }/* visited links */
A:active { colour: yellow }/* active links */
```

The Background Properties

The background properties set aspects of the background of an element that is, the surface onto which text is displayed. That background can be either transparent, a colour or an image. You can also set the position of an image, if and how often the image should be repeated on the screen and whether it should be fixed or scrolled relative to the canvas. Five of the properties set specific aspects of the background while the sixth the background property, is a shorthand method that lets you set all the first five properties at one time. The properties do not inherit. The following are the background properties:

- background-colour
- background-image
- background-repeat
- background-attachment
- background-position
- background

The Background-Colour Property

The background-colour property set the background colour of an element

- a colour
- the keyword transparent – this is the default

when a colour value is provided, the specified colour will be visible behind the text of the element. How much of the surface will actually get that colour depends on the type of element and on the amount of padding. It also depends on whether there is a background image in addition to the colour, as specified with the background-image property below.

Background colour in inline elements: For an inline element, the colour will be visible only behind the text itself and behind the padding around the element. Hence, if the element is broken across lines, the background colour will be visible behind the words and spaces at the end of the first line and also behind the words and spaces on the second line. For example the words in the middle of the sentence are displayed in white on black by putting them in an EM element with the colour: white and background-colour: black. Here is the code to produce this result:

```
<STYLE TYPE="text/css">
    EM {
        background-colour: black;
        colour: white
    }
</STYLE>
<BODY>
<P>This paragraph has <EM> few emphasized words</EM> in the middle.
</BODY>
```

This would result in:

This paragraph has few emphasized words in the middle.

Background colour in block elements: For block-level elements, the colour will occupy a rectangular region that includes the indent of the first line (if any) of the paragraph as well as any empty space at the end of the each line.

It will also occupy the padding around the block, if any. In such cases, it is a good idea to add a little padding to leave room between the letters and the edge of the background.

```
P.statement {
    Background: black;
    Colour: white;
}
```

Background colour in list items: For list item elements, the background is not applied to the label if the label is outside the text box. If the label is inside the text box, then the background will be hind the label as well.

The Background-Image Property

The background-image property lets you set an image as background for an element.

This property has two values:

- a URL
- the keyword none – this is the default

To specify an image as a background you enter the URL of the image as the <url> value. When specifying an image as the background, you should also specify a colour. When the document is displayed, the image will overlay the colour.

The Background-Repeat Property

The background-repeat property determines whether and how an image is repeated in the element. By default, an image is initially placed in the upper-left corner of the element. Repetition of an element begins from either this default position or from a new position that you set with the background-position property, which we discuss shortly. The following are examples of rules for each value of background-repeat show the result of each.

```
background-image: url(wave);/* (a) */
background-repeat: repeat;
background-image: url(wave);/* (b) */
background-repeat: repeat-x;
background-position: center;
```

```
background-image: url(wave);/* (c) */
background-repeat: repeat-y;
background-position: right;
background-image: url(wave);/* (d) */
background-repeat: no-repeat;
```

The Background-Attachment Property

The background-attachment property determines whether the image should be fixed of moveable on the canvas.

This property has two values:

- scroll – the image will scroll along with the content. This is the default.
- fixed – the image will be fixed with regard to the canvas.

A background image is visible behind the element to belongs. When the element scrolls, you usually want its background to move along.

The Background-Position Property

The background-position property has lets you override the default position of an image and specify the image's initial position, whether a single image or an image that is repeated. Essentially, you can set the position of a background image in either of three ways:

- percentages
- absolute positions
- keywords

Placing images using percentages: When you place an image using percentages, you tell the browser where the background image is relative to the size of the element. Here's an example of how this works. Suppose you have an element and you want to place an image in that element. Assume the element is BODY and your percentage values are 20% and 60% written as a rule like this:

```
BODY{
background-position: 20% 60%;
background-image: url(tile.png)
}
```

Placing images using absolute positions: When you give two length values instead of two percentages, the upper left corner of the image will be that far away from the upper left corner of the element. For example:

```
BLOCKQUOTE{
background-image: url(shape.png);
background-position: 1cm 5mm
}
```

will put the background image shape.png at 1cm from the left and 5mm from the top of the element. As with percentages, if you only give one value, the image will be centered vertically. That is, a value of 1cm is equivalent to 1cm 50%. Negative values are possible, if you want to put the image partially outside the element. Only the part that is inside the element will be visible, though.

Placing images using keywords: When placing an image using keywords, you use any combination of two keywords. One of three keywords – top, center, bottom – represents the horizontal (x-axis) dimension. And one of three keywords – left, center and right – represents the vertical (y-axis) dimension.

The Background Property

The background property is a shortcut means of setting all of the first five properties at the same time. Its values are all of the possible values of those five properties. You may set from one to five of the properties, in any order. However, this property will always set all five of the properties, regardless of whether you explicitly set values for all five. If you don't explicitly seta value for a specific property, the background property uses that property's initial value as the property's value. For example, here:

```
BODY { background: red }
```

Only the value of the background-colour property has seen set. The background property will assume the initial values for all of the other properties. In contrast:

```
BODY { background: url(chess.png} gray 50% repeat }
```

Sets values for four of the properties, but not that for the background-attachment property; hence, that property's value will be the initial value (which is scroll). In other words, the rule above is exactly equivalent to:

```
P {
background-image: url(chess.png);
background-colour: gray;
background-position: 50%;*/ = 50% 50% */
background-repeat: repeat;
background-attachment: scroll;/* implicit */
}
```

You can set separately the five aspects of a background – colour, image, repeat, scrolling and position – but we don't recommend this as a rule. The different aspects of a background are so tightly linked in how they work to produce effects that you could end up with some unexpected, even weird results. For example, setting a background-repeat value without setting the image at the same time may produce a very strange-looking background. The same is true with specifying an image without also setting the colour behind it. In both case, problems could arise when style sheets are cascaded. Cascading involves using more than one style sheet for your document; for example, yours (that is, the designer's) the browser's default style sheet and possibly one attached by the user. If you were to satisfy an image without also setting a colour behind it, the background might be composed of any combination of background from these three style sheets.

The Font-Family

The font-family property lets you specify the font families which will be used in your documents. Here is the formal definition of the property. Each CSS property has a "data sheet" like the above which describes, in a formal way, key characteristics of the property. Also, on the left there is a "button bar" which gives you an overview of how widely supported the property is.

The font-family property accepts a list of font families as value. The font families are separated with a comma and there are two different kinds:

- specific font families (e.g. Arial or Courier)
- generic font families (one of: serif, sans-serif, monospace, cursive, fantasy)

For example, a document that is to be displayed in Times might have a style sheet like this:

```
BODY { font-family: Times }
```

A document that is to be displayed in Garamond, if possible and Times if Garamond is not available would have a style sheet like this:

```
BODY { font-family: Garamond, Times }
```

where Garamond and Times are both family names.

It's important to remember the comma when specifying a list of font families. Some font family names have spaces in them (for example "New Century Schoolbook") and the space character is therefore not sufficient for separating different font families. In addition, you may use quotation marks around family names, as in this example rule:

BODY { font-family: "new century schoolbook", serif }

The quotation marks are necessary if the name contains character other than letters, digits, dashes and spaces. The value can also be the name of a generic font family – instead of or in addition to the specific font families. Use of generic font families is a partial solution to the problem of dealing with unavailable fonts. A browser may not have any of the fonts listed in the value. To make sure that something will be displayed in the worst case of no match, you can add a generic font family like serif to the list. In this case, if the browser does not have any of the given families, it will have to use any font it has that has serifs. Here are the generic names:

serif
sans-serif
monospace
cursive
fantasy

They correspond to the font category. Here's an example rule that includes a generic family name:

BODY { font-family: Garamond, Times,
"New Century Schoolbook", serif }

In this example, the browser will first check to see if it has Garamond. If it doesn't it will check for Times. If it doesn't have Times, it will search all the fonts it has until it finds one in a serif style (if any) that it can use to display the document.

The use of generic font families is encouraged. Each browser has a list of fonts that it can display. Those lists usually differ among browsers. However, all browsers must understand the previous five category names. Hence, when a requested font is not available on a particular browser and you told the browser which category the font is in, the browser is expected to substitute a font from the same category. Although the resulting translation may be a bit rough, it will usually be a font that tools at least somewhat like the one intended.

Font Metrics

Fonts are typically measured in a unit specific to the printing industry called the point. A point (abbreviated "pt") is the traditional printer's typographer's unit for specifying the size of fonts, the spacing between adjacent lines and the thickness of rules – among other things. It is still used a log, although some countries and some publishers now prefer to use the metric system (specifically the millimeter – mm – and centimeter – cm). CSS uses only the Anglo-American point. It does so because that point's value is in between the other two values. Also, it conforms to the point size used in PostScript printers, the most common type of printer.

To understand how type is sized, you first need some information about what makes up a letter. The x-height is the size of the body or main part of the letter and is approximately equal to the height of the x of the font. The ascender is that part of the lowercase letter that extends above the x-height. The descender is that part of the lowercase letter that extends below the x-height.

The size of type is usually obtained by measuring from roughly the top of its ascenders to the bottom of its descenders. (In the days of metal type, the font size was the size of the letter body). The measurement is expressed in points. The difference of a point is more noticeable in smaller sizes than in larger sizes. Text type, used for running text, is generally 14pt or less. Common sizes are 10 pt and 12 pt. Display type, type used for headings, is generally bigger than 14pt.

Length units

The point unit is the traditional measurement of length is typography, but not always the best for Web design. CSS accepts a range of different units which fall into three categories.

- absolute units – this category includes the point unit described above as well as other units (mm, cm, in, pc) which describe physical distances that can be measured with a measuring band.

- relative units – these units describe distances relative to other distances. The "em" unit, is the foremost member of this group which also includes the ";ex" unit.
- the pixel unit – the pixel unit as defined in CSS, forms a group of its own since it has some unique characteristics. On normal computer screens, the pixel unit will do what you expect. On printers and other high-resolution devices it will do what you hope!

The three groups their units are described in more detail below. Common for all the units is the way they are combined with numbers to form values.

Absolute units

An absolute unit is a unit of measure that specifies a fixed length – a length which can be measured with a measuring band. The units are:

- **millimeter:mm**
- **centimeter: command (1cm = 10 mm)**
- **inch: in (1in = 25.4mm)**
- **point: pt (72pt = 1 in)**
- **pica: pc (1pc = 12pt)**

The mm and command units come from the metric system of measurements. The other units are all defined as fraction of an inch. A pica is 12 pints. Six picas equal one inch. Absolute units have limited usefulness because they cannot be scaled. In general, they should only be used if you know the physical properties of the output medium. For example, if you write a style sheet for a document you know will only be printed on A4-sized paper, absolute units may be the right choice. In most cases, however, you will be better off using relative units.

Relative units

A relative unit is a unit of measure that specifies a value that is relative to the font size. There are two relative units in CSS: "em" and "ex". Usually the font size they refer to is the font size of the element itself. (The only exception is the font-size property, which we discuss later in the chapter where the value scales to the font size of the element's parent.) Relative units have the advantage over absolute unit sin that they scale automatically. When you choose a different font, all properties that were expressed in em or ex don't have to be changed. The em unit was heavily promoted earlier and we will not repeat all the good arguments for using the em unit here, but merely state the facts. In CSS, the em is exactly equal to the font size (i.e. the height of the font). For example, in a 12pt font, the em is 12pt wide, while in a 15pt font the em is 15pt wide.

The ex unit is also relative to the font size but in a different way. It is called "ex" because it is defined as the x-height. The em can be set explicitly in CSS1, but the ex is a characteristic of the font, so it cannot be set explicitly. This means you can determine how big it is only by inspecting the font. For example, Times Roman has a relatively large ex (x-height) compared to Baskerville. So even though a 12pt Times Roman and a 12 pt Baskerville both have an em of 12pt, their ex values will vary because their x-heights vary. Times Roman's will be somewhat larger than Baskerville's.

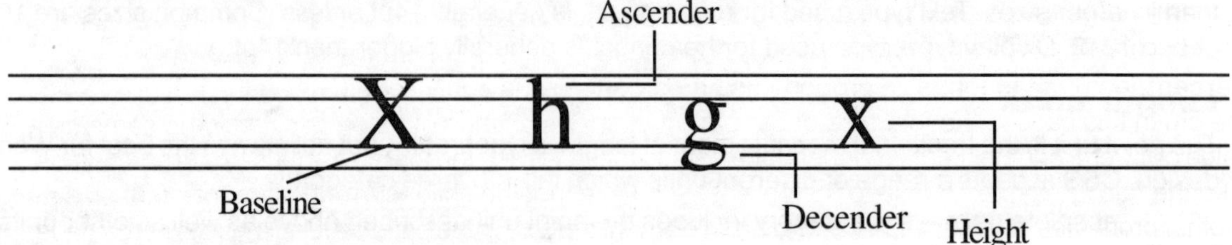

The pixel unit

The term pixel is derived from Picture Element. The pixel is the smallest element on a video display screen, such as a computer monitor or a television. It also applies to the output from certain types of printers, such as laser printers. The pixel unit in CSS is based on pixels, but is slightly more refined than the name implies. On computer screen, the pixel unit behaves as you expect it to. Consider these two rules:

```
H1 { border-width: 4px }
H2 { border-width: 3px }
```

In the example above, the width of the border surrounding H1 element is set to four pixels and the border around H2 elements is set to three pixels. On a normal computer screen, these values will be used as specified. That is the border around H1 elements will actually be four pixels wide and the border around H2 elements will be three pixels wide. Another seemingly obvious fact is that the H1 border will be exactly one pixel thicker than the H2 border.

Percentages as Values

Many properties that accept a number or a length as a value also accept a percentage, such as 50%, 33.3% and 100%. Although not a unit of measure, percentage offer similar advantages that relative units do, that is, they automatically scale. What the percentage is a percentage of – that is, what it is relative to – depends on the property. Usually, it is a percentage of the value that the property has in the parent element. For example, an H1 with a font size of 80% means the font is 80% of the font size of its parent (often the BODY element). The only exception is the line-height property. (Line heights is also known as interline spacing.)

Keywords as Values

Keywords are not units of measure, but some have connotations of being relative. In this case, they have the same advantage as relative units and percentages in that they can be scaled. For example the keywords bolder and lighter which are for the font-weight property are clearly relative: they are relative to the font weight of the parent. The same applies to larger and smaller for the font-wise property, which get to next.

The font-size property

The font-size property specifies the size of the font. This property has four possible types of values:

- **length**
- **percentage**
- **absolute size**
- **relative size**

The first two types are general types of values which can be used on many CSS properties. The last two types are sets of keywords which are particular to the font-size property.

Length: When a property accepts length values, all units described in the "Length units" section above can be used. This include absolute units (mm, command, in, pt, pc), relative units (em, ex) and the pixel unit (px). Here are some examples:

```
BODY { font-size: 12 pt }
PRE { font-size: 0.4mm }
H1.chapter { font-size: 16px }
```

Percentage: On the font-size property using percentages is equivalent to using the em unit. Since the em in this case is the font size of the parent element, a value of 120% is exactly the same as a value of 1.2em. another context-dependent method for specifying the font size is thus to use percentage. Like the relative to the relative to the parent's font size, not the child (current) element's font size as in other properties. A percentage value gives in element the size of the parent's font times the percentage.

Thus 120% gives a size that is 20% more than the size of the parent element, while 80% gives a size that is 80% of the parent element's size. For example, if the parent element size is 12pt, a 120% value will result in a size of 14.4pt, while an 80% value would result in a size of 9.6 pt. Here is an example:

H1 { font-size: 120% }

The example above is equivalent to the previous example using the em unit.

Absolute size: The "absolute-size" value is an index to a table of font sizes that is computed and kept by the browser. These sizes are expressed as keywords, as follows:

- **xx-small**
- **x-small**
- **small**
- **medium**
- **large**
- **x-large**
- **xx-large**

These sizes from an increasing range. For computer screens, the CSS2 specification suggests that each of them is 1.2 times bigger than the previous one. That is, if the "medium" version of a font is 10pt the "large" version would be 1.2 times that size of 12pt. Different media may use or need different scaling factors. When the browser is computing the table of size values. It takes into account the quality and availability of fonts.

Here's an example usage of the "absolute" value of the font-size property:

H1 { font-size: xx-small }

Relative size: The "relative-size" values lets you specify the size in a context-dependent manner. Recall from earlier in the chapter that relative values usually are relative to the font size of the element itself. The only exception is the font-size property. For that property the value scales in relation to the font size of the element's parent. The value has two keywords: larger and smaller. The keywords are interpreted relative to the table of font sizes mentioned in the previous section and the font size of the element's parent element. Specifying one of these keywords is a safe way to provide for context-dependent size changes. For example, suppose the parent element has a font size of "medium". If you then set the font-size property on the child element to larger, that is, larger than medium, the resultant size of the child will be large, which is the next value in the table of size values. Here's the code for doing this:

BODY { font-size: medium }
H2 { font-size: larger }
P { font-size: smaller }

BODY has a font size of medium H2, as a child element of BODY, will be one size larger than BODY; that is, large P will be one size smaller than BODY; that is, small.

Font-Style Property

The font-style property lets you specify the oblique or italic, style within the current type family. The property has three keywords values:

- **normal (also known as "roman", "regular", or sometimes "upright"). This is the default value**
- **Italic**
- **oblique**

Italic and oblique styles are similar, but not the same, Sans-serif families usually consider the two to be the same. Both are a variant that looks slanted to the right. For some families that is indeed how they are produced: by mechanically slanting the roman letters. Usually, however, that doesn't lead to beautiful letters and for most font the italic and roman forms are designed separately. Serif families usually distinguish between the two. The italic from looks very different from the oblique form. The two

Font		*Normal*	*Bold*	*Italics*
Arial	Sans serif	Aa	**Aa**	*Aa*
Times New Roman	Serif	Aa	**Aa**	*Aa*

differ in the shape of their serif and often look like completely different letter forms. Oblique refers to a version that looks like slanted roman.

Here an example rules using the font-styles property:

H1, H2, H3 { font-style: italic }
P { font-style: oblique }
H1 EM { font-style: oblique }

Font Variant Property

The font-variant property lets you specify a small-caps style within the current font family. This property has two values:

- **normal**
- **small-caps**

A small caps font style, despite its name does not really consist of small capital letters. It is a differently shaped letter entirely that resembles capital letters. They are a little smaller and have slightly different proportions. The first line of the figure shows a true small-caps font .the second line shows uppercase letters from a roman font that have been reduced by 80% to simulate small-caps. Notice that the stems of the small caps are the same width as the stems of the lowercase letters a around them while the reduced uppercase letters are too thin. The small caps are also wider than the reduced uppercase letters.

a Small letter
A Capital letter
A Small Caps letter

The value normal selects a font that is not a small-cap font, while small-caps selects a small-cap font. In the absence of a small-caps font, the browser may try to create one from an available normal font by scaling down and then stretching some uppercase letters. As a last resort, the browser may even use a normal font's uppercase letters, without reducing them.

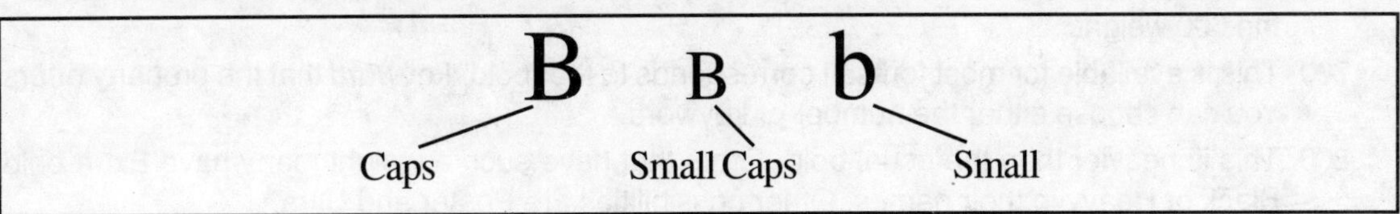

On the screen in low resolutions, as well as on low-grade printers, the difference between this adaptation and true small-caps will be hard to discern. On paper or a higher-quality output device, where the resolution is normally much higher, the difference will be noticeable.

The following is an example rule that sets the font variant property value:

H3 { font-variant: small-caps }

Font-Weight Property

The font-weight property specifies the weight of the desired font within the current family. This property has nine levels of weight:

- The nine values 100 to 900 from an ordered sequence, where each number indicates a weight that is at least as dark as it predecessor, 100 being the lightest
- normal is the same as 400
- bold is the same as 700

Two of the values – bolder and lighter – select a weight that is relative to the parent's weight. The font-weight property is a superset of what is available for most families. Very few families have the equivalent of all nine different weights. Hence, many of the CSS weight values will result in the same font. Numerical values are used because we felt that descriptive words, other than normal and bold, could be potentially confusing.

This is because there is no universal naming system for fonts. For examples, a font that you might consider bold could be described, depending on the font, as Roman, Regular, Book, Medium, Semi-Bold, Demi-Bold or a number of other names. These names are generally meaningful mainly within the family that uses them and are used to differentiate fonts within that family. Outside that family, a name may have a different meaning. So you cannot necessarily gauge the weight of a font from its name. In CSS1 the value normal (400) will have the weight that is normal for the specified font. What it means for another font may differ.

The following are description of the nine numerical values:

100 This is largest weight. Not many fonts have weights lighter than 400 ("normal"). Those that do, for which the browser would select the 100 value, will often have Thin, Light or Extra-light in their names. But font names are very inconsistent in their description so we suggest you don't rely on them. Let the browser figure out which weight a font is.

200 This is not usually any different from 100. However, it is included because some fonts, such as the Multiple Master fonts, have very many weights.

300 This is a little heavier than the 100 and 200.

400 This is available for all fonts. It corresponds to the "normal" keyword that the property offers. You can choose either the number or keyword.

500 This is little bolder than 400, but not by much. Some fonts have a weight called Medium that is just a little bolder than the normal weight: 500 would result in that weight. If there is no such font, 500 will be the same as 400.

600 This is a little bolder than 500. A weight name of Demi-bold or Semi-bold would correspond to the 600 weight.

700 This is available for most fonts. It corresponds to the "bold" keyword that the property offers. You can choose either the number or keyword.

800 This is heavier than the usual bold. Fonts that have such a weight many have Extra-bold, Black or Heavy in their names. Other possibilities are Poster and Ultra.

900 This is the very heaviest weight available in a font. Fonts with this weight often have rather strange names, such as Nord or Ultima. Or they may have cryptic numbers, often the case with Multiple Master fonts. Most fonts on the typical computer don't go beyond bold (700), so choosing either 800 or 900 will result in the same effect as choosing 700.

The bolder and lighter values select a weight that is relative to the parent's weight. For example, consider this style sheet:

```
P { font-weight: normal } /* same as 400 */
H1 { font-weight: 700 } /* same as "bold" */
STRONG { font-weight: bolder }
```

In this example, a STRONG element that appears in either the P or H1 element will be bolder than its parent. However, the weight of STRONG when it appears in the P element may be less than its weight in the H1 element.

The Font Property

The font property lets you to specify, in one action, all other font properties described above plus line-height. The font property is the first of the so-called shorthand properties we describe in detail. Shorthand properties let you set a group of properties which naturally belong together. The font properties described in this chapter (font-style, font-variant, font-weight, font-size, font-family) from one such group and the font property sets all of them in one statement. Also, the line-height property is set by the font property, but note that font-stretch and font-size-adjust are not included in this group since they were added in CSS2.

The font property is one of the most common CSS properties, and a simple example of its use was given in chapter. Here is another example:

```
H1 { font: 1.3em sans-serif }
```

The font property requires that you always specify a font size and a font family – all other values are optional. The example above is equivalent to writing:

```
H1 {
    font-style: normal;
    font-weight: normal;
    font-variant: normal;
    font-size: 1.3em;
    line-height: normal;
    font-family: sans-serif;
}
```

In principle, all values that are legal on the individual properties are also legal on the font property. There are restrictions, however, on the order the values must come in. First comes the font-style, font-weight and font-variant values, in any order. These are optional and will be set to their initial value. Finally comes the font-family value which is required.

The syntax resembles a shorthand common in traditional typography, for example "12/14 Times bold italic." Here is an example of how to write a rule that reflects this shorthand:

```
P { font; italic bold normal 112pt/14pt Times, serif }
```

For demonstration purposes, we set every property, including font-variant, which has its initial value of "normal". Specifying initial values is not necessary. The browser will automatically use the initial values for those properties not explicitly set. For example:

```
P { font: italic bold large palatino, serif }
```

Sets all properties except font-variant and line-height whose initial values will be used. In other words, the above is equivalent to:

```
P {
    font-style: italic;
    font-weight: bold;
    font-variant: normal; /* default */
    font-size: large;
    line-height: normal; /* default */
    font-family: palantino, serif;
}
```

Here are some more examples:

```
P { font: italic 112pt/14pt bodoni, bamboo, serif }
P { font: normal small-caps 120%/120% fantasy }
P { font: x-large/100% "new century schoolbook", serif }
```

The keyword "message-box" refers to the font used in user interface dialog boxes. The rule above will set all six properties to the values used in dialog boxes. The system font keywords are:

caption	The font used for captioned controls (e.g. buttons, drop-downs etc.)
icon	The font used to label icons.
menu	The font used in menus (e.g. dropdown menus and menu lists).
message-box	The font used in dialog boxes.
small-caption	The font used for labeling small controls.
status-bar	The font used in window status bars.

System fonts may only be set as a whole; that is, the font family, size, weight, style, etc., are all set at the same time. These values may then be altered individually. For example:

```
BODY {
    font: message-box;
    Font-size: x-large;
```

Font Stretch Property

The font-stretch property selects a normal, condensed or extended face from a font family. The property accepts two keywords which make the text wider or narrower than the text in the parent element. Also, nine keywords which express stretching in absolute terms are allowed. They have the following order, from narrowest to widest: ultra-condensed, extra-condensed, condensed, semi-condensed, normal, semi-expanded, expanded, extra-expanded, ultra-expanded.

ABCDEG	Normal
ABCDEF	Expanded

It's up to the font designer to decide exactly how wide, say the extra expanded font is. Browsers can also synthesize condensed or expanded fonts from normal fonts and the example below show how a browser might interpret the keyword values.

Font Stretch Property

The font-stretch property selects a normal, condensed or extended face from a font family. The property accepts two keywords which make the text wider or narrower than the text in the parent element. Also, nine keywords which express stretching in absolute terms are allowed. They have the following order, from narrowest to widest: ultra-condensed, extra-condensed, condensed, semi-condensed, normal, semi-expanded, expanded, extra-expanded, ultra-expanded.

ABCDEG	Normal
ABCDEF	Expanded

It's up to the font designer to decide exactly how wide, say the extra expanded font is. Browsers can also synthesize condensed or expanded fonts from normal fonts and the example below show how a browser might interpret the keyword values.

Numbers as Values

Some properties accept numbers without unit identifiers as values. The font-size-adjust property is one example and line-height is another. The syntax of number values are very simple: the number can be a whole number (0, 1, 2…._ a fractional number (0.5, 2.57. 1.04…….) or a negative number (-1, -3.14, -0.25….) if the integer part of the fractional number is zero, it can be omitted (.25, -, .61).

Text Decoration Property

The text-decoration property does not specify a font property, though it seems to fit best in this chapter because it affects type. The property is used to add underlining, overlining, strike-out or a blinking effect to the text.

The value is either none, meaning no decoration, or any combination of these:

underline an underline is added below the text

overline an overline is added above the text

line-through a horizontal line is inserted through the text (also known as strike-out)

blink the text is made to blink

In many browsers, underlining is used with the A element to mark the status of hyperlinks. The default style sheet for those browsers includes a rule like this:

A:link, A:visited, A:active {text-decoration: underline}

You cannot specify the exact position and thickness of the decorations. Many fonts come with indications of the preferred thicknesses of an overline, underline and line-through and their distances from the baseline and the browser will try to use those thickness and distance values. Otherwise, it will compute appropriate values based on the size of the font. The color of the lines will be the same as the color of the text.

You have a similar lack of precise control over the blink decoration. The blinking text will be shown in its own colors about half of the time; how it looks the other half is not specified. It may be invisible or it may be shown in a different color so that the two colors show alternately. Most browsers will blink at a rate of approximately half a second on and half a second off. Not all browsers can blink and of course, blink will have no effect when the document is printed.

The text-decoration property is not inherited. However, a decoration on a parent will continue in child elements. The effect of this continuation differs from the effect that would result if we were to give the child elements its own decoration. For example, suppose you added this rule to your style sheet:

EM { text-decoration: underline }

In the following, the underline decoration will affect all of the EM element, event he child element, STRONG:

Some <EM>very, <STRONG>very</STRONG>important things </EM> resulted from this effect.

The result would like this:

Some, <u>very, very important things</u> resulted from this effort.

The color of the decoration (if any) also will continue across child elements. Thus even if STRONG had a different color value, its underline would still have been black.

Text Transform Property

The text transform property, like the text-decoration property, does not specify a font. However, it too seems to fit best in this chapter because it affects the case of text.

This property has four values:

- **capitalize**
- **uppercase**
- **lowercase**
- **none**

The capitalize value capitalizes The First Letter Of Each Word. To Give An Effect Like This. Uppercase converts everything to UPPERCASE LETTERS, like these; lowercase does the opposite. None neutralizes an inherited value. These effects are often used in headings and titles but seldom in running text. Here are some example usages:

H1 { text-transform: uppercase }
H1 { text-transform: capitalize }

The rules for converting from uppercase to lowercase and vice versa depend on the language used and on browser. For example, the French typically remove accents when the write in uppercase so that, for example, telephone becomes TELEPHONE. The Dutch have a special letter that is usually written "ij" although it is really a single letter. So a word like ijstijd would be capitalized as IJstijd.

The property also is useful for converting acronyms, like NAATO, BASIC, PIN and NASA to small-caps, as in, NATO, BASIC, PIN and NASA. Acronyms in standard uppercase letters look too large. So in the next example we use small-caps instead. We use font-variant to change the font to small-caps. But we don't use the uppercase version of small-caps, since our intent is to make the acronym look smaller. So we use the lowercase version of small-caps (text-transform: lowercase) to achieve the right effect.

There is often confusion about the effects of text-transform's uppercase and font-variant's small caps. However, as stated earlier in the chapter selecting a small-caps font doesn't change the characters. A text transform on the other hand, does. Font-variant selects a font, but text-transform doesn't. There is still a connection between them, though. The fact that small-caps relies on the availability of a small-caps font also means that it can fail, if there is no such font available. For example, if the style sheet reads;

```
P {
    font-family: "zapf Chancery", cursive;
    font-variant: small-caps;
}
```

The small-caps will fail, since the Zapf Chancery family doesn't have a small-caps variant. The browser may be able to synthesize one by taking the capitals and reducing them in height, but if it is not able to do that, it will show the P in all capitals instead, exactly as if the rule had read text-transform: uppercase. In other words, uppercase is a fallback strategy for browsers that don't have access to small-caps fonts.

Dealing with Space

Extra space in and around elements on a page can enhance your presentation and help get your message across to the reader. Along with influencing color and fonts, influencing spacing has been high on the wishlists of Web page designers. You can effect the space inside a block-level element by changing the amount of space between letters, words, lines and / or paragraphs as well as by varying the alignment of text. Six properties help you influence space inside paragraphs:

- **text-align**
- **text-indent**
- **line-height**
- **word-spacing**
- **letter-spacing**
- **vertical-align**

Among these six properties, text-align, text-indent and line-height are used most often, because they are the primary means of expressing the character of the text and of safeguarding readability. The other three, word-spacing, letter-spacing and vertical-align, are usually used only to achieve special, localized effects.

Text Align Property

The text-align property sets the way the lines are adjusted horizontally between the left and right margins of the element. This property have five values:

Left	Lines are aligned at the left margin; the right margin is ragged (uneven). Sometimes called left-justified.
Right	Lines are aligned at the right margins; the left margin is ragged. Sometimes called right-justified.
Center	Lines are individually centered in the middle of the box; both the right and left margins are ragged.

Justified Lines are aligned on both the left and right margins; text is spread out between the margins as evenly as possible. Sometimes called fully justified.

String Used to align table columns.

Here's an example rule for changing the alignment of a P element from the default to center:

P { text-align: center }

The text-align property is inherited, so you can set the alignment of the whole document by using the BODY element as follows, where we change the alignment from the default to justify:

BODY { text-align: justify }

Revision

1. CSS predefines 16 colour names; ***aqua*** (a light greenish blue, sometimes called cyan), ***black***, ***blue***, ***fuchsia*** (light purple/pink) ***gray***, ***green***, ***lime*** (light green) ***maroon*** (dark red), ***navy*** (dark blue), ***olive***, ***purple***, ***red***, ***silver*** (light gray) ***teal*** (blue-green) ***yellow***, ***white***.
2. Computer monitors commonly use the RGB – red, green, blue – colour model to display colour.
3. The colour property sets the colour of the text of an element.
4. Setting the colour of a border can be done in two ways.
5. Until CSS, the most common use of colour in Web browsers (other than in images) was to draw attention to hyperlinks.
6. The size of type is usually obtained by measuring from roughly the top of its ascenders to the bottom of its descenders.
7. Fonts are typically measured in a unit specific to the printing industry called the point.
8. The font-family property lets you specify the font families which will be used in your documents.
9. A relative unit is a unit of measure that specifies a value that is relative to the font size.

Multiple Choice Questions

1. RGB stands for Red, Green and :
   (a) Blue (b) Badge
   (c) Black
2. CSS has how many defined colours :
   (a) 15 (b) 16
   (c) 14
3. Fonts are measured in :
   (a) Pints (b) Inches
   (c) Points
4. Spacing can be done by :
   (a) Margins (b) Spacer
   (c) Space
5. The most common alignments are, Left, Right, Center and :
   (a) Justify (b) Vertical
   (c) Horizontal
6. The border-style property sets the __________ of the border.
   (a) Appearance (b) Display
   (c) Properties

Answers

1. a 2. b 3. c 4. a 5. a
6. a

Subjective Questions

1. What is CSS?
2. How would you put Style Sheets to the document?
3. What do you mean by Overriding Inheritance?
4. Which properties are not inherited?
5. Describe the common tasks of Fonts, Margins, Links?
6. How does the CSS handle colours?
7. Write short notes on the following:

   Font Metrics
   Font-size property
   Font weight property
   Text Decoration Property
   Relative units
   Font-variant property
   Font stretch property
   Text Transform Property

4 Extensible Markup Language (XML)

Introduction

XML is a language defined by the World Wide Web Consortium the body that sets the standards for the Web. XML supersedes other markup languages such as Hypertext Markup Language (HTML). In HTML all the HTML elements you can use are predefined and there are not enough of them. In fact, XML is a meta markup language because it lets you create your own markup language.

Features of XML

XML is popular for many reasons and few of them are listed here.

Easy Data Exchange

In XML, data and markup are stored as text that you can configure. If you like, you can use XML editors to create XML documents, but if something goes wrong, you can examine or modify the document directly because its all just text. The data also is not encoded in some way that has been patented or copyrighted, so it's more accessible. You might think that binary formats would be more efficient because they can store data more compactly, but that's not the way things have worked out. Even large amounts of data are not necessarily stored efficiently; for instance, Microsoft Excel routinely creates even larger files that are five times as long as the corresponding text. As we'll see, XML provides a very efficient way of storing most data. In addition, when you standardize markup languages, many different people can use them. I'll take a look at that next.

Customizing Markup Languages

You can create customized markup languages using XML, and that represents its extraordinary power. When you and a number of other people agree on a markup language, you can create customized browsers or applications to handle that language. Hundreds of such languages already are being standardized now, including these:

- Banking Industry Technology Secretariat (BITS)
- Financial Exchange (IFX)
- Bank Internet Payment System (BIPS)
- Telecommunications Interchange Markup (TIM)
- Schools Interoperability Framework (SIF)
- Common Business Library (CBL)
- Electronic Business XML Initiative (ebXML)
- Product Data Markup Language (PDML)
- Financial Information eXchange protocol (FIX)
- The Text Encoding Initiative (TEI)

Some customized markup languages, such as Chemical Markup Language (CML), let you represent complex molecules graphically, as we'll see later in this chapter. Likewise, you can imagine how useful a language would be that creates graphical building plans for architects when you open a document in a browser. Not only can you create custom markup languages, but you also can extend them using XML. If someone creates a markup language based on XML, you can add the extensions you want easily.

Self-Describing Data

The data in XML documents is self-describing. Take a look at this document:

```
<?xml version=”1.0” encoding=”UTF-8”?>
<DOCUMENT>
    <GREETING>
        Hello From XML
    <\GREETING>
    <MESSAGE>
        Welcome to the wild and woolly world of XML.
    <\MESSAGE>
<\DOCUMENT>
```

Based solely on the names we've given to each XML element here, you can figure out what's going on: This document has a greeting and a message to impart. Even if you came back to this document years later, you could figure out what's going on. This means that XML documents are, to a large extent, self-documenting.

Structured and Integrated Data

Another powerful aspect of XML is that it lets you specify not only data, but also the structure of that data and how various elements are integrated into other elements. This is important when you're dealing with complex and important data. For example, you could represent a long bank statement in HTML, but in XML you actually can build in the semantic rules that specify the structure of the document so that the document can be checked to make sure it's setup correctly.

Take a look at this XML document:

```
<?xml version=”1.0”?>
<SCHOOL>
    <CLASS type=”seminar”>
        <CLASS_TITLE>XML In The Real World</CLASS_TITLE>
        <CLASS_NUMBER>6.031</CLASS_NUMBER>
        <SUBJECT>XML</SUBJECT>
        <START_DATE>6/1/2002</START_DATE>
        <STUDENTS>
            <STUDENT status=”attending”>
                <FIRST_NAME>Edward</FIRST_NAME>
                <LAST_NAME>Samson</LAST_NAME>
            </STUDENT>
            <STUDENT status=”withdrawn”>
                <FIRST_NAME>Ernestine</FIRST_NAME>
                <LAST_NAME>Johnson</LAST_NAME>
            </STUDENT>
        </STUDENT>
    </CLASS>
</SCHOOL>
```

Here's I've set up an XML seminar and added two students to it. Here emphasis is on the correctness of documents is storing in XML. In XML, browsers are supposed to check your document; if there's a problem, they are not supposed to proceed any further. They should let you know about the problem, but that's as far as they're supposed to go.

Well-formed XML Documents

To be wellformed an XML document must follow the syntax rules set up for XML by W3C in the XML 1.0 specifications. Each element also must nest inside any enclosing elements properly. For example, this document is not well-formed because the </GREETING> closing tag comes after the opening <MESSAGE> tag for the next element:

```
<?xml version="1.0" encoding="UTF-8"?>
<DOCUMENT>
    <GREETING>
        Hello From XML
    <MESSAGE>
    <\GREETING>
        Welcome to the wild and woolly world of XML.
    <\MESSAGE>
<\DOCUMENT>
```

Valid XML Documents

Most XML browsers will check your document to see whether it is well formed. Some of them also can check whether it's valid. An XML document is valid if there is a Document Type Definition (DTD) associated with it and if the document complies with that DTD. A document's DTD specifies the correct syntax of the document. DTDs can be stored in a separate file or in the document itself, using a <!DOCTYPE> element. Here's an example in which I add a <!DOCTYPE> element to the greeting XML document we developed earlier:

```
<?xml version="1.0" encoding="UTF-8"?>
<?xml-stylesheet type="text/css" href="greeting.css"?>
<!DOCTYPE DOCUMENT [
    <!ELEMENT DOCUMENT (GREETING, MESSAGE)>
    <!ELEMENT GREETING (#PCDATA)>
    <!ELEMENT MESSAGE (#PCDATA)>
]>
<DOCUMENT>
    <GREETING>
        Hello From XML
    <\GREETING>
    <MESSAGE>
        Welcome to the wild and woolly world of XML.
    <\MESSAGE>
<\DOCUMENT>
```

DTD indicates that you can have <GREETING> and <MESSAGE> elements inside a <DOCUMENT> element, that the <DOCUMENT> element is the root element and that the <GREETING> and <MESSAGE> elements can hold text. Most XML browsers will check XML documents for well-formedness, but only a few will check for validity.

Parsing XML Yourself

Let's say that you have this XML document, greeting.xml, which we developed earlier in this chapter.

```
<?xml version="1.0" encoding="UTF-8"?>
<DOCUMENT>
    <GREETING>
        Hello From XML
    <\GREETING>
    <MESSAGE>
        Welcome to the wild and woolly world of XML.
```

```
        <MESSAGE>
    <\DOCUMENT>
```

Now say that you want to extract the greeting Hello From XML from this XML document. One way of doing that is by using XML data islands in Internet Explorer and then using a scripting language, such as JavaScript to extract and display the next content of the <GREETING> element. Here's how that looks in a Web page:

```
<HTML>
    <HEAD>
        <TITLE>
            Finding Element Values in an XML Document
        </TITLE>

        <XML ID="firstXML" SRC="greeting.xml"></XML>

        <SCRIPT LANGUAGE="JavaScript">
            function getData()
            {
                xmldoc= document.all("firstXML").XMLDocument;

                nodeDoc = xmldoc.documentElement;
                nodeGreeting = nodeDoc.firstChild;

                outputMessage = "Greeting: " +
                    nodeGreeting.firstChild.nodeValue;
                message.innerHTML=outputMessage;
            }
        </SCRIPT>
    </HEAD>

    <BODY>
        <CENTER>
            <H1>
                Finding Element Values in an XML Document
            </H1>
                <div id="MESSAGE"></div>
                <P>
                <INPUT TYPE="BUTTON" VALUE="Get the Greeting"
                    ONCLICK="getData()">
        </CENTER>
    </BODY>
</HTML>
```

This Web page displays a button with the caption "Get The Greeting,". When you click the button, the JavaScript code in this page reads in greeting.xml, extracts the text from the <GREETING> element and displays that text, as you also can see in figure. In this way, you can see how you can create applications that handle XML documents in a customized way – even customized XML browsers.

Although JavaScript is useful for lightweight XML uses, most XML programming is done in Java today. Here's an example Java program, read XML.java using XML4J probably the most widely used XML praser which is free from IBM's Alpha Works. This program also reads greeting.xml and extracts the text content of the <GREETING> element:

```
import org.apache.xerces.parsers.DOMParseer;
import org.w3c.dom.Document;
import org.w3c.dom.Element;
```

```
import org.w3c.dom.Node;
import org.w3c.dom.Text;

public class readXML

    static public void main(String[ ] argv)
    {
        try {
            DOMParser parser = new DOMParser();
            Parser.parse("greeting.xml");
            Document doc = parser.getDocuemnt();

            For (Node node = doc.getDocumentElement().getFirstChild();
                node != null; node = node.getNextSibling()) {

                if (node instanceof Element) {
                    if (node.getNodeName().equals("GREETING")) {
                        StringBuffer buffer = new StringBuffer();
                        For (Node subnode = node.getFirstChild();
                            subnode != null; subnode =
                                subnode.getNextSibling()){
                            if (subnode instanceof Text) {
                                buffer.append(subnode.getNodeValue());
                            }
                        }
                        System.out.printin(buffer.toString());
                    }
                }
            }
        } catch (Exception   e) {
            e.printStackTrace();
        }
    }
}
```

Structure of a XML document

XML uses its start tags and end tags as containers. Together, the start tags, the content and the end tag all form a single element. Elements are building bricks out of which an XML document is assembled. Each XML document must have only one root element, and other elements must be perfectly nested inside that element. This means that if an element contains other elements, those elements must be completely enclosed within that element.

See the figure here which shows the document in a tree-like structure with the root element at the top of the tree. All the elements that are inside this element are neatly contained within each other. An XML document must contain one and only one root element, and there must not be any elements that are either partially or completely outside, before or after, that element.

To make it easier to refer to the relationships between elements and to elements with respect to other elements, you could say that an element is the parent of the elements that it contains. The elements that are inside an element are called its children. Elements that share the same parent element are called siblings.

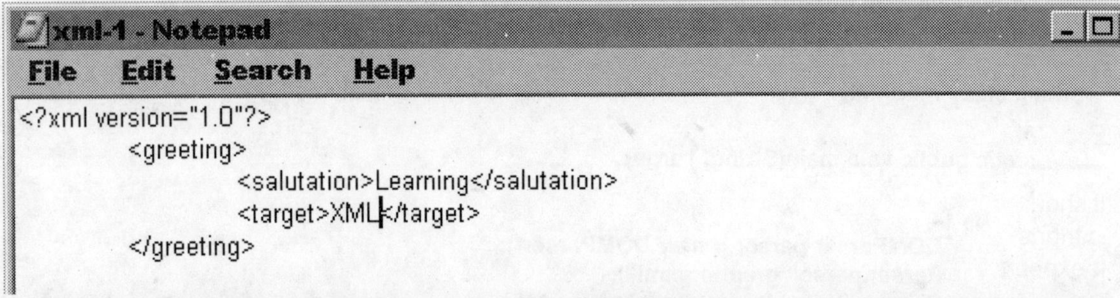

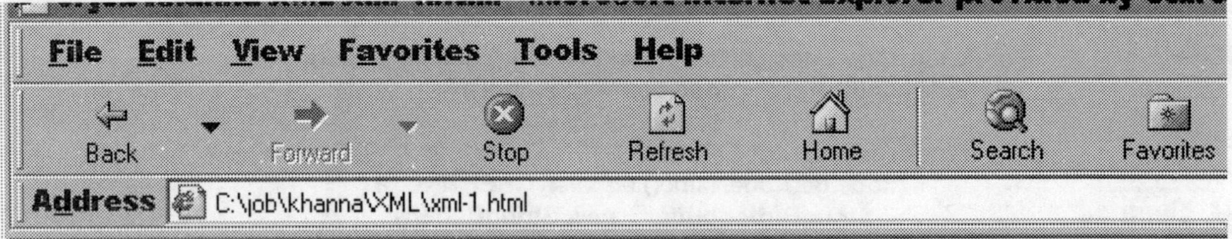

Nesting

It is the process of embedding one object or construct within another as shown above. Every child element resides completely within its parent element. This can be represented as follows:

```
<DOCUMENT>
    <PARENT1>
        <CHILD1></
CHILD1>
        <CHILD2></
CHILD2>
    </PARENT1>
</DOCUMENT>
```

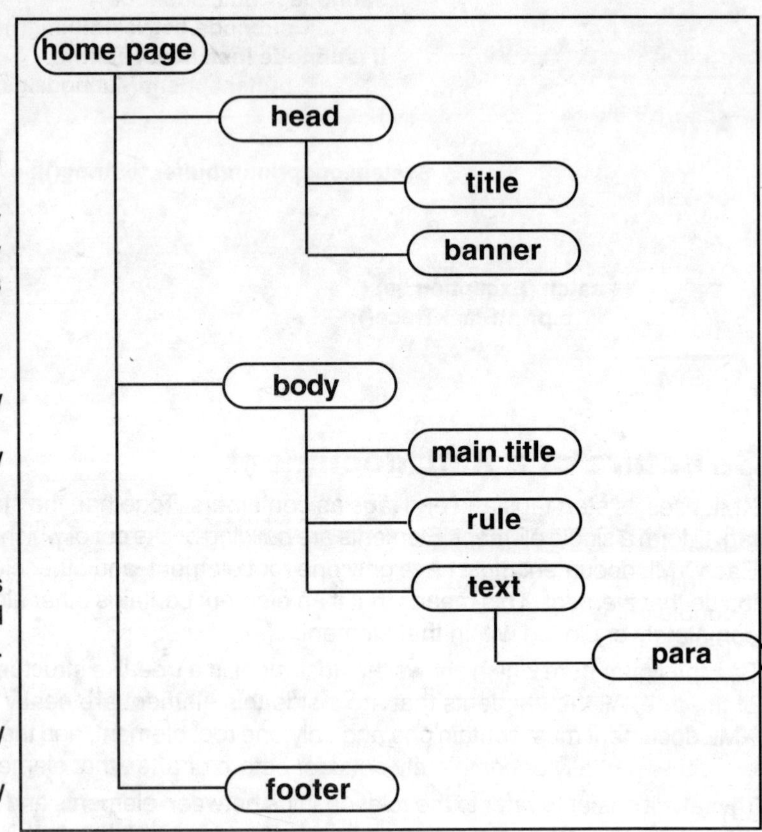

The following code, however, does not contain properly nested elements and will cause an error.

```
<DOCUMENT>
    <PARENT1>
        <CHILD1></
CHILD1>
        <CHILD2></
CHILD2>
</DOCUMENT>
    </PARENT1>
```

XML Element Tags

Following are the parts of XML's element tags.

Symbol	Description	Symbol	Description
<	Start tag open delimiter	</	End tag open delimiter
foo	Element name	>	Tag close delimiter
/>	Empty tag close delimiter		

It should be remembered that whereas HTML simply relies on recognizing preprogrammed tags, XML is triggered by these specific parts of the element tags, and the XML processor's behaviour and what it expects to see next are directly controlled by the named symbols.

Element Markup

Instead of XML's tags being markers that indicate where a style should change or a new line should begin, XML's element markup is composed of 3 parts: a start tag, the contents and the end tag. The start tag and end tag should be treated as wrappers, and when you think of an element, you should have a mental picture of piece of text with both tags in place.

Symbol	Name	Description
foo	Start tag	At the start of an element, the opening tag
text	Content	In the middle of an element, its content
</foo>	End tag	At the end of an element, the closing tag

Attribute Markup

Attributes are used to attach information to the information contained in an element. The general form for using an attribute is:

<!element.name property="value">

or

<!element.name property='value'>

The technical description of the markup of this attribute specifications are given below:

Symbol	Description	Symbol	Description
<	Start tag open delimiter	element.name	Element name
property	Attribute name	=	Value indicator
"	Literal string delimiter	'	Alternative literal string delimiter
value	Value of the attribute	>	Start tag close delimiter

Note that an attribute value must be enclosed in quotation marks. You can use either single quotes or double quotes, but you cannot mix the two in the same specification.

There is one attribute that is reserved for XML's own use—the xml.lang attribute. This attribute is reserved to identify the human language in which the element was written. The value of the attribute is one of the ISO 639 country codes, some of the most common languages are shown below:

Code	Language	Code	Language
ar	Arabic	ch	Chinese
de	German	en	English
es	Spanish	fr	French
gr	Greek	it	Italian
ja	Japanese	nl	Dutch
pt	Portuguese	ru	Russian

When there are several version of a language, such as British and American English, the language code can be followed by a hyphen (-) and one of the ISO 3166 country codes, given below.

Code	*Country*	*Code*	*Country*
AT	Austia	BE	Belgium
CA	Canada	CN	China
DE	Germany	DK	Denmark
EN	England	ES	Spain
FR	France	GR	Greece
IT	Italy	JA	Japan
NL	The Netherlands	PT	Portugal
RU	Russia	US	United States

These codes are just suggestive and not complete.

Structures in XML

A data object is an XML document if it is well formed, as defined in this specification. A well formed XML document has both a logical and a physical structure. Physically, the document is composed of units called entities. An entity may refer to other entities to cause their inclusion in the document. A document begins in a root or document entity. Logically, the document is composed of declarations, elements, comments, character references, and processing instructions, all of which are indicated in the document markup.

Well Formed XML Documents

A textual object is a well formed XML document if:

1. Taken as a whole, it matches the production labeled document.
2. It meets all the well formedness constraints given in this specification.
3. Each of the parsed entities which is referenced directly or indirectly within the document is well formed.

Matching the document production implies that:

1. It contains one or more elements.
2. There is exactly one element, called the root, or document element, no part of which appears in the content of any other element. For all other elements, if the start tag is in the content of another element, the end tag is in the content of the same element. More simple stated, the elements, delimited by start- and end-tags, nest properly within each other.

As a consequence of this, for each non-root element c in the document, there is one other element p in the document such that c is in the content of p, but is not in the content of any other element that is in the content of p. p is referred to as the parent of c, and c as the child of p.

Logical Structures

Each XML document contains one or more elements, the boundaries of which are either delimited by start-tags, or, for empty elements, by an empty-element tag. Each element has a type, identified by name, sometimes called its Generic Identifier (GI), and may have a set of attribute specifications. Each attribute specification has a name and a value.

```
element::= EmptyElemTag
     STage content ETag  [ WFC: Element Type Match ]
                         [ VC: Element Valid ]
```

This specification does not contain the semantics, use, or (beyond syntax) names of the element types and attributes, except that names beginning with a match to (('X' 'x') ('M' 'm') ('L' 'l')) are reserved for standardization in this or future versions of this specification.

Well-Formedness Constraint: Element Type Match: The Name in an element's end-tag must match the element type in the start-tag.

Validity Constratint: Element Valid: An element is valid if there is a declaration matching elementdec1 where the Name matches the element type, and one of the following holds:

1. The declaration matches EMPTY and the element has no content.
2. The declaration matches children and the sequence of child elements belongs to the language generated by the regular expression in the content model, with optional white space (characters matching the nonterminal S) between each pair of child elements.
3. The declaration matches Mixed and the content consists of character data and child elements whose types match names in the content model.
4. The declaration matches ANY, and the types of any child elements have been declared.

Physical Structures

An XML document may consist of one or many storage units. These are called entities; they all have content and are all (except for the document entity and external DTD subset) identified by name. Each XML document has one entity called the document entity, which serves as the starting point for the XML processor and may contain the whole document.

Entities may be either parsed or unparsed. A parsed entity's contents are referred to as its replacement text; this text is considered an integral part of the document.

An unparsed entity is a resource whose contents may or may not be text, and if text, may not be XML. Each unparsed entity has an associated notation, identified by name. Beyond a requirement that an XML processor make the identifiers for the entity and notation available to the application, XML places no constraints on the contents of unparsed entities.

Parsed entities are invoked by name, using entity references; unparsed entities by name, given in the value of ENTITY or ENTITIES attributes. General entities are entities for use within the document content. In this specification, general entities are sometimes referred to with the unqualified term entity when this leads to no ambiguity. Parameter entities are parsed entities for use within the DTD. These two types of entities use different forms of reference and are recognized in different contexts. Furthermore, they occupy different namespaces; a parameter entity and a general entity with the same name are two distinct entities.

Markup and Character Data

XML documents are made up of markup and character data. Binary data might contribute to XML documents some day, but there is no provision for enclosing binary data in a document made up of markup and character data yet; until there is, you refer to external binary data with entity references.

The markup in a document gives it its structure. Markup includes start tags end tags, empty element tags, entity references, character references, comments, CDATA section delimiters document type declarations and processing instructions. So what's the character data in an XML document? All the text in a document that is not markup is character data.

Here's quick example using markup and character data that we've already seen:

```
<?xml version="1.0" encoding="UTF-8"?>
<DOCUMENT>
    <GREETING>
        Hello From XML
```

```
        <GREETING>
        <MESSAGE>
            Welcome to the wild and true world of XML.
        <\MESSAGE>
    <\DOCUMENT>
```

Tags begin with < and end with >, so it's easy to see that the markup here consists of tags, such as <?xml version="1.0" encoding="UTF-8"?>, <DOCUMENT>, and so on. The text Hello From XML and Welcome to the wild and true world of XML is the character data.

However, markup does not need to begin and end with < and >. Markup also can start with & and end with ;- in the case of general entity references – or can start with % and end with ;- in the case of parameter entity references, which are used in DTDs. Using entity references, some of the markup in a document can become character data when you process that document. For example, the markup > is a general entity reference that is turned into a > when parsed and the markup < is turned into a < when parsed. Here's an example:

```
    <?xml version="1.0" encoding="UTF-8"?>
    <DOCUMENT>
        <GREETING>
            This text is inside the &lt;GREETING&gt; element.
        <\GREETING>
    <\DOCUMENT>
```

You can see this XML document in Internet Explorer in figure given below, where you see that the markup > was turned into a > and that the markup < was turned into a <. Because some markup can turn into character data when parsed, the character data that results after everything has been parsed – and markup that should be replaced by character data has been replaced – has a special name: parsed character data.

WhiteSpace

If you're ever concerned about exactly what characters are legal in XML documents, you'll find them listed in the XML 1.0 specification under the production named Char. It's worth nothing that spaces, carriage returns, line feeds and tabs are all treated as whitespace in XML. Take a look this document.

```
<?xml version="1.0" encoding="UTF-8"?>
<DOCUMENT>
<GREETING>
Hello From XML
<\GREETING>
<MESSAGE>
Welcome to the wild and true world of XML.
<\MESSAGE>
<\DOCUMENT>
```

Practically speaking, that document is equivalent to this one:

```
<?xml version="1.0" encoding="UTF-8"?>
<DOCUMENT><GREETING>Hello From XML<\GREETING>
<MESSAGE>Welcome to the wild and true world of XML.<\MESSAGE><\DOCUMENT>
```

It's also worth nothing that the XML recommendation specifies that XML documents use the UNIX convention for line endings, which means that lines are ended with a linefeed character only (ASCII code 10). In DOS files, lines are ended with carriage-return linefeed pairs (ASCII codes 13 and 10), but when parsed, that's treated simply as a single linefeed (ASCII code 10). That gets us started what can go into XML documents: markup and character data. It's now time to move to the next step up and begin working on the actual structure of XML documents, starting with the prolog.

The XML Declaration

An XML document can (and should, according to W3C) start with an XML declaration, which can indicate that the document is written in XML. If you use an XML declaration, it should be the first line in the document. Nothing should come before the XML declaration. Here's an example:

```
<?xml version = "1.0" standalone="yes"? endcoding="UTF-8"?>
```

the XML declaration uses the <?xml?> element. In earlier drafts of XML that was <?XML?> but it was made lowercase in the final recommendation; it's an error to use uppercase.

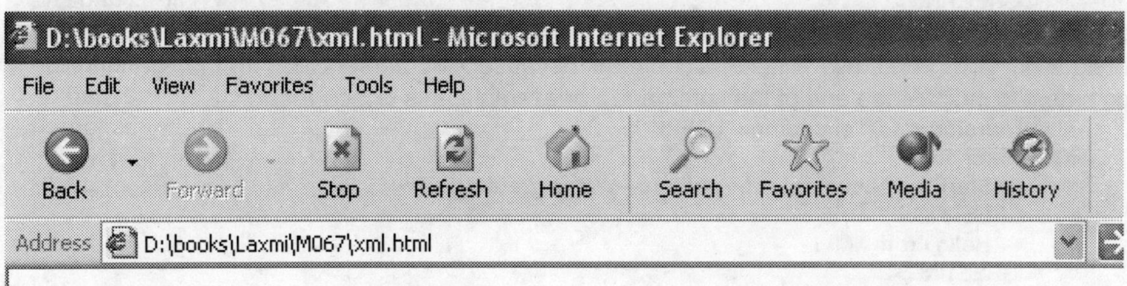

You can use three possible attributes in the XML declaration:

version This is the XML version; currently, only 1.0 is possible here. This attribute is required if you use an XML declaration.

encoding This is the language encoding for the document. As discussed in chapter 1, the default here is UTF-8. You also can use Unicode, UCS 2 or UCS-4 and many other character sets such as ISO character sets. This attribute is optional.

standalone Set this to "yes" if the document does not refer to any external entities; otherwise, use "no". This attribute is optional.

Comments

XML comments are very much like HTML comments. You can use comments to include explanatory notes in your document that are ignored by XML parsers; comments may appear anywhere in a document outside other markup. As in HTML you start a comment with <!.... and end it with…>. Here's an example:

```
<?xml version="1.0" encoding="UTF-8"?>
<DOCUMENT>
    <!—Start the document off with a greeting.—>
    <GREETING>
    <!—Here's the greeting's text,—>
        Hello From XML!
    <\GREETING>
<\DOCUMENT>
```

You should follow a few rules when adding comments to an XML document. Comments must not come before an XML declaration; for example, this is incorrect:

```
<!—Here's my document—>
<?xml version="1.0" encoding="UTF-8"?>
<DOCUMENT>
    <GREETING>
```

```
            Hello From XML
        <\GREETING>
    <\DOCUMENT>
```

You also can't put a comment inside markup, like this:

```
    <?xml version="1.0" encoding="UTF-8"?>
    <DOCUMENT <!—Start the document—>>
        <GREETING>
            Hello From XML!
        <\GREETING>
    <\DOCUMENT>
```

In addition, you cannot use – inside a comment because XML parsers look for that string inside a comment to indicate the end of the comment. For example, this is incorrect:

```
    <?xml version="1.0" encoding="UTF-8"?>
    <DOCUMENT>
        <!—Start the document off—politely—with a greeting.— >
        <GREETING>
            Hello From XML!
        <\GREETING>
    <\DOCUMENT>
```

You can use comments to remove parts of documents as long as the enclosed parts do not themselves contain any comments. For example, here I'm commenting out the <MESSAGE> element:

```
    <?xml version="1.0" encoding="UTF-8"?>
    <DOCUMENT>
        <GREETING>
            Hello From XML
        <\GREETING>
    <!—
        <MESSAGE>
            Welcome to the wild and wooly world of XML.
        </message>
    —>
    <\DOCUMENT>
```

Here's how a parser treats this document:

```
    <?xml version="1.0" encoding="UTF-8"?>
    <DOCUMENT>
        <GREETING>
            Hello From XML
        <\GREETING>
    <\DOCUMENT>
```

Processing Instructions

As their name implies, processing instructions are instructions to the x processor. These instructions start with <? And end with ?>. The only restriction here is that you can't use <?xml?> (or <?xml?>, which is also reserved). Processing instructions must be understood by the XML processor, so they're processor-dependent not build into the XML recommendation. A very common and well-understood processing instruction (although like other processing instructions, not a part of the XML 1.0 recommendation) is <?xml-stylesheet?>, which connects a style sheet with the document. Here's an example:

```
    <?xml version = "1.0" standalone="yes"?>
    <?xml-stylesheet type="text/css"  href="greeting.css"?>
    <DOCUMENT>
```

```
        <GREETING>
            Hello From XML
        <\GREETING>
        <MESSAGE>
            Welcome to the wild and wooly world of XML.
        </message>
    <\DOCUMENT>
```

XML processors, such as Internet Explorer 5 or Netscape Navigator 6, understand <?xml-stylesheet?> already. I've now take a look at everything that a prolog can contain (except DTDs): XML declarations, comments, processing instructions and whitespace. It's time to take a look at the actual structure of an XML document, as created with tags and elements.

Tags and Elements

You give structure to an XML document using markup, which consists of elements. In turn, an XML element consists of a start tag and an end tag, except in the case of elements that are defined to be empty which consist of only one tag. A start tag (also called an opening tag) starts with < and ends with >. End tags (also called closing tags) begin with </ and end with >.

Tag Names

The XML specification is very specific about tag names; you can start a tag name with a letter, an underscore or a colon. The next characters may be letters, digits, underscores, hyphens, periods and colons (but no whitespace).

Avoid Colors in Tag Names: Although the XML 1.0 recommendation does not say so, you should definitely avoid using colons in tag names because you use a colon when specifying namespaces in XML. Here are some allowable XML tags:

```
<DOCUMENT>
<document>
<_Record>
<customer>
<PRODUCT>
```

Note that because XML PROCESSORS ARE CASE-SENSITIVE, THE <DOCUMENT> tag is not the same as a <document> tag. (in fact, you can even have <DOCUMENT> and <DOCUMENT>-and even <DoCuMeNt> - as different tags in the same document, but I really recommend against it.) Here are the corresponding closing tags:

```
</DOCUMENT>
</document>
</_Record>
</customer>
</PRODUCT>
```

These are some tags that x considers illegal:

```
<2001DOCUMENT>
<.document>
<Record Number>
<customer*name>
<PRODUCT (ID)>
```

Using start and end tags, you can create elements as in this example, which has three elements, the <DOCUMENT>, <GREETING> and <MESSAGE> elements; the <DOCUMENTS> element contains the <GREETING> and <MESSAGE> elements:

```
<?xml version = "1.0" standalone="yes"?>
<DOCUMENT>
```

```
<GREETING>
    Hello From XML
</GREETING>
<MESSAGE>
    Welcome to the wild and true world of XML.
<\MESSAGE>
<\DOCUMENT>
```

You also can create elements without using end tags if the elements are explicitly declared to be empty.

Empty Elements

Empty elements have only one tag, not a start and end tag. You may be familiar with empty elements from HTML; for example, the HTML <IMG>, <LI>, <HR> and <BR> elements are empty, which is to say that they do not enclose any content (either character data or markup). Empty elements are represented with only one tag (in HTML there is no closing </IMG>, <LI>, <HR> and <BR> tags). In XML, you can declare elements to the empty in the document's DTD.

In XML you close an empty element with />. For example, if the <GREETING> element is empty, it might look like this in an XML document:

```
<?xml version = "1.0" standalone="yes"?>
<DOCUMENT>
    <GREETING TEXT = "Hello From XML" />
<\DOCUMENT>
```

This usage might seem a light strange at first, but this is XML's way of making sure that an XML processor isn't left searching for a nonexistent closing tag. In fact, in XHTML, which is a derivation of HTML in XML, the <IMG>, <LI>, <HR> and <BR> tags are actually used as <IMG />, <LI />, <HR / > and
 (except that XHTML tags use lowercase letters). The additional / doesn't seem to give the major browsers any trouble.

The Root Element

Each well-formed XML document must contain one element that contains all the other elements. This containing element is called the root element! The root element is a very important part of XML documents, especially when you look at them from a programming point of view, because you parse XML documents starting with the root element. In order.xml developed at the start of this chapter, the root element is the <DOCUMENT> element (although you can give the root element any name):

```
<?xml version = "1.0" standalone="yes"?>
<DOCUMENT>
<CUSTOMER>
<NAME>
<LAST_NAME>Smith</LAST_NAME>
<FIRST_NAME>Sam</FIRST_NAME>
</NAME>
<DATE>October 15, 2001</DATE>
<ORDERS>
        <ITEM>
            <PRODUCT>Tomatoes</PRODUCT>
            <NUMBER>8</NUMBER>
            <PRICE>$1.25</PRICE>
        </ITEM>
        .
        .
        .
```

```
        <ITEM>
            <PRODUCT>Lettuce</PRODUCT>
            <NUMBER>6</NUMBER>
            <PRICE>$11.50</PRICE>
        </ITEM>
    </ORDERS>
</CUSTOMER>
</DOCUMENT
```

Attributes

Attributes in XML are much like attributes in HTML – they're name-value pairs that let you specify additional data in start and empty tags. To assign a value to an attribute, you use an equal sign. For example, I'm assigning values to the STATUS attributes of the <CUSTOMER> elements in this XML to indicate the status of a customer's credit:

```
<?xml version = “1.0” standalone=”yes”?>
<DOCUMENT>
<CUSTOMER STATUS=”Good Credit”>
<NAME>
<LAST_NAME>Smith</LAST_NAME>
<FIRST_NAME>Sam</FIRST_NAME>
</NAME>
<DATE>October 15, 2001</DATE>
<ORDERS>
        <ITEM>
            <PRODUCT>Tomatoes</PRODUCT>
            <NUMBER>8</NUMBER>
            <PRICE>$1.25</PRICE>
        </ITEM>
        <ITEM>
            <PRODUCT>Oranges</PRODUCT>
            <NUMBER>24</NUMBER>
            <PRICE>$4.98</PRICE>
        </ITEM>
    </ORDERS>
</CUSTOMER>
<CUSTOMER STATUS=”Lousy credit”>
    <NAME>
<LAST_NAME>Jones</LAST_NAME>
<FIRST_NAME>Polly</FIRST_NAME>
</NAME>
<DATE>October 20, 2001</DATE>
<ORDERS>
        <ITEM>
            <PRODUCT>Bread</PRODUCT>
            <NUMBER>12</NUMBER>
            <PRICE>$14.95</PRICE>
        </ITEM>
        <ITEM>
            <PRODUCT>Apples</PRODUCT>
            <NUMBER>6</NUMBER>
            <PRICE>$1.50</PRICE>
        </ITEM>
    </ORDERS>
```

```
</CUSTOMER>
<CUSTOMER STATUS="Good credit">
     <NAME>
<LAST_NAME>Weber</LAST_NAME>
<FIRST_NAME>Bill</FIRST_NAME>
</NAME>
<DATE>October 25, 2001</DATE>
<ORDERS>
          <ITEM>
               <PRODUCT>Asparagus</PRODUCT>
               <NUMBER>12</NUMBER>
               <PRICE>$2.95</PRICE>
          </ITEM>
          <ITEM>
               <PRODUCT>Lettuce</PRODUCT>
               <NUMBER>6</NUMBER>
               <PRICE>$11.50</PRICE>
          </ITEM>
     </ORDERS>
</CUSTOMER>
</DOCUMENT>
```

An XML processor can read the attributes and their values and you can put that data to work in your own applications. A lot of debate occurs over when you should store data using attributes and when you should store data using elements. There simply is no hard and fast rule, but here are a couple guidelines that I find useful.

First, too many attributes definitely make documents hard to read. For example, take a look at this element:

```
<CUSTOMER>
<NAME>
<LAST_NAME>Smith</LAST_NAME>
<FIRST_NAME>Sam</FIRST_NAME>
</NAME>
<DATE>October 15, 2001</DATE>
<ORDERS>
     <ITEM>
          <PRODUCT>Tomatoes</PRODUCT>
          <NUMBER>8</NUMBER>
          <PRICE>$1.25</PRICE>
     </ITEM>
     </ORDERS>
</CUSTOMER>
```

It's fairly clear what going on here, even if it is a little involved. However, if you try to convert all this data to attributes, you end up with something like this:

```
<CUSTOMER LAST_NAME="Smith" FIRST_NAME="Sam"
DATE="October 15, 2001" PURCHASE="Tomatoes"
PRICE="$1.25" NUMBER="8" />
```

Clearly, this will be a mess if you have a few such elements.

Another point is that you really can't specify document structure using attributes. For example, the example we've already seen in this chapter stores multiple ordered items per customer. However, attributes names must be unique, so it's much tougher to store data like this using attributes:

```
<CUSTOMER>
```

```
<NAME>
<LAST_NAME>Smith</LAST_NAME>
<FIRST_NAME>Sam</FIRST_NAME>
</NAME>
<DATE>October 15, 2001</DATE>
<ORDERS>
    <ITEM>
        <PRODUCT>Tomatoes</PRODUCT>
        <NUMBER>8</NUMBER>
        <PRICE>$1.25</PRICE>
    </ITEM>
    <ITEM>
        <PRODUCT>Oranges</PRODUCT>
        <NUMBER>24</NUMBER>
        <PRICE>$4.98</PRICE>
    </ITEM>
</ORDERS>
</CUSTOMER>
```

The upshot is that deciding whether to store your data in attributes or to create new elements is really a matter of taste until you get beyond a few attributes per tag. If you find yourself using (not just defining, but using more than four attributes in a tag, consider breaking up the tag into a number of enclosed tags. Doing so will make the document structure much easier to work with the edit later.

You should follow specific rules when creating attributes and those include correctly setting attribute names and specifying attribute values.

Attribute Names

According to the XML 1.0 specification, attribute names must follow the same rules as those for tag names, which means that you can start an attribute name with a letter, an underscore or a colon. The next chapters may be letters, digits, underscores, hyphens, periods and colons (but no whitespace because you separate attribute name-value pairs with whitespace).

Take a look at these examples showing legal attribute names:

```
<circle origin_x="10.0 origin_y="20.0" radius="10.0" />
<image3 src="image1.jpg">
<pen color="red" width="5">
<book pages="1231" >
```

Here are some illegal attribute names:

```
<circle 1origin_x="10.0 1origin_y="20.0" 1radius="10.0" />
<image src="image1.jpg">
<pen color@="red" width@="5">
<book pages (excluding front matter)="1231" >
```

Attribute Values

Because markup is always text, attributes are also text. Even if you're assigning a number to an attribute, you treat that number as a text string and enclose it in quotes, like this:

```
<circle origin_x="10.0 origin_y="20.0" radius="10.0" />
```

Among other things, this means that XML processors will return attribute values as text strings. If you want to treat them as numbers, you'll have to make sure that you translate them to numbers as the programming language you're using allows.

In XML, you must enclose attribute values in quotation marks. Usually, you use double quotes, but consider the case in which the attribute value itself contains double quotes – you can't just surround such a value with double quotes because an XML processor will get confused as to where your text ends. In such a case, you can use single quotes to surround the text, like this:

```
<quotation text="He said, "Not that!" ' />
```

what if the attribute value contains both single and double quotes? In that case, you can use the XML – defined entity ' for a single quote and " for double quotes, like this:

```
<person height="5apos;6&quot;" />
```

Assigning Values to Attributes

If you're going to use an attribute, you must assign it a value. Not doing so is a violation of well-formedness – that is, you cannot have "standalone" attributes in XML, such as the BORDER attribute in HTML tables, which need not be assigned a value.

Creating Document Type Declarations

You define the syntax and structure of elements using a document type definition (DTD) and you declare that definition in a document using a document type declaration. We've seen that you use <!DOCTYPE> to create a document type declaration. This element can take many different forms, as you see here:

- **<!DOCTYPE rootname [DTD]>**
- **<!DOCTYPE rootname SYSTEM URL>**
- **<!DOCTYPE rootname SYSTEM URL [DTD]>**
- **<!DOCTYPE rootname PUBLIC identifier URL>**
- **<!DOCTYPE rootname PUBLIC identifier URL [DTD]>**

To use a DTD, you need a document type declaration, which means that you need <!DOCTYPE>. The <!DOCTYPE> declaration is part of a document's prolog (also called prologue).

```
<?xml version = "1.0" standalone="yes"?>
<!DOCTYPE DOCUMENT [
.
.
.

]>
<DOCUMENT>
<CUSTOMER>
<NAME>
<LAST_NAME>Smith</LAST_NAME>
<FIRST_NAME>Sam</FIRST_NAME>
</NAME>
<DATE>October 15, 2001</DATE>
<ORDERS>
        <ITEM>
                <PRODUCT>Tomatoes</PRODUCT>
                <NUMBER>8</NUMBER>
                <PRICE>$1.25</PRICE>
        </ITEM>
        .
        .
        .
<ITEM>
                <PRODUCT>Asparagus</PRODUCT>
                <NUMBER>12</NUMBER>
                <PRICE>$2.95</PRICE>
```

```
            </ITEM>
            <ITEM>
                <PRODUCT>Lettuce</PRODUCT>
                <NUMBER>6</NUMBER>
                <PRICE>$11.50</PRICE>
            </ITEM>
        </ORDERS>
    </CUSTOMER>
    </DOCUMENT
```

Now it's up to us to supply the actual document type definition, the DTD, that's part of this <!DOCTYPE> declaration.

Creating Document Type Definitions

To introduce DTDs, I'll start with a DTD that's internal to the document whose syntax it specifies. In this case, the DTD itself goes inside the square brackets in <!DOCTYPE> (note that I've set the standalone attribute to "yes" here because this document doesn't rely on any external resources):

```
<?xml version="1.0" standalone="yes"?>
<!DOCTYPE DOCUMENT [
<!ELEMENT DOCUMENT (CUSTOMER)*>
<!ELEMENT CUSTOMER (NAME,DATE,ORDERS)>
<!ELEMENT NAME (LAST_NAME,FIRST_NAME)>
<!ELEMENT LAST_NAME (#PCDATA)>
<!ELEMENT FIRST_NAME (#PCDATA)>
<!ELEMENT DATE (#PCDATA)>
<!ELEMENT ORDERS (ITEM)*>
<!ELEMENT ITEM (PRODUCT,NUMBER,PRICE)>
<!ELEMENT PRODUCT (#PCDATA)>
<!ELEMENT NUMBER (PCDATA)>
<!ELEMENT PRICE (#PCDATA)>
]>
<DOCUMENT>
<CUSTOMER>
<NAME>
<LAST_NAME>Smith</LAST_NAME>
<FIRST_NAME>Sam</FIRST_NAME>
</NAME>
<DATE>October 15, 2001</DATE>
<ORDERS>
        <ITEM>
            <PRODUCT>Tomatoes</PRODUCT>
            <NUMBER>8</NUMBER>
            <PRICE>$1.25</PRICE>
        </ITEM>
        .
        .
        .
        <ITEM>
            <PRODUCT>Asparagus</PRODUCT>
            <NUMBER>12</NUMBER>
            <PRICE>$2.95</PRICE>
        </ITEM>
        <ITEM>
            <PRODUCT>Lettuce</PRODUCT>
```

```
            <NUMBER>6</NUMBER>
            <PRICE>$11.50</PRICE>
        </ITEM>
    </ORDERS>
</CUSTOMER>
</DOCUMENT>
```

When a DTD is in place – and we'll see how to create this DTD in the following sections – you have a valid document. Having gotten <!DOCTYPE> in place, we're ready to start creating the DTD, starting with <!ELEMENT>.

Element Declaration

To declare the syntax of an element in a DTD, you use <!ELEMENT> like this: <!ELEMENT NAME CONTENT_MODEL>. Here, NAME is the name of the element that you're declaring; CONTENT_MODEL can be set to EMPTY or ANY or it can hold mixed content (other elements as well as parsed character data) or child elements.

Here are a few examples. Note the expressions starting with % and ending with ;. Those expressions are parameter entity references, much like general entity references except that you use them in DTDs not the body of the document (we'll see parameter entities in the next chapter):

```
<!ELEMENT direction (left, right, top?)>
<!ELEMENT CHAPTER (INTRODUCTION, (P I QUOTE I NOTE)*, DIV*)>
<!ELEMENT HR EMPTY>
<!ELEMENT p (#PCDATA I i)* >
<!ELEMENT %title; %content; >
<!ELEMENT DOCUMENT ANY>
```

We're going to see how to create<!ELEMENT? declarations like these in this and the next chapter. I'll start by declaring the root element of the example document for this chapter, order.xml:

```
<?xml version="1.0" standalone="yes"?>
<!DOCTYPE DOCUMENT [
<!ELEMENT DOCUMENT ANY>
]>
<DOCUMENT>
</DOCUMENT>
```

Notice that I'm specifying a content model of ANY here; see the next section for the details on this keyword.

ANY

When you declarer an element with the content model of ANY, that means that the declared element can contain any type of content – any element in the document, as well as parsed character data. (Effectively, this means that the contents of elements that you declare with the ANY content model are not checked by XML validators. Here is how you specify a content model of ANY:

```
<?xml version="1.0" standalone="yes"?>
<!DOCTYPE DOCUMENT [
<!ELEMENT DOCUMENT ANY>
]>
<DOCUMENT>
</DOCUMENT>
```

However, giving an element the content model ANY is often not a good idea because it removes syntax checking. It's usually far better to specify an actual content model and I'll start doing that with a child list of elements.

Child Element Lists

Besides using the content model of ANY, you can specify that the element you're declaring contain another element by giving the name of that element in parentheses, like this:

```
<?xml version="1.0" standalone="yes"?>
<!DOCTYPE DOCUMENT [
<!ELEMENT DOCUMENT (CUSTOMER)*>
]>
<DOCUMENT>
.
.
.
</DOCUMENT>
```

In this case, I'm indicating that the root element, <DOCUMENT> can contain any number (including zero) of <CUSTOMER> elements. Because the <DOCUMENT> element can contain any number of <CUSTOMER> elements, I can now add a <CUSTOMER. Element to the document, like this:

```
<?xml version="1.0" standalone="yes"?>
<!DOCTYPE DOCUMENT [
<!ELEMENT DOCUMENT ANY>
]>
<DOCUMENT>
    <CUSTOMER>
    .
    .
    .
    </CUSTOMER>
</DOCUMENT>
```

However, this is not a valid document because I haven't declared the <CUSTOMER> element yet. I'll do that next.

#PCDATA

Say that we want to let the <CUSTOMER> element store some plain text – in particular, say that we want to store the name of a customer. All nonmarkup text is referred to as parsed character data in a DTD and it is abbreviated as #PCDATA in element declarations. Parsed character data explicitly means text that does not contain markup, just simple character data.

The parsed character data is where you store the actual content of the document as plain text. Note, however, that this is the only way to specify the content of the document using DTDs – you can't say anything more about the actual type of content.

For example, even though you might be storing numbers, that data is only plain text as far as DTDs are concerned. This lack of precision is one of the reasons that XML schemas, the alternative to DTDs were developed. With schemas, you can specify much more about the type of data you're storing, such as whether it's in integer, floating point or even date format and XML processors can check to make sure the data matches the format that it's supposed to be expressed in.

Here's how I declare the <CUSTOMER> element so that it can contain PCDATA (and only PCDATA):

```
<?xml version="1.0" standalone="yes"?>
<!DOCTYPE DOCUMENT [
<!ELEMENT DOCUMENT (CUSTOMER)*>
<!ELEMENT CUSTOMER (#PCDATA)>
]>
<DOCUMENT>
    <CUSTOMER>
```

```
.
.
.
</CUSTOMER>
</DOCUMENT>
```

Now I can add text to a <CUSTOMER> element in the document like this:

```
<?xml version="1.0" standalone="yes"?>
<!DOCTYPE DOCUMENT [
<!ELEMENT DOCUMENT (CUSTOMER)*>
<!ELEMENT CUSTOMER (#PCDATA)>
]>
<DOCUMENT>
    <CUSTOMER>
    Sam Smith
    </CUSTOMER>
</DOCUMENT>
```

Note that elements that have been declared to hold PCDATA can hold only PCDATA; you cannot, for example, place another element in the <CUSTOMER> element the way that it has been declared now – this document is not valid:

```
<?xml version="1.0" standalone="yes"?>
<!DOCTYPE DOCUMENT [
<!ELEMENT DOCUMENT (CUSTOMER)*>
<!ELEMENT CUSTOMER (#PCDATA)>
]>
<DOCUMENT>
    <CUSTOMER>
    Sam Smith
    <CREDIT_RATING>
        Lousy
    </CREDIT_RATING>
    </CUSTOMER>
</DOCUMENT>
```

The content model that support both PCDATA and other elements inside an element inside an element is called the mixed content model. There's another thing to note here now that we're dealing with multiple declarations – the order in which you declare elements doesn't matter, so this DTD, where I've declared the <DOCUMENT> element after the <CUSTOMER> element, works just as well:

```
<?xml version="1.0" standalone="yes"?>
<!DOCTYPE DOCUMENT [
<!ELEMENT CUSTOMER (#PCDATA)>
<!ELEMENT DOCUMENT (CUSTOMER)*>
]>
<DOCUMENT>
    <CUSTOMER>
    .
.
.
</CUSTOMER>
</DOCUMENT>
```

Note that although the order of element declarations is not supposed to matter – and in practice, that's the way I've always seen it – some XML processors may demand that you declare an element before using it in another declaration. It's also possible to declare elements in such a way that they can contain multiple children. In fact, you an specify the exact types of child elements that an elements can enclose and in what order those child elements must appear. I'll take a look at that now.

Dealing with Multiple Children

When you want to declare an element that can contain multiple children, you have several options. DTDs use a syntax to deal with multiple children that is much like working with regular expressions in languages such as Perl, in case you're familiar with that. Here's the syntax that you can use (here, a and b are child elements of the elements you're declaring):

a+ -One or more occurrences of a.

a* - Zero or more occurrences of a.

a? a or nothing.

a, b a followed by b.

a I b a or b, but not both.

(expression) – Surrounding an expression with parentheses means that it's treated as a unit and may have the suffix operator ? (Question Mark), *(Asterik Sign), or +(Plus Sign). If you're not familiar with this kind of syntax, it's not much use asking why things are set up this way; this syntax has been around a long time and W3C adopted it for DTDs because many people were familiar with it. If this books totally strange to you, it's just one of the skills you'll have to master when writing DTDs – but, fortunately, it soon becomes second nature.

Using Plus (+) Operator

If you must specify that the <DOCUMENT> element can contain only between 12 and 15 <CUSTOMER> elements, you'll have a problem when working with DTDs because the DTD syntax won't allow you to do that without getting very complex. However, you can specify that the <DOCUMENT> element must contain one or more <CUSTOMER> elements like this, using the + operator:

```
<?xml version="1.0" standalone="yes"?>
<!DOCTYPE DOCUMENT [
<!ELEMENT DOCUMENT (CUSTOMER)*>
<!ELEMENT CUSTOMER (#PCDATA)>
]>
<DOCUMENT>
    <CUSTOMER>
        Sam Smith
    </CUSTOMER>
    <CUSTOMER>
        Fred Smith
    </CUSTOMER>
</DOCUMENT>
```

In this case, the XML processor now know that you want the <DOCUMENT> element to contain one or more <CUSTOMER> elements, which makes sense if you want a useful document that actually contains some data. In this way, we've been able to specify the syntax of the <DOCUMENT> element in some more detail.

Using * (Asterisk) Operator

Besides specifying one or more child elements, you can also declare elements so that they can enclose zero or more of a particular child element. This is useful if you want to allow an element to have a particular child element or any number of such elements, but you don't want to force it to have that particular child element. For example, a <CHAPTER> element might be capable of containing a <FOOT-NOTE> element or even several <FOOTNOTE> elements, but you wouldn't necessarily want to force all <CHAPTER> elements to have <FOOTNOTE> elements. Using the * operator, you can do that.

The * operator means that the indicated child element can appear any number of times in the declared element, (including zero times). Here's how I indicate that the <DOCUMENT> element can contain any number of <CUSTOMER> elements:

```
<?xml version="1.0" standalone="yes"?>
<!DOCTYPE DOCUMENT [
<!ELEMENT DOCUMENT (CUSTOMER)*>
<!ELEMENT CUSTOMER (#PCDATA)>
]>
<DOCUMENT>
    <CUSTOMER>
        Sam Smith
    </CUSTOMER>
    <CUSTOMER>
        Fred Smith
    </CUSTOMER>
</DOCUMENT>
```

Using Question Mark (?) Operator

Besides using + to specify one or more occurrences of a particular child element and * to specify zero or more occurrences of a child element, you can also use ? to specify zero or one occurrences of a child element. In other words, using ? indicates that a particular child element may be present in the element you're declaring, but it need not be. For example, a <CHAPTER> element might be capable of containing one <OPENING_QUOTATION> element, but you wouldn't necessarily way to force all <CHAPTER> elements to have an <OPENING_QUOTATION> element. Using the ? operator, you can do that.

Here's an example: in this case, I'm allowing the <DOCUMENT> element to contain only zero or one <CUSTOMER> element (rather a limited clientele):

```
<?xml version="1.0" standalone="yes"?>
<!DOCTYPE DOCUMENT [
<!ELEMENT DOCUMENT (CUSTOMER)*>
<!ELEMENT CUSTOMER (#PCDATA)>
]>
<DOCUMENT>
    <CUSTOMER>
        Sam Smith
    </CUSTOMER>
</DOCUMENT>
```

We've advanced a little in DTD power now by allowing multiple child elements, but so far, we've allowed only child elements of the same type in any one declared element – but that's about to change.

What Does XML Look Like?

So what does XML look like and how does it work? Here's an example that mimics the HTML page just introduced:

```
<?xml version="1.0" encoding="UTF-8"?>
<DOCUMENT>
    <GREETING>
```

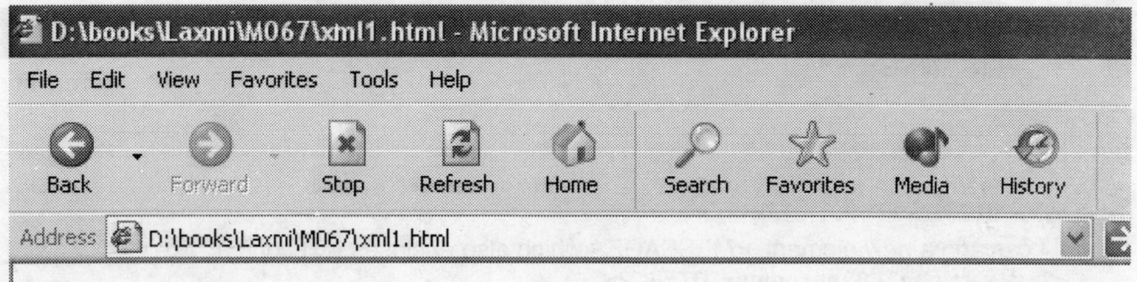


```
        Hello From XML
    <\GREETING>
    <MESSAGE>
        Welcome to the wild and woolly world of XML.
    <\MESSAGE>
<\DOCUMENT>
```

Here's how this one works: I start with the XML processing instruction <?xml version="1.0" encoding="UTF-8"?>>(all XML processing instructions start with <? And end with?>), which indicates that's I'm using XML version 1.0 (the only version currently defined) and the UTF-8 character encoding, an 8-bit condensed version of Unicode. Also, when I add new sections of code, they'll be highlighted with shading to point out the actual lines I'm discussing.

```
<DOCUMENT>
    <GREETING>
        Hello From XML
    <\GREETING>
    <MESSAGE>
        Welcome to the wild and woolly world of XML.
    <\MESSAGE>
<\DOCUMENT>
```

Next, I create a new tag named <DOCUMENT>. As we'll see in the next chapter you can use any name for a tag, not just DOCUMENT as long as the name starts with a letter or an underscore (_) and as long as the following characters consist of letters, digits, underscores, dots (.) or hyphen (-) but no spaces. In XML, tags always start with < and end with >.

XML documents are made up of XML elements. Much like in HTML, you create XML elements content (if any), such as text or other elements and ending with the matching closing tag that starts with </ , such as <\DOCUMENT>. (There are additional rules we'll see in the next chapter if the element doesn't contain any content.) It's necessary to enclose the entire document, except for processing instructions, in one element, called the root element – that's the <DOCUMENT> element here:

```
<?xml version="1.0" encoding="UTF-8"?>
<DOCUMENT>
.
.
.
<\DOCUMENT>
```

Now I'll add to this XML document a new element that I made up <GREETING> which encloses text content (in this case, Hello From XML), like this:

```
<?xml version="1.0" encoding="UTF-8"?>
<DOCUMENT>
```

```
<GREETING>
    Hello From XML
<\GREETING>
.
.
.
.
<\DOCUMENT>
```

Next, I can add a new element, <MESSAGE> which also encloses text content, like this:

```
<?xml version="1.0" encoding="UTF-8"?>
<DOCUMENT>
    <GREETING>
        Hello From XML
    <\GREETING>
    <MESSAGE>
        Welcome to the wild and true world of XML.
    <\MESSAGE>
<\DOCUMENT>
```

Now the <DOCUMENT> root element contains two elements: <GREETING> and <MESSAGE>. Each of the <GREETING> and <MESSAGE> elements also hold text themselves. In this way, I've created a new XML document.

Note the similarly of this document to the HTML page we saw earlier. Note also, however, that in the HTML document, all the tags were predefined and a Web browser knew how to handle them. Here, we've just created the tags <DOCUMENT>, <GREETING> and <MESSAGE> from thin air. How can we use an XML document like this one? What would a browser make of these new tags?

What Does XML Look Like in a Browser?

It turns out that a browser such as Microsoft Internet Explorer version 5 or later lets you display raw XML documents directly. For example, if I saved the XML document we just crate in a document named greeting.xml and then opened that document in the Internet Explorer, you'd see something like what appears in figure on the next page.

You can see our complete XML document in figure but it's nothing like the image you see in figure; there is no particular formatting at all. So, now that we've created our own markup elements how do you tell a browser how to display them?

Many people who are new to XML find the claim that you can use XML to create new markup languages very frustrating – after all, what then? It turns out that it's up to you to assign meaning to the new elements you create and you can do that in two main ways. First, you can use a style sheet to indicate to a browser how you wan the content of the elements you've created to be formatted. The second way is to use a programming language, such as JAVA or JavaScript to handle the XML document in programming code. We'll see both ways throughout this book and I'll take a quick look at them in this chapter as well. I'll start by adding a style sheet to the XML document we've already created.

There are two main ways of specifying styles when you format XML documents: with cascading style sheets (CSS) and with the Extensible style Sheets Language (XSL). We'll see both in this book; here, I'll apply a CSS style sheet by using the XML processing instruction <?xml-stylesheet type="text/css" href="greeting.css"?> which tells the browser that the type of the style sheet I'll be using is CSS that its name is greeting.css:

```
<?xml version="1.0" encoding="UTF-8"?>
<?xml-stylesheet type="text/css"  href="greeting.css"?>
```

```
<DOCUMENT>
    <GREETING>
        Hello From XML
    <\GREETING>
    <MESSAGE>
        Welcome to the wild and true world of XML.
    <\MESSAGE>
<\DOCUMENT>
```

Here's what the contents of the file greeting.css itself looks like. In this case, I'm customizing the <GREETING> element to display its content in red, centered in the browser and in 36-point font. The <MESSAGE> element has been customized to display its text in black 18-point font. The display: block part indicates that I want the content of these elements to be displayed in a block, which translates here to being displayed on individual lines:

```
GREETING {display: block; font-size: 36pt; color: #FF0000; text-align: center}
MESSAGE (display: block; font-size: 18pt; color: #000000}
```

You can see the results in two browsers that support XML.

XSL—Putting Styles in XML

CSS is the short form of Cascading Style Sheet whereas XSLFO is the XSL Formatting Objects. They both have different approach. CSS describes directly how to render documents onscreen or on paper. CSS is the mechanism you use to tell the browser how to style the elements. It also deals with fonts, colours, text identification, and so on.

On the other hand XSLT style sheet converts the XML title element in two HTML elements: paragraph and italicized text.

Two versions of CSS have released so far: CSS1 and CSS2. CSS2 builds on CSS1 but it improves XML support, adds new styling options, and supports alternate media such as paper printing. Microsoft Internet Explorer supports CSS1.

The various style sheet options supported by CSS are:

Comments

They are enclosed within /* and */. Very much similar to C language. They are ignored by the browser. For example:

```
/*an illustration */
```

Selector

CSS rules are associated with elements through selectors. To select an element, use its name. For example,

```
article {
    font-family: Arial, Times New Roman,
"Helvetica", serif;
    font-size: 12pt;
    margin: 6px;
}
```

To select several items separate them with commas. You can select an element depending on its ancestor, list the two elements separated by space. For example,

```
article title {
```

```
    font-size: larger;
    font-weight: bold;
}
```

Properties

They are enclosed in curly brackets. Each property has a name followed by a colon and one or more values (separated by commas). A semicolon terminates the property. For example:

```
article {
    font-family: Arial, Times New Roman,
"Helvetica", serif;
    font-size: 12pt;
    margin: 6px;
}
```

Flow Objects

They are used to render the screen or print a page. The document flows from the top to the bottom of the canvas. Anything in the flow, characters, words, paragraphs, images is a flow object. Let us learn about a few of them.

Boxes

The simplest form of flow object is a box. A box is a rectangular area on the screen or on paper. Every element is rendered in a box.

CSS supports several types of boxes. The important ones are:

- Block box, which a line break before and after it. It is the most important element in organizing a document on the screen.
- Inline box, which appears in a single line within a block box. There are no line breaks before or after it.
- Anonymous box, which is created automatically by the browser when needed.

The display property describes how an element is displayed. The following example selects the element as block box:

```
display: block;
```

Various other properties of the box are:

Margins

You can have the margin on all sides of the box, i.e., margin-top, margin-right, margin-bottom, and margin-left. These can be applied as this:

```
margin-right: 15px;
```

You can give here the percentage, relative weightage, etc., too.

```
margin-right: 15px;
margin-top: 50%
margin-left: 10px;
margin-bottom: auto;
```

Or you can combine all of them to give one instruction of margin, as follows:

```
margin: 15px;
```

Padding

It is similar to margins, but the value auto is not acceptable here.

```
padding: 0.3in;
```

Border width

Similar to margins, you can have border-top, border-right, border-bottom and border-left, which control the width of each border independently. Acceptable values are thin, medium, or an absolute length.

```
border-width: thin;
```

Border shorthand

The border-top, border-right, border-bottom and border-left are the shorthand properties for all pervious properties. They list the width, style and colour of the border:.

```
border: thin solid silver;
```

Revision

1. CSS predefines 16 colour names; ***aqua*** (a light greenish blue, sometimes called cyan), ***black***, ***blue***, ***fuchsia*** (light purple/pink) ***gray***, ***green***, ***lime*** (light green) ***maroon*** (dark red), ***navy*** (dark blue), ***olive***, ***purple***, ***red***, ***silver*** (light gray) ***teal*** (blue-green) ***yellow***, ***white***.
2. Computer monitors commonly use the RGB – red, green, blue – colour model to display colour.
3. The colour property sets the colour of the text of an element.
4. Setting the colour of a border can be done in two ways.
5. Until CSS, the most common use of colour in Web browsers (other than in images) was to draw attention to hyperlinks.
6. The size of type is usually obtained by measuring from roughly the top of its ascenders to the bottom of its descenders.
7. Fonts are typically measured in a unit specific to the printing industry called the point.
8. The font-family property lets you specify the font families which will be used in your documents.
9. A relative unit is a unit of measure that specifies a value that is relative to the font size.

Multiple Choice Questions

1. RGB stands for Red, Green and :
   (a) Blue (b) Badge
   (c) Black
2. CSS has how many defined colours :
   (a) 15 (b) 16
   (c) 14
3. Fonts are measured in :
   (a) Pints (b) Inches
   (c) Points
4. Spacing can be done by :
   (a) Margins (b) Spacer
   (c) Space

5. The most common alignments are, Left, Right, Center and :
   (a) Justify (b) Vertical
   (c) Horizontal
6. The border-style property sets the __________ of the border.
   (a) Appearance (b) Display
   (c) Properties

Answers

1. a 2. b 3. c 4. a 5. a
6. a

Subjective Questions

1. What is CSS?
2. How would you put Style Sheets to the document?
3. What do you mean by Overriding Inheritance?
4. Which properties are not inherited?
5. Describe the common tasks of Fonts, Margins, Links?
6. How does the CSS handle colours?
7. Write short notes on the following:
   Font Metrics
   Relative units
   Font-size property
   Font-variant property
   Font weight property
   Font stretch property
   Text Decoration Property
   Text Transform Property

5 JavaScript

Introduction

Like many other programming languages JavaScript is a programming language, which is mainly used with Web Pages. This language is very useful when providing interactivity to a Web page. There are lots of JavaScript programs available over the net which can be copied and made use by making alterations desired by you. A JavaScript is a program which is included on an HTML page. On the HTML page you enclose the JavaScript program within <SCRIPT> and end it up with </SCRIPT>. When you run it with a browser, it knows that this is a program written under JavaScript it runs it to give you the result instead of just giving the text of the script. The <SCRIPT> tag is most often found within the <HEAD> section of the HTML page, though you can, if you wish, have scripts under the <BODY> section.

What you can do with JavaScript

There are lots of thing which you can do with JavaScript. For example, to enhance your Web page besides making it more interactive. It allows you to have Web pages users feedback by providing buttons on the Web page. You can create forms which the users of the Web page can fill to provide you with the information needed by you. You can even do calculations on these forms using CGI. It also allows you to create new windows, display alert boxes and messages on the Web pages.

JAVA and JavaScript

Let me make it very clear once and for all, JAVA and JavaScript have only name in common and nothing else. They are poles apart. JAVA is basically a programming language which is enhancement of C++ having a distinctive feature of being able to run on various machines and operating systems like Windows NT, Mac OS or even Unix.

JavaScript as Object Oriented Language

An object is some kind of a thing. A computer, a mouse, etc., are all objects. To JavaScript they are all objects it deals with in Web browsers such as windows, forms and elements of the form such as buttons and check boxes.

Events

Events are actions that the user performs while visiting your page. Submitting a form and moving a mouse over an image are two examples of events. JavaScript deals with events with commands called event handlers. There are as many as 12 JavaScript event handlers, as listed below. An action by the user on the page triggers an event handler in your script.

Event Handlers

Event	***What it handles***
onAbort	The user aborted loading the page
onBlur	The user left the object

onChange	The user changed the object
onClick	The user clicked on an object
onError	The script encountered an error
onFocus	The user made an object active
onLoad	The object finished loading
onMouseOver	The cursor moved over an object
onMouseOut	The cursor moved off an object
onSelect	The user selected the contents of an object
onSubmit	The user submitted a form
onUnload	The user left the window

Values and Variables

You can give values to your variables in JavaScript as you would give in any other programming language. A string value is a word or words enclosed in quotes. Variables contain values. Various types of values which can be given are:

Type	***Description***	***Example***
Number	Any numeric value	1233654
String	Characters inside quote marks	"Hello"
Boolean	True or False	False
Null	Empty and meaningless	
Object	Any value associated with the object	
Function	Value returned by a function	

JavaScript is case sensitive so to it **MyName** and **MyName** does not mean the same. You cannot have spaces in between the variables.

Operators

These are the symbols used to work with variables. They are simple arithmetic symbols. Various operators used in JavaScript are:

Operator	***What it does***
a + b (Numeric)	It adds a to b
a + b (String)	It adds text of b to the end of text a
a - b	It subtracts b from a
a * b	It multiplies a with b
a / b	It divides a by b
a % b	It modules of a and b
a++, ++a	It adds one to a, same as a = a + 1
a--, --a	It subtracts one from a, same as a = a - 1
-a	It reverses the sign of a

Assignments

By using assignments you can assign a value to a variable. There are various assignment operators in use in JavaScript. These are:

Assignment	*What it does*
a = b	It sets a to the value of b
a += b	It is same as a = a + b
a -= b	It is same as a = a - b
a *= b	It is same as a = a * b
a /= b	It is same as a = a/b
a %= b	It is same as a = a % b

Comparisons

It is often required to compare the value of one variable with another, or the value of a variable against a literal value, i.e., a value typed into the expression.

Comparisons	*What it does*
a = = b	It returns true if a and b are equal
a ! = b	It returns true if a and b are not equal
a > b	It returns true if a is greater than b
a > = b	It returns true is a is greater than or equal to b
a < b	It returns true if a is less than b
a < = b	It returns true if a is less than or equal to b
a && b	It returns true if both a and b are true
a I I b	It returns true if either a or b is true
! a	It returns true if a is false

Reserved Word

Following are the reserved words of JavaScript so they cannot be used otherwise.

break	continue	else
false	for	function
if	in	new
null	return	this
true	var	void
while	with	

Reserved by Java

Following words are used by Java language, and are reserved in JavaScript to avoid possible confusion.

abstract	boolean	byte	case
catch	char	class	const
default	do	double	extends
final	finally	float	goto
implements	import	instanceOf	int
interface	long	native	package
private	protected	public	short
static	super	switch	synchronized
throw	throws	transient	try

Words to be Avoided

The following word though not reserved but as they part of the JavaScript language, you should not use them as function as variable names.

alert	Anchor	Area	Array
assign	blur	Boolean	Button
Checkbox	clearTimeout	close	closed
confirm	Date	defaultStatus	document
Document	Element	escape	eval
FileUpload	focus	Form	Frame
frames	Function	getClass	Hidden
history	History	Image	isNan
Java	JavaArray	JavaClass	JavaObject
JavaPackage	length	Link	location
Location	Math	MimeType	name
navigate	navigator	Navigator	netscape
Number	Object	onBlur	onError
onFocus	onLoad	onUnload	open
opener	Option	Packages	parent
parseFloat	parseInt	Password	Plugin
prompt	prototype	Radio	ref
Reset	scroll	Select	self
setTimeout	status	String	Submit
sun	taint	Text	Textarea
top	toString	unescape	untaint
valueOf	window	Window	

Browsers to use

You can use any browser to create web document. It is important that your computer has more than one browser in it. It is important since it lets you view the document through more than one browser. It gives your document more versatility and makes you sure that your document would run all the browsers. The two most popular browsers are Netscape Communicator (used to be called Navigator) and Microsoft Internet Explorer. Both are quite popular and can be both used in your computer.

Netscape Communicator

More web designer use this software than any other. It is also called Netscape. Developed by some of the same engineers who created popular software called Mosaic. It offers more nonstandard features that make its enhanced pages much more attractive to the eye.

Microsoft Internet Explorer

Anything with Microsoft tag is bound to create hype. It is fast becoming popular. It is no doubt a viable competitor Netscape.

Software Requirement

JavaScript can run on most of the platforms available in the market. In this book, I restrict myself to

Windows versions, may it be 95, 98, or even ME. You would need a machine with minimum of 32MB of RAM and a 486 processor. Also you would need Netscape Communicator 4.5 or later and Microsoft Internet Explorer 4 or later and the language mentioned in the book is JavaScript 1.3.

However, as mentioned above it is better to have both the browsers in the machine. The following table shows the various versions which can be used with various browsers.

JavaScript and Browsers

Browser	***JavaScript version***
Netscape Navigator 2.0x	1.0
Netscape Navigator 3.0x	1.1
Netscape Navigator 4.0x	1.2
Microsoft Internet Explorer 3.0x	1.0
Microsoft Internet Explorer 4.0x	1.2

Starting with JavaScript

Now that we have learned the initial introduction of JavaScript, let us straight way start writing our small programs in JavaScript. As usual the first program will be "Learning JavaScript".

Writing first JavaScript program

For creating the text file, you can use any of the text program, I am using here Notepad. As you can see I have created a new file for this. As you would notice that I have given the name of the file as New1 with html extension and run it under Netscape. Both text and the result in browser are shown on the next page. So you how you have been able to run your first JavaScript program and see the result too. Let me now tell what has gone into it.

<SCRIPT>	This is the opening script tag which tells the browser to expect JavaScript instead of HTML.
document.write	This is your first line of the program for execution which has been taken up by the browser in the document window and shown the result.
"Learning JavaScript"	I have given this, you could have given anything else of your choice.
</SCRIPT>	This shows the end of the JavaScript program.

Rest of the commands are from HTML, I would not explain them here, since I presume that you must be knowing them.

Putting Comments

As you would have done in any other program which you would have written, you need to put comments

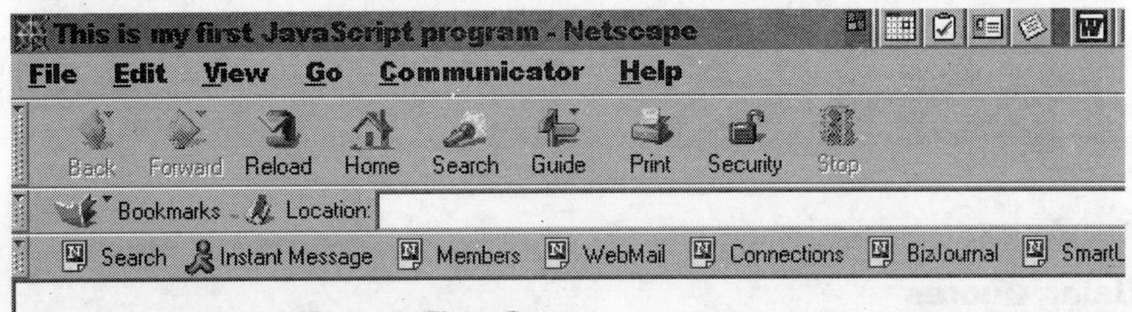

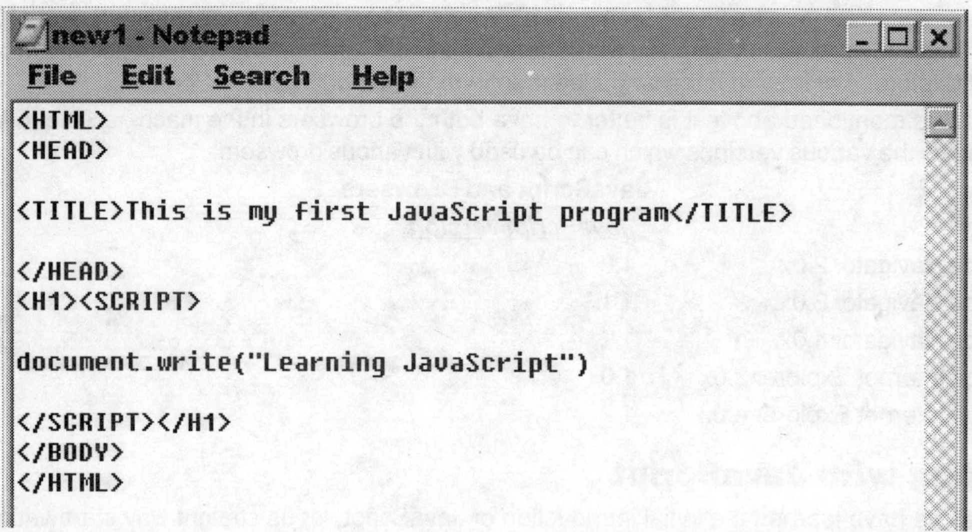

in the programs. Each programming language allows you to do the same. Some of them give you the option of REM or /* or any other similar command. JavaScript also gives you the option in the form of /*. So anything written after this is not executed till it comes across another command as */ which breaks the comment option and starts executing the further lines. In the above example, you will see that I have put "This is a comment line and is not to be executed" within the scope of /* and */.

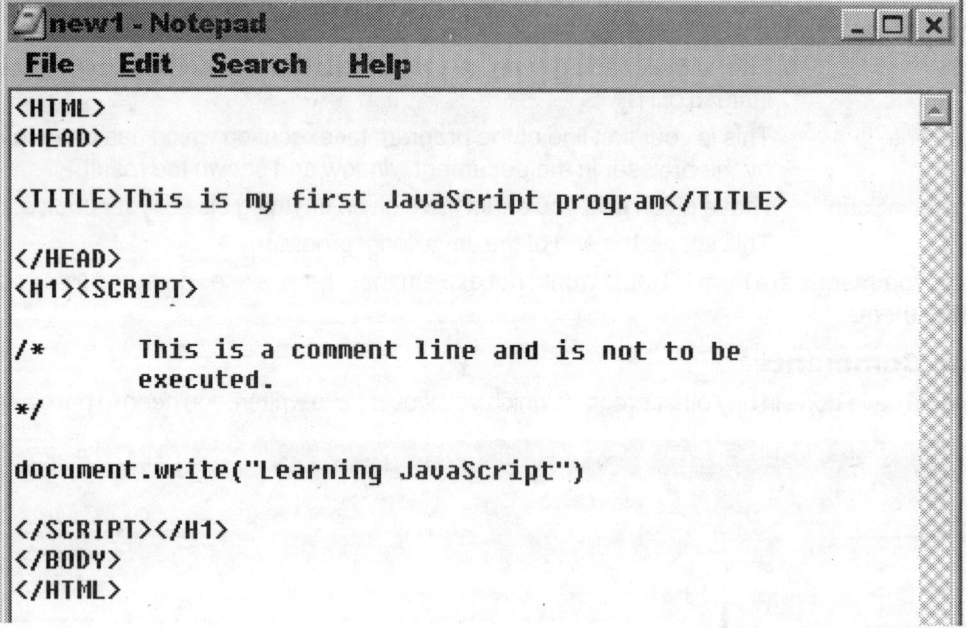

The result is not required to be shown because it will remain the same despite these lines.

Using Quotes

As you have seen in the above example, I have used quotes for displaying the text. What would happen if I have to display the quotes themselves.

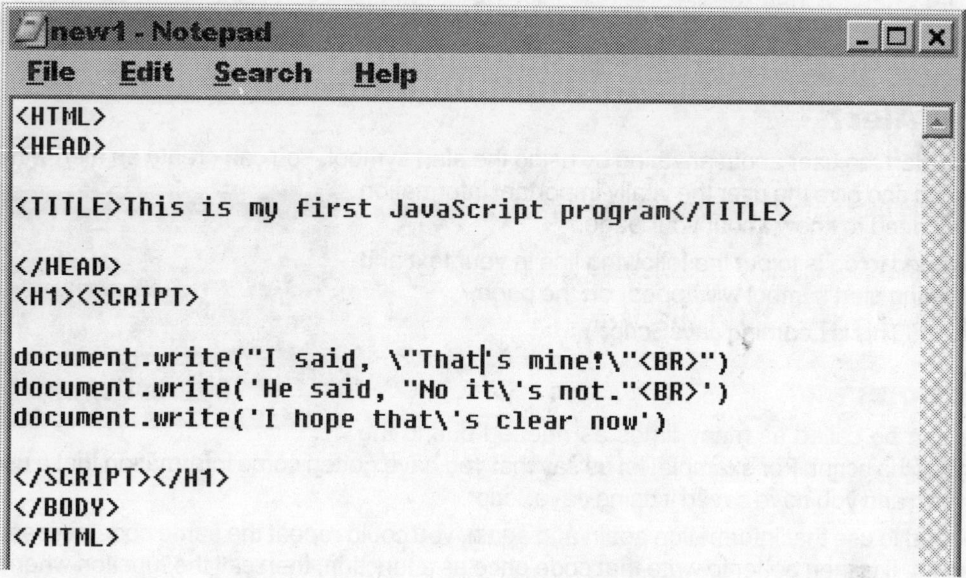

Let me explain this with the help of another example. The text of the program is shown below while the output of it is given on the next page.

If you are still not clear about the various quotes, have a look at the table given below and everything would be clear to you.

Character	***Meaning***
\'	single quote
\"	double quote
\\	backslash
\n	new line
\r	carriage return
\f	form feed

\t horizontal tab
\b backspace

Using Alert

You can alert the user about anything by using the alert symbol. You can create an alert window that will pop up and give the user the vitally important information that they need to know about your page.

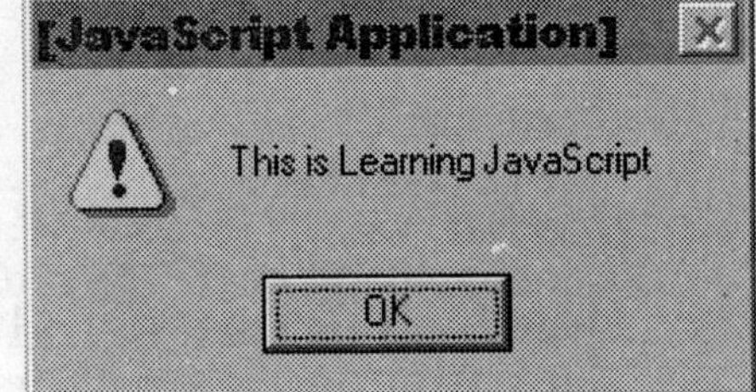

All you need to do is to put the following line in your text and the following alert symbol will appear on the page.

```
alert("This is Learning JavaScript")
```

Functions

These can be called as many times as needed during the running of the script. For example, let us say that you have gotten some information that a user typed into a form and you have saved it using JavaScript.

If you need to use that information again and again, you could repeat the same code over and over in your script. It is then better to write that code once as a function, then call the function whenever you need it.

A function consists of the word function followed by the function name. There are always parentheses after the function name, followed by an opening brace.

The statements that make up the function go on the following lines, then the function is closed by another brace.

Here is how a function looks like:

```
function saySomething() {
    alert("Whatever you want to say")
}
```

Notice that the line with alert is indented. That makes it easier to read your code. All of the statements between the first brace and the last one are part of the function.

Let us see this with a live example. On the next page, you would see a JavaScript program which has been written to have 3 buttons each one of them having a new text underneath it. This text can be displayed by clicking at the corresponding button.

The text displayed by the various buttons is also shown on the next page.

Next we will use the conversion functions to find out the values of various variables. Most common among them is the eval function which is used to evaluate the value.

eval Function

This can be used to convert a string expression to a numeric value. Supposing we have to find out the value of the string 5 + 6 = . You can use this function to find out the value as 11. Not only this you can use the various mathematical operations too in this.

Let us find out the value of following expressions using JavaScript.

```
5 * 6
6 / 5
```

I have used these examples on the next page to find out the values, see the text and the result on the next page which have been given as 30 and 1.2 respectively.

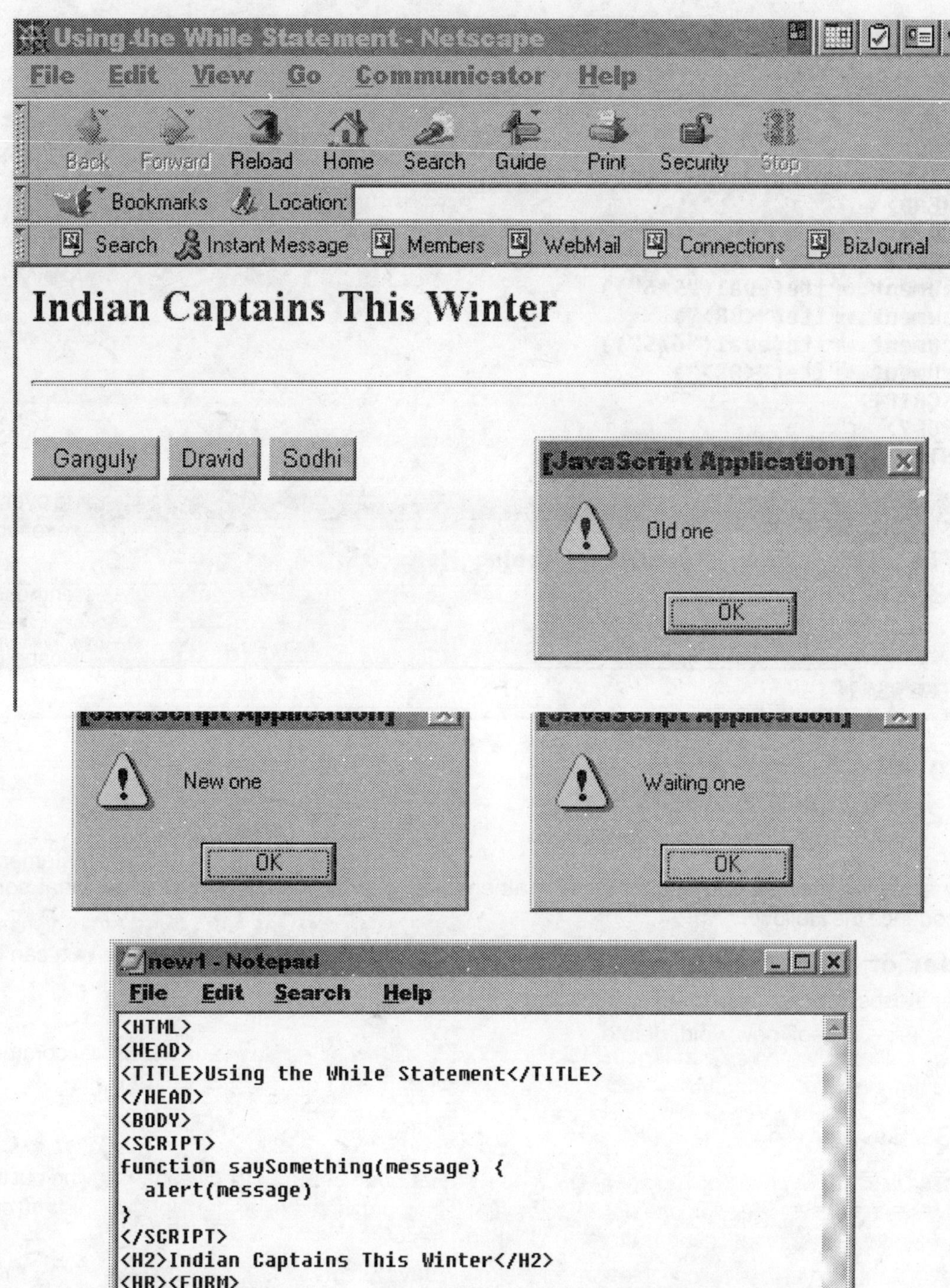
Using the While Statement - Netscape
File Edit View Go Communicator Help
Back Forward Reload Home Search Guide Print Security Stop
Bookmarks Location:
Search Instant Message Members WebMail Connections BizJournal
Indian Captains This Winter
Ganguly Dravid Sodhi
[JavaScript Application]
Old one
OK
New one
OK
Waiting one
OK
new1 - Notepad
File Edit Search Help
<HTML>
<HEAD>
<TITLE>Using the While Statement</TITLE>
</HEAD>
<BODY>
<SCRIPT>
function saySomething(message) {
alert(message)
}
</SCRIPT>
<H2>Indian Captains This Winter</H2>
<HR><FORM>
<INPUT TYPE="button" VALUE="Ganguly" onClick="saySomething
('Old one')">
<INPUT TYPE="button" VALUE="Dravid" onClick="saySomething
('New one')">
<INPUT TYPE="button" VALUE="Sodhi" onClick="saySomething
('Waiting one')">

new1 - Notepad

File Edit Search Help

```
<HTML>
<HEAD>
<TITLE>Learning JavaScript</TITLE>
</HEAD>
<BODY>
<SCRIPT>
document.write(eval("5*6"))
document.write("<BR>")
document.write(eval("6/5"))
document.write("<BR>")
</SCRIPT>
</BODY>
</HTML>
```

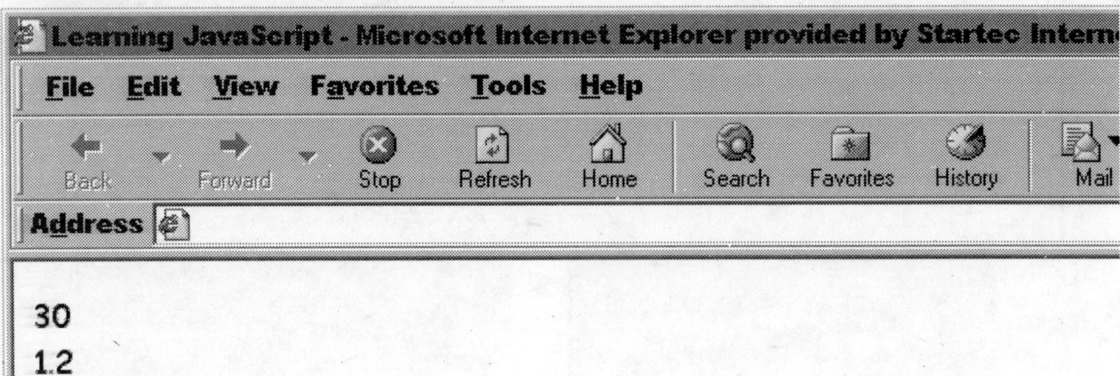

Now since we are going to use all these mathematical expressions, we must know what sort of precedence they follow.

Order of precedence

Parentheses, function call, or array subscript

!, ~, -, ++, –, typeof, new, void, delete

*, /, %

+, –

<<, >>, >>>

<, <=, >, >=

= =, !=, = = =, != =

&

^

|

&&

| |

?:

=, +=, −=, /=, %=, <<=, >>=, &=, ^=, |=

The comma (,) operator

Now keeping this in mind, you can go ahead and perform the mathematical and other functions in JavaScript.

Using Statements in JavaScript

Statements form the integral part of any programming language. You will find statements in most of the programming language starting with BASIC. JavaScript also provides a complete range of programming statements. In this chapter we will go through these statements as supported by JavaScript programming.

Assignment Statement

As in any other programming language, assignment is used to assign value to the variables. The assignment statement actually updates the value of the variable based upon the expression.

For example, a simple assignment statement

```
a = b + 5
```

will assign the value of b to a and increment it with 5.

Data Declaration Statement

This statement is similar to assignment, where the assignment assigns the value based upon the expression, here the value is directly assigned to the variable.

For example, following is simple data declaration

```
x = 20
```

which will assign the value 20 to the variable x.

If Statement

This statement is used when you have to give more than one option. For example, "If there is rain, the play will not take place", has two options, if there is rain, there will be no play. However, if there is no rain, there will be play.

In JavaScript this can be defined as:

```
if ("it rains") {
    there is no play
} else {
    there is play
}
```

This means that if the first condition is true, then the first set of statement will be executed. If false, then the second statement will take place. The braces enclosing the statements in the if and else clauses can be omitted if only one statement is enclosed.

Switch Statement

It is quite similar to the switch statement of C++. The syntax of the statement is as follows:

```
switch (expression) {
    case value1:
        statements
        break
```

```
    .
    .
case valuen:
    statements
    break
default:
    statements
}
```

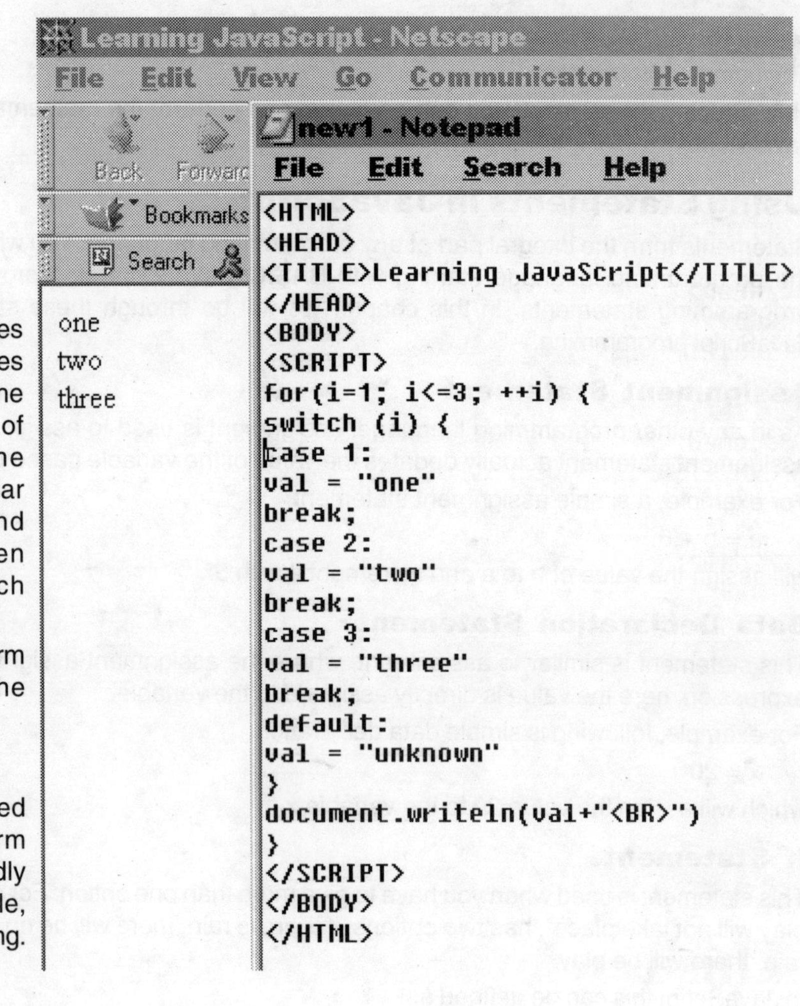

The switch statement evaluates the expression and determines if any of the values match the expression's value. If one of them matches, then the statements for that particular case are executed, and statement execution then continues after the switch statement.

Let us now put this in the form of an example and see the result.

While Statement

The while statement is used where you have to perform certain statements repeatedly depending upon the variable, whose value keeps on changing. It can be written as:

```
while (condition) {
    statement
}
```

This can be shown by an interesting example, where the value of i is decreasing and the heading changes accordingly. See the text of the program and its result in the browser on the next page.

For Statement

This statement is similar to the while statement and is used as the following:

```
for (initializationStatement; condition; updateStatement) {
    statements
}
```

This means that the statement will be executed till the condition in the for is met. For example, variable like For i = 1 to 10, do the condition1. A small example of the same can be:

```
x = new array(3, 6, 9, 12)
    sum = 0
for (i = 0;i < x.length;++i) {
```

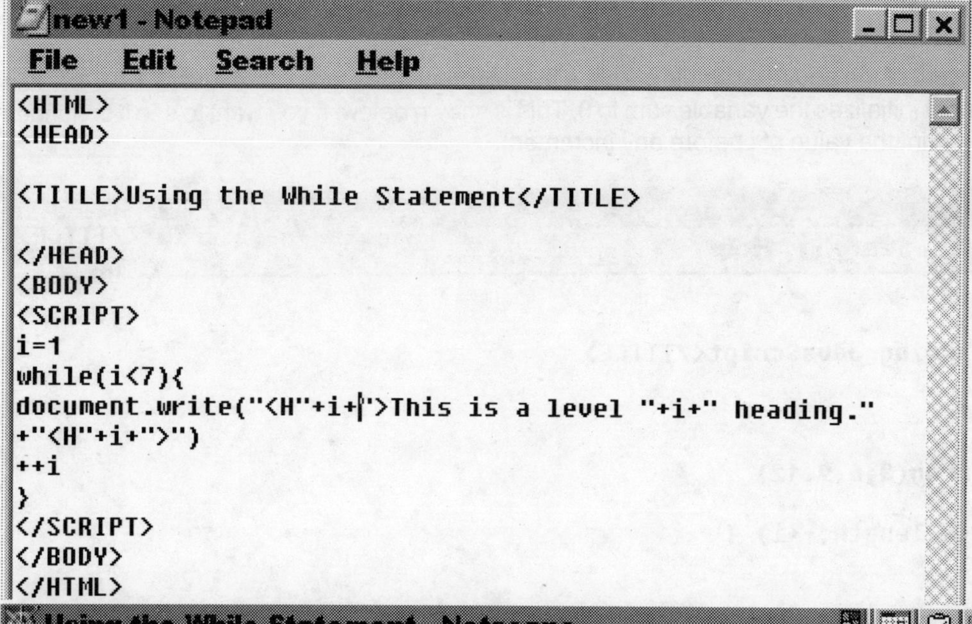


```
<HTML>
<HEAD>

<TITLE>Using the While Statement</TITLE>

</HEAD>
<BODY>
<SCRIPT>
i=1
while(i<7){
document.write("<H"+i+">This is a level "+i+" heading."
+"<H"+i+">")
++i
}
</SCRIPT>
</BODY>
</HTML>
```

This is a level 1 heading.

This is a level 2 heading.

This is a level 3 heading.

This is a level 4 heading.

This is a level 5 heading.

This is a level 6 heading.

```
        sum += x[i]
    }
```

The first statement of the above example create a five-element array with the values 3, 6, 9, 12. The second statement initializes the variable sum to 0. This is shown below. If you want to see it continues mode, you can print the value of i before any increment.

new1 - Notepad

File Edit Search Help

```
<HTML>
<HEAD>
<TITLE>Learning JavaScript</TITLE>
</HEAD>
<BODY>
<SCRIPT>
x = new Array(3,6,9,12)
sum = 0
for(i=0;i<x.length;++i) {
sum += x[i]
}
document.write(sum)
</SCRIPT>
</BODY>
</HTML>
```

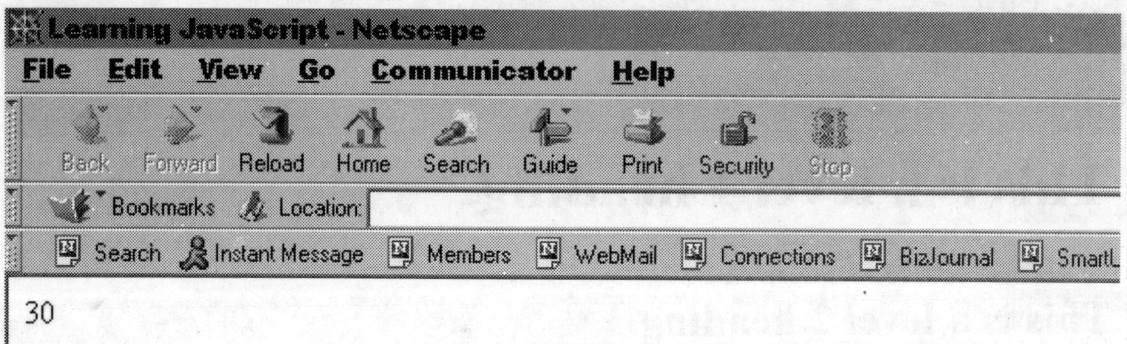

Do While

Instead of For, you can use the option of do while too. This can be represented as the following:

```
    do {
    statements
    while (condition);
```

When applied to an example, this can be written as following:

```
    i = 0
    do {
    ++i
    document.writeln(i+"<BR>")
    } while (i < 7);
```

This you can see from the figure below and see the result too, as shown.

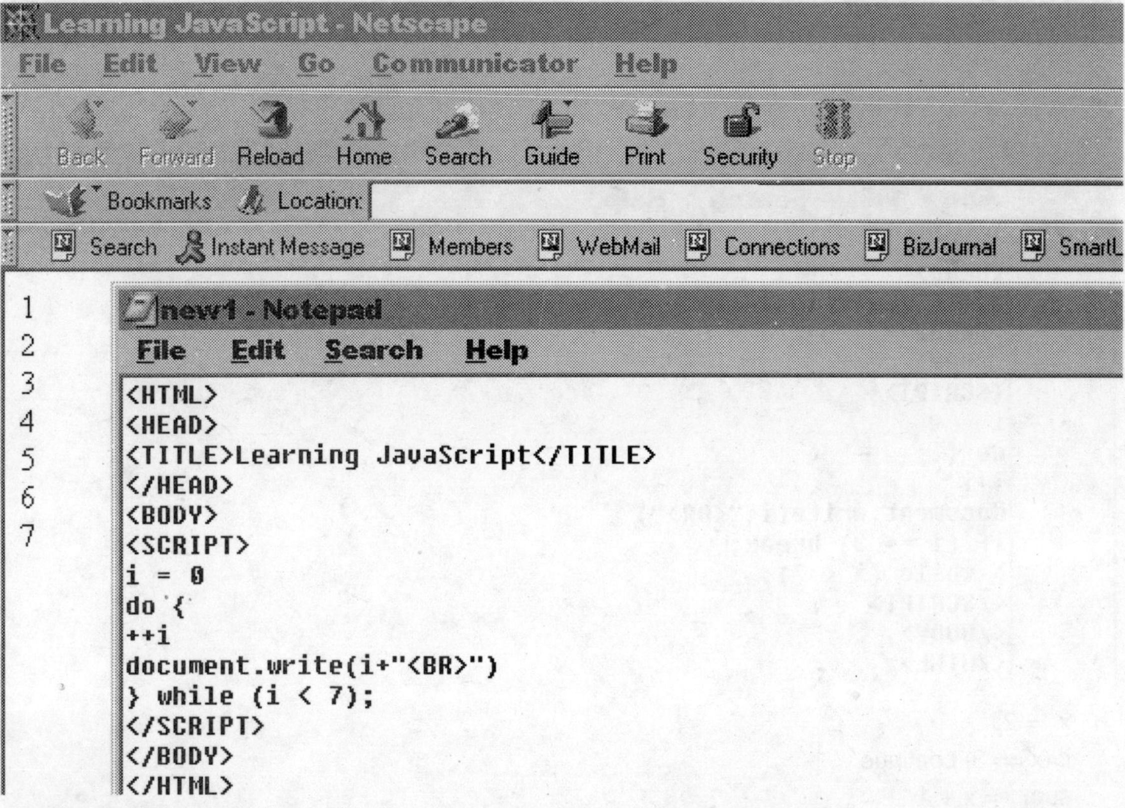

Label Statement

This is used to assign a label to the statement. For example the foilowing:

```
NewLabel:
    x = 3 * y
```

causes the above statement to be labeled with the label NewLabel. Labels are used to identify statements.

Break Statement

It is used to break an onging execution. It consists of the word break, followed by an optional label. When the break statement is encountered, the loop is immediately terminated.

Let us put a break in the above example and see the result which is shown on the next page.

Continue Statement

This statement is quite similar to the break statement. It affects the execution of the for, do while, or while statement in which it is contained. It however, does not completely terminates the loop's execution but only terminates the execution of the statements in the loop's current iteration.

For example, see the following example,

```
x = 1
sum =0
while(x<7) {
```

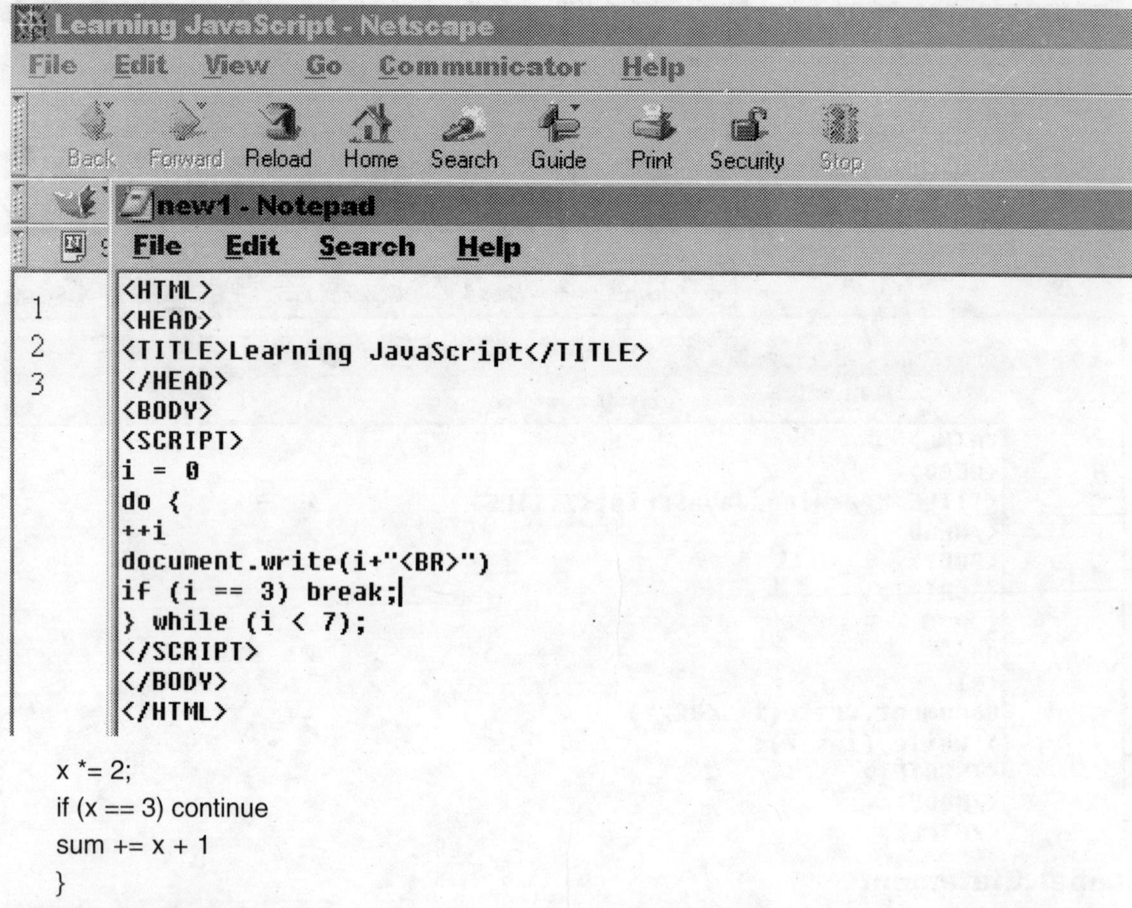


```
x *= 2;
if (x == 3) continue
sum += x + 1
}
```

Return Statement

This statement is used to return a value which has been processed as a result of the processing of a function. The syntax of the return statement is:

```
return expression
```

It can be better explained with the help of an example:

```
function factorial(n) {
    var sum = 1
    for(i=1;i<n;++i)
    sum *= i
    return sum
}
```

Here the sum is calculated as the product of all integers from 1 to n.

With Statement

This statement helps you in eliminating retyping the name of an object that is to be referenced in a series of references and method invocations. The following syntax will explain it better.

```
with(variableName) {
    statements
```

```
}
```

An example for the above can be as follows:

```
with(document) {
        write("<H1>Learning JavaScript</H1>")
        write("<P>")
        write("Using the with Statement")
        write("<P>")
}
```

See the result of this example below and the text too.

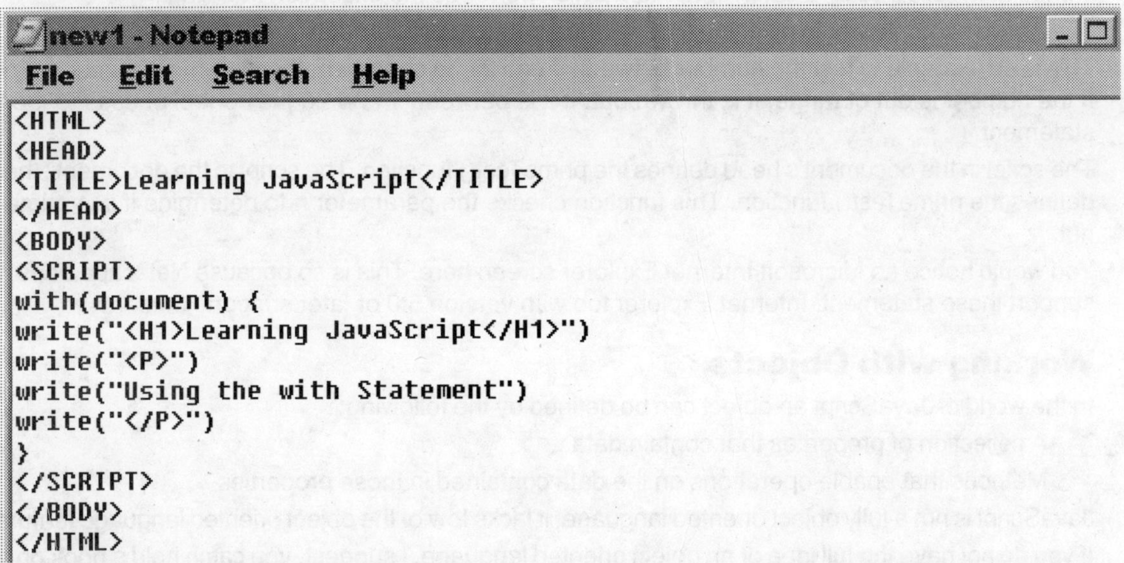
new1 - Notepad

File Edit Search Help

```
<HTML>
<HEAD>
<TITLE>Learning JavaScript</TITLE>
</HEAD>
<BODY>
<SCRIPT>
with(document) {
write("<H1>Learning JavaScript</H1>")
write("<P>")
write("Using the with Statement")
write("</P>")
}
</SCRIPT>
</BODY>
</HTML>
```

For in

Earlier we had learned the For statement. This statement is similar but it repeatedly executes a set of statements. Instead of iterating the statements based on the loop condition, it executes the statements for all properties that are defined for an object.

The syntax for this is like this:

```
for (variableName in objectName) {
    statements
}
```

The For In loop executes the statements enclosed with the braces one time for each property defined in objectName.

Throw Statement

The throw statement is used to throw an exception.

The syntax is:

```
throw expression
```

The value of expression is used to identify the type of error that occurred.

Try & Catch Statement

The try statement and the catch statement work together to support exception handling.

Take a look at the following syntax to understand it better.

```
try {
    statement(s) where an exception may be thrown
}
catch(errorVarible) {
    statement(s) that handle the exception
}
```

The try statement surrounds the statements for which exception handling is to be performed. It is immediately followed by a catch statement which performs the exception handling.

The above example would give you a better idea about the statements.

This is an example to test the number between 0 and 21, to check whether it is prime number or not. If the number is out of range, it is thrown out. If it is between 1 and 20 the control passes to the for statement.

The script in the document's head defines the primeTest() function. The script in the document's head defines the primeTest() function. This function checks the parameter n to determine if it is prime or not.

You would notice an Microsoft Internet Explorer screen here. This is so because Netscape does not support these statement. Internet Explorer too with version 5.0 or later supports these.

Working with Objects

In the world of JavaScript an object can be defined by the following:

A collection of properties that contain data

Methods that enable operations on the data contained in those properties

JavaScript is not a fully object oriented language, it lacks few of the object oriented language features.

If you do not have the full idea of an object oriented language, I suggest, you catch hold a book on the same subject and get some more information. If you want you can refer to my book, Learning C++, which also explains Object Oriented Programming (OOPs).

Following are the properties of JavaScript which make it capable to run object oriented based programming.

new1 - Notepad

File Edit Search Help

```
<HTML>
<HEAD>
<TITLE>Learning JavaScript</TITLE>
</HEAD>
<BODY>
<SCRIPT>
function primeTest(n) {
document.write("Testing "+n+": ")
try {
if(n < 1 || n > 20) throw "It's out of range"
for(var i = 2; i < n; ++i)
if(n %i == 0) throw "It's divisible by  " + i
document.writeln("It's prime.<BR>")
}
catch (exception) {
document.writeln(exception+" .<BR>")
}
}
</SCRIPT>
<BODY>
<SCRIPT>
for(i = 0; i <= 21; ++i) {
primeTest(i)
}
</SCRIPT>
</BODY>
</HTML>
```

Object Types

JavaScript supports both predefined and user-defined object types. However, JavaScript does not provide capabilities for type enforcement. An object of any may be assigned to any variable.

Object Instantiation

Object types are instantiated using the new operator to create specific object types.

Object Composition

Object types may be defined in terms of other predefined or user-defined object types.

Modularity

JavaScript code may be defined in terms of other predefined or user-defined object types.

Object reuse

JavaScript code may be defined in a modular fashion, but JavaScript does not provide any features that enforce modular software development.

Information Hiding

JavaScript does not provide any capabilities to support information hiding.

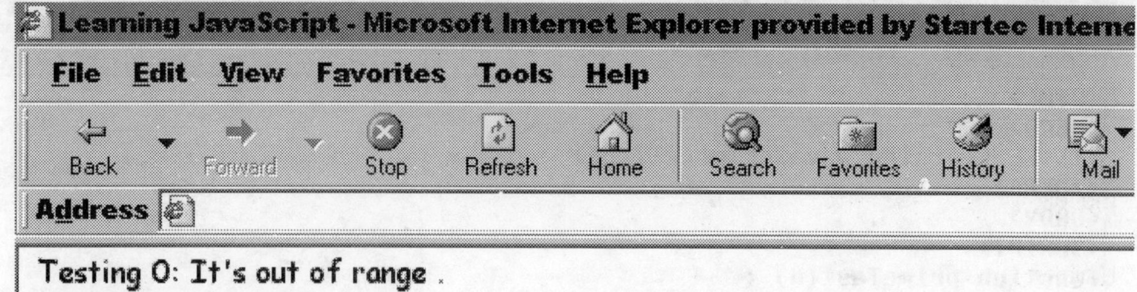

Encapsulation

Because JavaScript lacks information hiding capabilities, it cannot be used to develop encapsulated object types. Any method or property that is defined for a type is always directly accessible.

Inheritance

JavaScript does not provide any language features that support inheritance between object types.

Classification

Because JavaScript does not support inheritance, it cannot be used to develop a hierarchy of object types.

Polymorphism

JavaScript supports polymorphism using the arguments array for function definitions.

Properties

An object's properties are accessed by combining the object's name and its property name as follows:

objectName.propertyName

Using this for example we can change the background color to white from any other colour by using the following:

document.bgColor="white"

I think it is better if I use this a program and allow you to see the result.

In the following program, I have used three buttons, clicking on any one of them changes the background colour to the name of the colour mentioned on the button, for example, red to red, blue to blue, etc. The web page in Netscape would like the one shown.

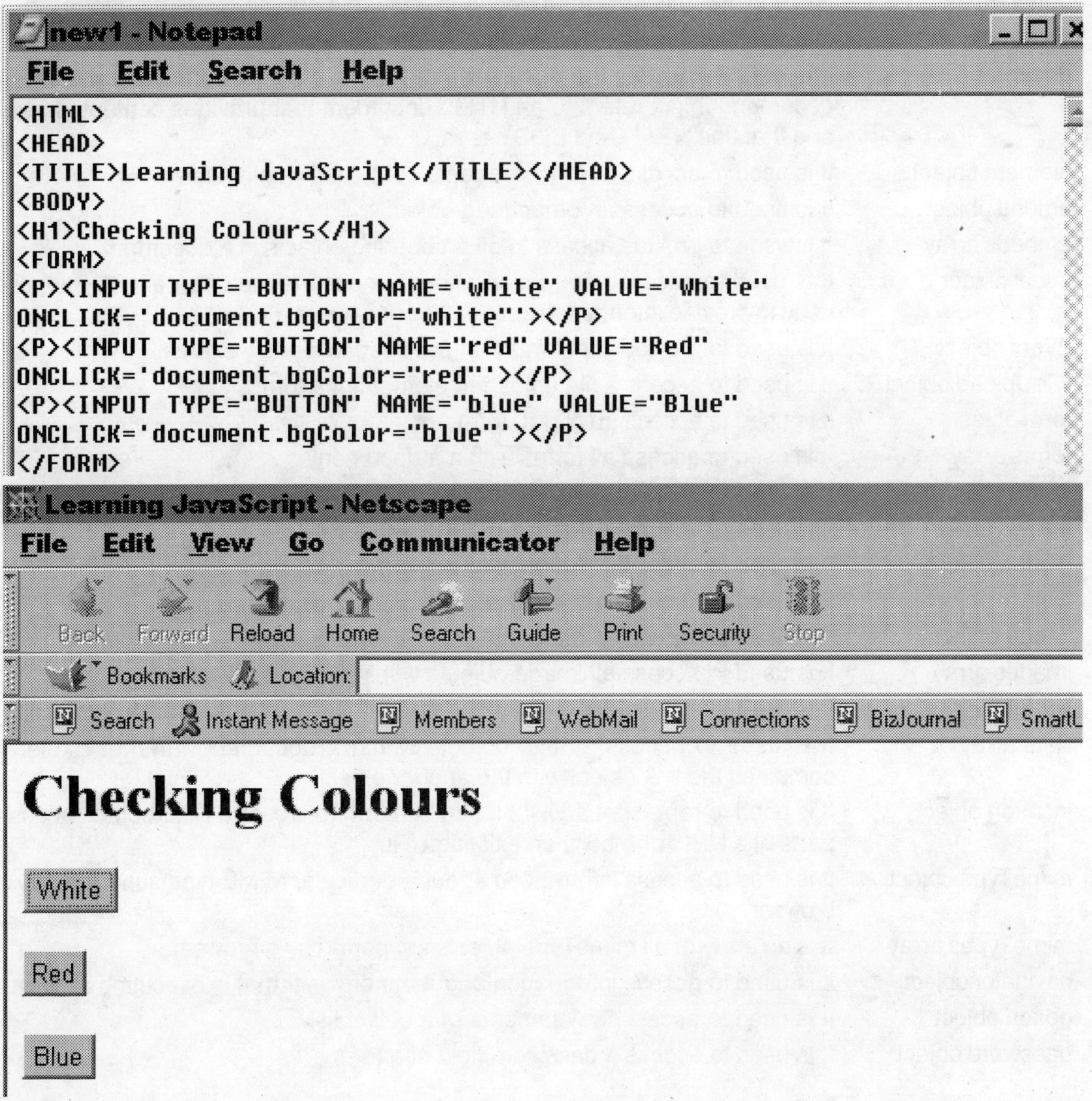

Browser Objects

When a Web page is loaded by a JavaScript capable browser, the browser creates a number of JavaScript objects that provide access to the Web page and the HTML elements it contains. These objects are used to update and interact with the loaded Web page.

Various objects and their uses are mentioned below:

Object	***Use***
anchor object	It is used to access the target of a hypertext link.
anchor array	It is used to access all anchor objects within a document.
applet object	It is used to access a Java applet.
applets array	It is used to access all applets in a document.
area object	It is used to access an area within a client-side image map.
button object	It is used to access a form button that is not a submit or reset button.
checkbox object	It is used to access a checkbox of a form.
document object	It is used to access the document that is currently loaded into a window. A document object refers to an HTML document that provides content, that is, one that has HEAD and BODY tags.
element object	It is used to access all form elements contained within a form.
embed object	It is used to access an embedded object.
embeds array	It is used to provide access to all embedded objects in a document.
event object	It is used to access information about the occurrence of an event. It is also used to provide information about a specific event.
Event object	It is used to provide constants that are used to identify events.
FileUpload object	It is used to access a file uploaded element of a form.
form object	It is used to access an HTML form.
forms array	It is used to access all forms within a document.
frame object	It is used to access an HTML frame.
frames array	It is used to access all frames within a window.
hidden object	It is used to access a hidden field of a form.
history object	It is used to maintain a history of the URLs accessed within a window.
image object	It is used to access an image that is embedded in an HTML document.
images array	It is used to access all image objects within a document.
link object	It is used to access a text- or image-based source anchor of a hypertext link.
links array	It is used to access all link objects within a document, Internet Explorer combines the link object with the anchor object.
location object	It is used to represent a URL. It can be used to create a URL object, access parts of a URL, or modify an existing URL.
mimeType object	It is used to access information about a particular MIME type supported by a browser.
mimeTypes array	It is an array of all mimeType objects supported by a browser.
navigator object	It is used to access information about the browser that is executing a script.
option object	It is used to access the elements of a select list.
password object	It is used to access a password field of a form.

plugup object	It is used to access information about a particular browser plug-in.
plugin array	It is an array of all plug-ins supported by a browser.
radio object	It is used to access a set of radio buttons of a form or to access an individual button within the set.
reset button	It is used to access a reset button on a form.
screen object	It is used to access information about the size and colour depth of a user's screen.
select object	It is used to access a select list of a form.
submit button	It is used to access a submit button on a form.
text object	It is used to access a text field of a form.
textarea object	It is used to access a text area field of a form.
window object	It is used to access a browser window or a frame within a window. The window object is assumed to exist and does not require the "window" prefix when referring to its properties and methods.

Date Object

This object type provides a common set of methods for working with dates and times. Instances of the Date object type may be created with any of the following constructors.

Constructor	***Description***
Date()	It creates a Date instance with the current date and time.
Date(dateString)	It creates a Date instance with the date specified in the dateString parameter. The format of the dateString is "month day year hours:minutes:seconds".
Date(milliseconds)	It creates a Date instance with the specified number of milliseconds since midnight January 1, 1970.
Date(year, month, day, hours, minutes, seconds, milliseconds)	It creates a Date instance with the date specified by the year, month, day, hours, minutes, seconds, and milliseconds integers. The year and month parameters must be supplied. If other parameters are included, then all preceding parameters must be supplied.

However, Date object type can be used as following:

Method	***Description***
getDate() getUTCDate() setDate() setUTCDate()	It returns/sets the day of the month of the Date object.
getDay() getUTCDay()	It returns the day of the week of the Date object
getHours() getUTCHours() setHours() setUTCHours()	It returns/sets the hour of the Date object.
getMilliseconds()	It returns/sets the millisecond value of the Date object.

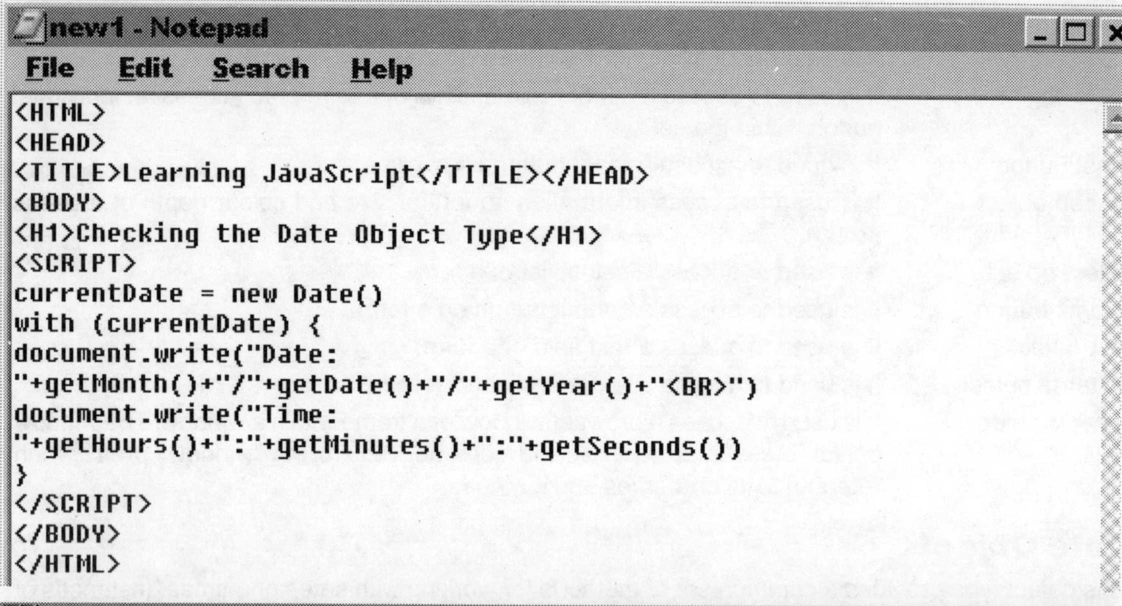

getUTCMilliseconds()	
setMilliseconds()	
setUTCMilliseconds()	
getMinutes()	It returns/sets the minutes of the Date object.
getUTCMinutes()	
setMinutes()	
setUTCMinutes()	
getMonth()	It returns/sets the month of the Date object.
getUTCMonth()	
setMonth()	
setUTCMonth()	

getSeconds() getUTCSeconds() setSeconds() setUTCseconds()	It returns/sets the seconds of the Date object.
getTime() setTime()	It returns/sets the time of the Date object.
getTimeZoneOffset()	It returns the time zone offset (in minutes) of the Date object.
getYear() getFullYear() getUTCFullYear() setYear() setFullYear() setUTCFullYear()	It returns/sets the year of the Date object. The full year method used 4-digit values.
toGMTString()	It converts a date to a string in Internet GMT (Greenwich Mean Time) format.
toLocaleString()	It converts a date to a string in locale format, which means the format commonly used in the geographical region in which the user is located.
toString()	It returns a string value of a Date object.
valueOf()	It returns the number of milliseconds since midnight January 1, 1970.
toUTCString()	It returns a string that represents the time in UTC.

The date object type can be best explained with the help of an example, which is shown on the next page.

Math Object

This is not a single object but a library of all the mathematical constants and functions needed by you. Most of the mathematical operations like Sin, Cos; Logarithm of 2 and 10; round off; square root; you name it and it is here. You will get all the results here. As usual I give you an example to illustrate the various math functions, as shown on next to next page.

Various math properties and methods which are used here are given below:

Math Methods

Method	***Description***
abs(x)	It returns the absolute value of x
acos(x)	It returns the arc cosine of x in radians
asin(x)	It returns the arc sine of x in radians
atan(x)	It returns the arc tangent of x in radians
atan2(x,y)	It returns the angle of the polar coordinates corresponding to (x,y)
ceil(x)	It returns the least integer that is greater than or equal to x
cos(x)	It returns the cosine of x
exp(x)	It returns value of e^x
floor(x)	It returns the greater integer that is less than or equal to x
log(x)	It returns the natural logarithm of x
max(x,y)	It returns the greater of x and y
min(x,y)	It returns the lesser of x and y

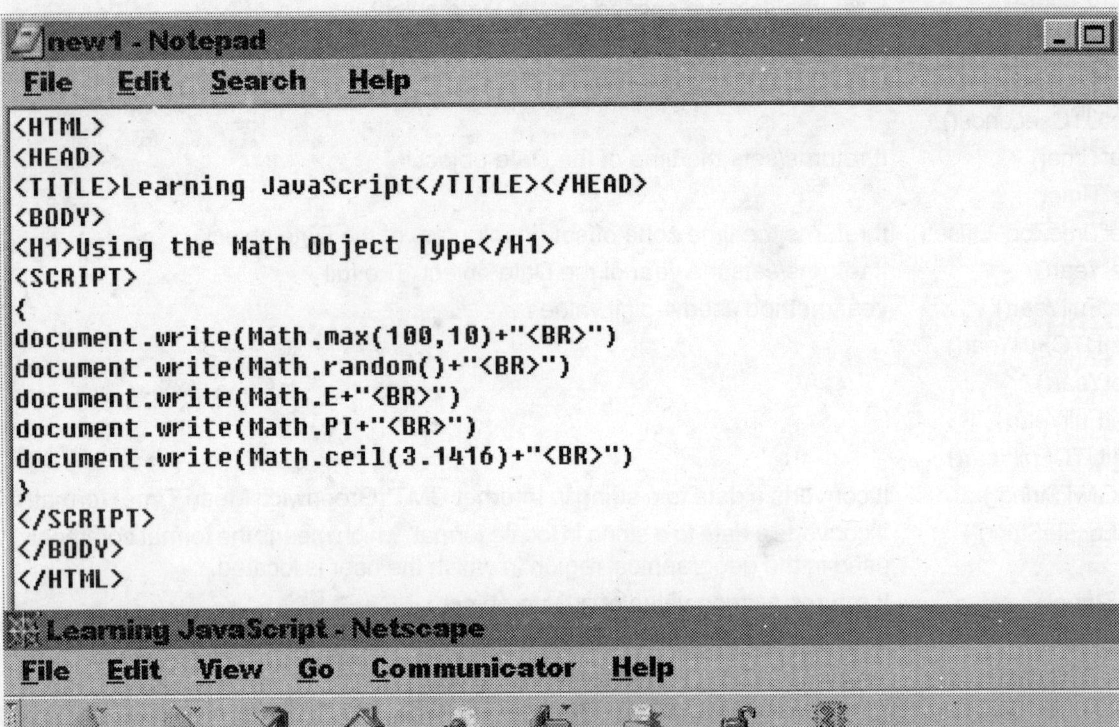

pow(x,y)	It returns the value of x^y
random	It returns a random number between 0 and 1
round(x)	It returns the value of x rounded off to the closet integer
sin(x)	It returns the value of sine of x
sqrt(x)	It returns the square root of x
tan(x)	It returns the tangent of x

Math Properties

Property	***Description***
E	Euler's constant

LN2	The natural logarithm of 2
LN10	The natural logarithm of 10
LOG2E	The base 2 logarithm of e
LOG10E	The base 10 logarithm of e
PI	The constant pi (π)
SQRT1_2	The square root of _
SQRT2	The square toot of 2

String Object

This object allows strings to be accessed as objects. It provides a set of methods for manipulating strings.

As usual I am showing here an example, with both text and its output in the browser. Text is shown below and the output on the next page. While you take a note of the example also note down the following methods which are allowed in the String as methods.

```
new1 - Notepad
File   Edit   Search   Help
<HEAD>
<TITLE>Learning JavaScript</TITLE></HEAD>
<BODY>
<H1>Using the String Object Type</H1>
<SCRIPT>
function displayLine(text) {
document.write(text+"<BR>")
}
x = new String("This is part of Learning JavaScript.")
displayLine('x = '+x)
displayLine('x.charAt(1) = '+x.charAt(1))
displayLine('x.charCodeAt(1) = '+x.charCodeAt(1))
displayLine('x.indexOf("is") = '+x.indexOf("is"))
displayLine('x.lastIndexOf("is") = '+x.lastIndexOf("is"))
displayLine('x.substring(18,25) = '+x.substring(21,31))
displayLine('x.toLowerCase() = '+x.toLowerCase())
displayLine('x.toUpperCase() = '+x.toUpperCase())
split = x.split(" ")
for(i=0; i<split.length; ++i)
displayLine('split['+i+'] = '+split[i])
</SCRIPT>
</BODY>
</HTML>
```

charAt(index)

It returns a string that consists of the character at the specified index of the string to which the method is applied.

charCodeAt(index)

It returns the Unicode encoding of the character at the specified index.

Using the String Object Type

x = This is part of Learning JavaScript.
x.charAt(1) = h
x.charCodeAt(1) = 104
x.indexOf("is") = 2
x.lastIndexOf("is") = 5
x.substring(18,25) = ing JavaSc
x.toLowerCase() = this is part of learning javascript.
x.toUpperCase() = THIS IS PART OF LEARNING JAVASCRIPT.
split[0] = This
split[1] = is
split[2] = part
split[3] = of
split[4] = Learning
split[5] = JavaScript.

fromCharCode(codes)

It creates a string from a comma-separated sequence of character codes.

indexOf(pattern)

It returns the index of the first string specified by he pattern parameter that is contained in a string. It returns -1 if the pattern is not contained in the string.

indexOf(pattern,startIndex)

It is same as the previous method except that searching starts at the position specified by startIndex.

lastIndexOf(pattern)

It returns the index of the last string specified by the pattern parameter that is contained in a string. It returns -1 if the pattern is not contained in the string.

lastIndexOf(pattern,startIndex)

It is same as the previous method except that searching starts at the position specified by startindex.

split(separator)

It separates a string into an array of substrings based upon the separator.

substring(startIndex)

It returns the substring of a string beginning at startindex.

substring(startlindex,endIndex)

It returns the substring of a string beginning at startIndex and ending at endIndex.

toLowerCase()

It returns a copy of the string converted to lowercase.

toString()

It returns the string value of the object.

toUpperCase()

It returns a copy of the string converted to uppercase.

valueOf()

It returns the string value of the object.

Defining Objects

In JavaScript you can create your own objects and use them whenever required. The process is called Defining Objects. Here I will take a specific example of a table creation and then its use. A table can be defined using the various function table(rows, columns) method. As you will see in the listing of the table which is being create, you can define not only the rows and columns of the table, but, can also define the border, colour, etc., too.

Handling Events in JavaScript

Events are nothing but the actions which can be performed by you using a hyperlink in JavaScript. This link can be in the form of a button, click of a mouse, or even entering of the data. The code which performs the processing is known as the event handler.

Few actions which can be associated with the event handling:

- Display a dialog box when a user moves the mouse over a link
- Validate the data a user has just entered into a form
- Load and display an animation sequence when a user clicks a button
- Interact with Java applets and browser plug-ins

I give below a list of all the events handled by both Internet Explorer and Netscape Navigator. As in the case of objects, I will take a few of them later with examples.

HTML Element	*HTML Tags*	*JavaScript Event*	*Description*
all elements	Various	mouseMove	The mouse is moved.
area	<AREA>	dblClick	The user double-clicks an area of an image map.
		mouseOut	The mouse is moved from within an image map area to outside of that area.
		mouseOver	The mouse is moved over an area of a client-side image map.
button	<INPUT	blur	A button loses the input.

	TYPE =	click	A button is clicked.
	"button">	focus	A button gains the input focus.
		mouseDown	A user presses the left mouse button over a button.
		mouseUp	A user releases the left mouse button over a button.
checkbox	<INPUT	blur	A checkbox loses the input focus.
	TYPE =	click	A checkbox is clicked.
	"checkbox"	focus	A checkbox gains the input focus.
document	<BODY>..	click	The use clicks in the body of a document.
body	</BODY>	dblClick	The user double-clicks in the body of a document.
		keyDown	The user presses a key.
		keyPress	The user presses and releases a key.
		keyUp	The user releases a key.
		mouseDown	The user presses the mouse button.
		mouseUp	The user releases the mouse button.
file uploaded	<INPUT	blur	A file upload from element loses the input focus.
	TYPE =	change	A user selects a file to be uploaded.
	"file"	focus	A file upload from element gains the input focus.
form	<FORM> ..	submit	A form is submitted by the user.
	</FORM>	reset	A form is reset by the user.
image	<IMG>	abort	The loading of an image is aborted as a result of a user action.
		error	An error occurs during the loading of an image.
		keyDown	The user presses a key.
		keyPress	The user presses and releases a key.
		keyUp	The user releases a key.
		load	An image is loaded and displayed.
link	<A>...</A>	click	The mouse is clicked on a link.
		dblClick	The mouse is double-clicked on a link.
		keyDown	The user presses a key.
		keyPress	The user presses and releases a key.
		keyUp	The user releases a key.
		mouseDown	The mouse button is pressed.
		mouseOut	The mouse is moved from within a link to outside of that link.
		mouseOver	The mouse is moved over a link.
		mouseUp	The mouse button is released.
password	<INPUT	blur	A password field loses the input focus.
	TYPE =	focus	A password field gains the input focus.
	"password"		
radio button	<INPUT	blur	A radio button loses the input focus.
	TYPE =	click	A radio button is clicked.
	"radio"	focus	A radio button gains the input focus.
reset	<INPUT	blur	A reset button loses the input focus.

	TYPE = "reset"	click	A reset button is clicked.
		focus	A reset button gains the input focus.
selection	<SELECT>.. </SELECT>	blur	A selection element loses the current input focus.
		change	A selection element is modified and loses the current input focus.
		focus	A selection element receives the current input focus.
submit	<INPUT TYPE = "submit">	blur	A submit button loses the input.
		click	A submit button is clicked.
		focus	A submit button gains the input focus.
text area	<TEXTAREA>.. </TEXTAREA>	blur	A text area loses the current input focus.
		change	A text area is modified and loses the current input focus.
		focus	A text area receives the current input focus.
		keyDown	The user presses a key.
		keyPress	The user presses and releases a key.
		keyUp	The user releases a key.
text field	<INPUT TYPE = "text">	blur	A text field loses the current input focus.
		change	A text field is modified and loses the current input focus.
		focus	A text field receives the current input focus.
		select	Text is selected within a text field.
window, frame set,	<BCDY>... </BODY>	blur	A window loses the current input focus.
		dragDrop	The user drops an object onto the window.
	<FRAMESET> ... </FRAMESET>	error	An error occurs when a window is loaded.
		focus	A window receives the current input focus.
	<FRAME> .. </FRAME>	load	The loading of a window is completed.
		move	A window is moved.
		resize	A window is resized.
		unload	The user exists a window.

As promised I am showing here the working of a very simple event, called onMouseOver. See the use of it in the script and its result in the browser and the result obtained by moving over the mouse.

Event Handling Attributes

Following is the list of attributes which connect the various events handled by JavaScript.

Event Handling Attributes	*Codes to be Executed*
onAbort	The loading of an image is aborted as the result of a user action.
onBlur	A document, window, frame set, or form element loses the current input focus.
onChange	A text field, text area, file upload field, or selection is modified and loses the current input focus.
onClick	A link, client-side image map area, or form element is clicked.
onDblClick	A link, client-side image map area, or document is clicked.
onDragDrop	A dragged object is dropped in a window or frame.

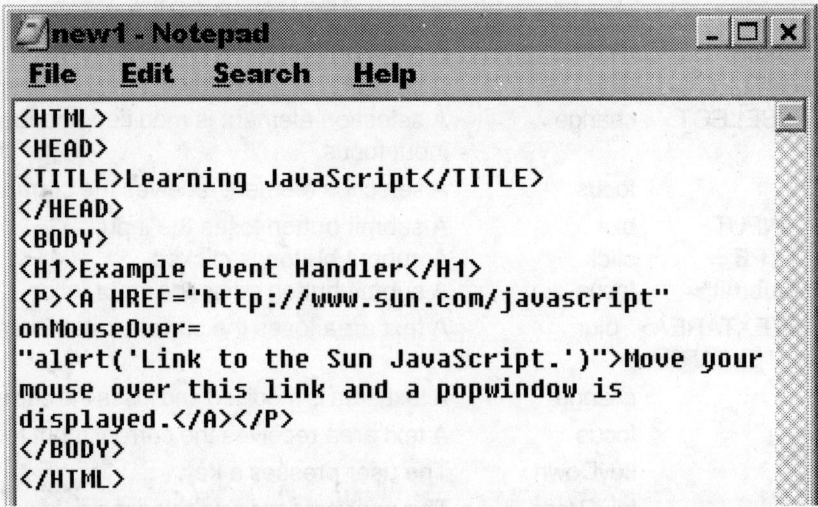

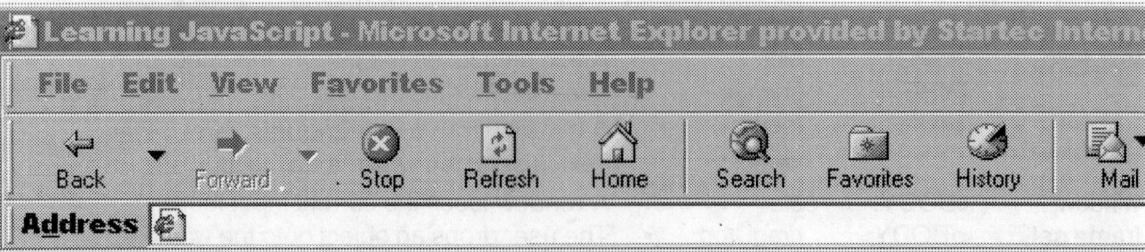

onError	An error occurs during the loading of an image, window, or frame.
onFocus	A document, window, frame set, or form element receives the current input focus.
onKeyDown	The user presses a key.
onKeyPress	The user presses a and releases a key.
onKeyUp	The user releases a key.
onLoad	An image, document, or frame set is loaded.

onMouseDown	The user presses a mouse button.
onMouseMove	The user moves the mouse.
on MouseOut	The mouse is moved out of a link or an area of a client-side image map.
onMouseOver	The mouse is moved over a link or an area of a client-side image map.
onMouseUp	The user releases a mouse button.
onMove	The user moves a window or frame.
onReset	A user resets a form by clicking on the form's reset button.
onResize	The user resizes a window or frame.
onSelect	Text is selected in a text field or text area.
onSubmit	A form is submitted.
onUnload	The user exists a document or frame set.

Supposing in the previous example of moving the mouse over the text, I add the option of counting the number of times you move over the link.

This is illustrated in the example given on the next page.

Window Events

These events produce a display; they also apply to layout documents that replace the document body with a frame set. A typical example of this is given below, where you will get Hello There! when you open the site. The listing of the program is listed below and the alert display too.

Form Events

You will read more about the Forms later. But, here I would like to give the various options of the form events. Following is the list of the activities associated with the forms.

Event-Handling	***Form Element***	***AttributeEvent***
on Blur	All form elements	A form element loses the input focus.
onChange	File upload field, select field, text field, or text area	A form element is modified and loses the current input focus.
onClick	Button, submit button, reset button, radio button, or checkbox	A button or checkbox is clicked

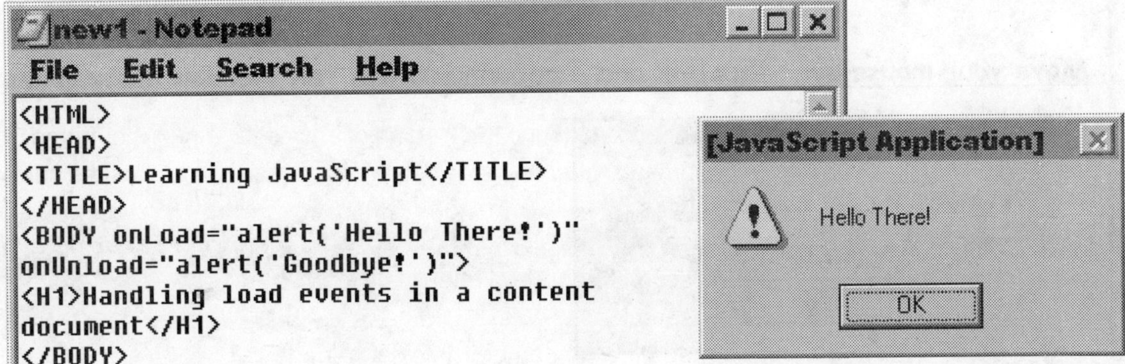

onFocus	All form elements	A form element receives the input focus.
onKeyDown onKeyUp, onKeyPress	Text area	A key is pressed, released, or pressed and released
onMouseDown onMouseUp	Button	A mouse button is pressed or released
onReset	Overall form	The form is reset by the user.
onSelect	Text field or text area	Text is selected within the text field or text area.
onSubmit	Overall form	The form is submitted by the user.

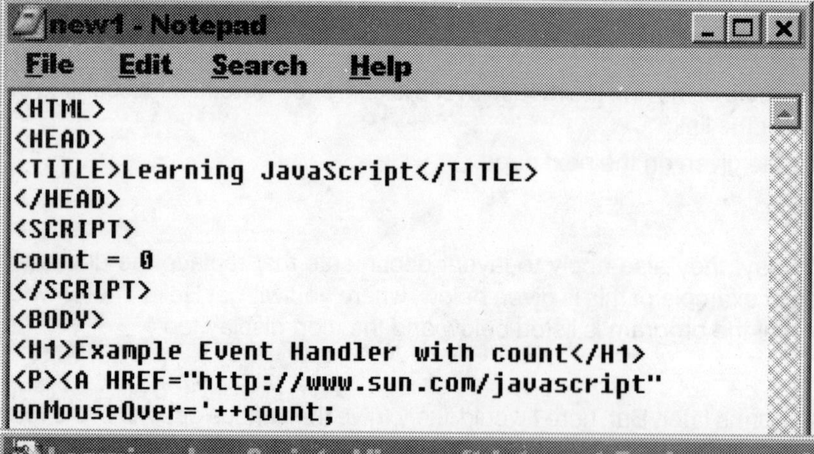

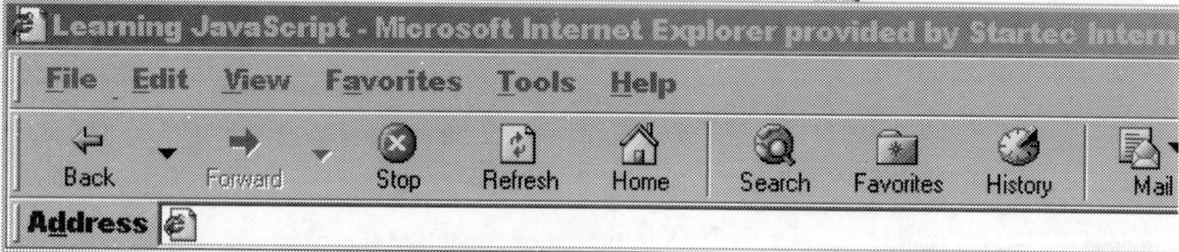

As usual now I present here an example of a form which would use the various options discussed here.

Event Object

This object is to provide additional information about events. It uses object's properties to provide this information. Let me give here, in the form of a table, how this is handled by the two main browsers, i.e., Internet Explorer and Netscape Navigator.

Property	*Browser*	*Description*
altKey, ctrlKey, and shiftKey	I	Set to true or false to indicate whether the Alt,, Control, or Shift keys were pressed when the event occurred.
button	I	Identifies the mouse button that was pressed when the event occurred. Values are: 0 No button 1 Left button 2 Right button 4 Middle button
cancelBubble	I	Set to true of false to cancel or enable event bubbling.
clientX and clientY	I	The cursor's horizontal and vertical position in pixels, relative to the Web page in which the event occurred.
data	N	An array of strings containing the URLs of objects dropped as the result of a DragDrop event.
formElement and toElement	I	Specifies the HTML element being moved from or to.
height and width	N	The height and width of a window or frame (in pixels).
keyCode	I	Identifies the Unicode key code associated with a key press.
layerX and layerY	N	The cursor's horizontal and vertical position in pixels, relative to the layer in which the event occurred. When used with the resize event, layerX and layerY specify the width and height of the object to which the event is targetted.
modifiers	N	Identifies the modifier keys (ALT_MASK, CONTROL_MASK, SHIFT_MASK, and META_MASK) associated with a mouse or key event.
offsetX and offsetY	I	The cursor's horizontal and vertical in pixels, relative to the container in which the event occurred.
pageX and pageY	N	The cursor's horizontal and vertical position in pixels, relative to the screen.
reason	I	Used to indicate the status of data transfer for data source objects.
returnValue	I	Set to true or false to indicate the event handler's return value. Same as returning true or false from the event handler.
screenX and screenY	N/I	The cursor's horizontal and vertical position in pixels, relative to the screen.
srcElement	I	The object to which the event was originally sent.
srcFilter	I	Specifies the filter object that caused an onfilterchange event.

target	N	The object to which the event was originally sent.
type	N/I	The type of event that occurred.
which	N	Either the mouse button (left=1, middle=2, and right=3) that was pressed or the ASCII value of a pressed key.
x and y	I	The cursor's horizontal and vertical position in pixels, relative to the document in which the event occurred.

N: Netscape Navigator   I: Microsoft Internet Explorer

Event Simulation

When an event simulation method is invoked, the object to which it refers acts as if the event is taking place. For example, button objects have the click() method that, when invoked, causes the button's event handler to be invoked.

See the following example, which has two buttons. On pressing the button, it will give you the name of the button you have pressed.

A complete listing of the program is given below and its outputs on the next page:

So from the site if you click Button 1, you will get the message as **Button 1 was clicked!**. Also if you click Button 2, you will get the message as **Button 2 was clicked!**.

Working with Forms

Forms play an important part in the Web page development. It is through forms that you get the important information from the user. You ask the user to fill up the form and you have all the information wanted by you.

new1 - Notepad

File Edit Search Help

```
<HTML>
<HEAD>
<TITLE>Learning JavaScript</TITLE>
<SCRIPT>
function button1Clicked() {
alert("Button 1 was clicked!")
}
function button2Clicked() {
alert("Button 2 was clicked!")
}
</SCRIPT>
</HEAD>
<BODY>
<FORM NAME="test">
<INPUT TYPE="BUTTON" NAME="button1" VALUE="Button 1"
ONCLICK="button1Clicked()">
<INPUT TYPE="BUTTON" NAME="button2" VALUE="Button 2"
ONCLICK="button2Clicked()">
</FORM>
</BODY>
</HTML>
```

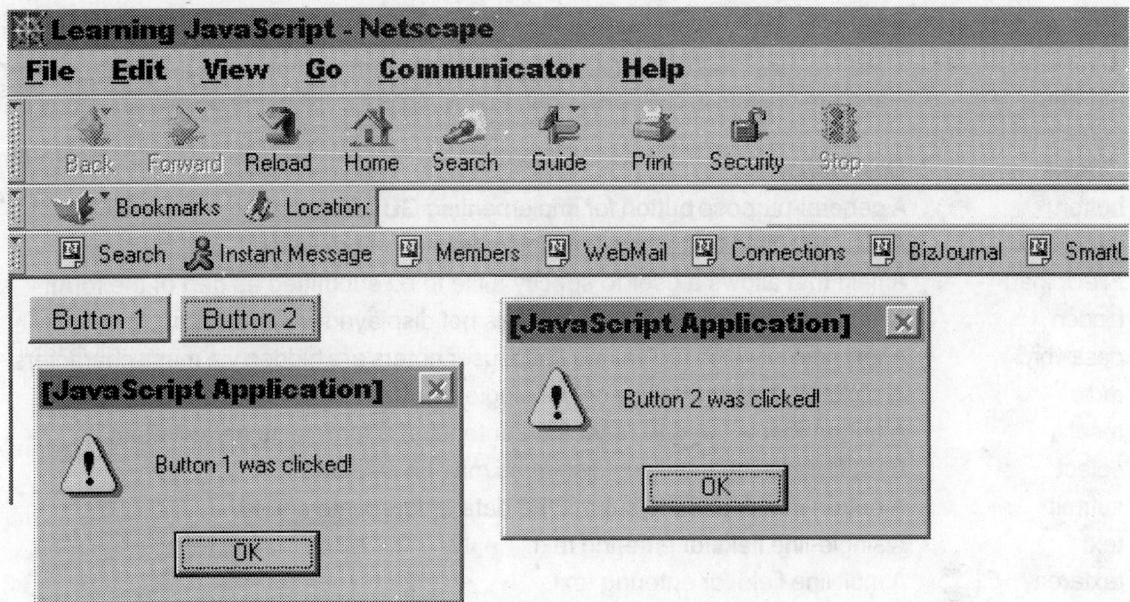

JavaScript provides a number of features that can be used to enhance the forms that you develop for your particular Web applications. These features allow you to validate form data before it is submitted to your server and to exercise greater control of the interaction between your forms and Web users. Following is the list of properties of the form object of JavaScript. These properties provide access to a form's attributes and allow you to work with a form's fields and GUI controls.

Property	***Description***
action	It provides access to the HTML action attributes of the <form> tag.
button	It is an object representing a button GUI control.
checkbox	It is an object representing a checkbox field
elements	It is an array containing all the fields and GUI controls included in a form.
encoding	It provides access to the HTML enctype attribute of the <form> tag.
FileUpload	It is an object representing a file-upload from field.
hidden	It is an object representing a hidden form field.
length	It provides access to the length of the elements array.
method	It provides access to the HTML method attribute of the <form> tag.
name	It identifies the name of the form.
password	It is an object representing a password field.
radio	It is an object representing a radio button field.
reset	It is an object representing a reset button.
select	It is an object representing a selection list.
submit	It is an object representing a submit button.
target	It provides access to the HTML target attribute of the <form> tag.
text	It is an object representing a text field.
textarea	It is an object representing a text area field.

Form Elements

A form may contain a wide range of fields and GUI controls. These form components are referred to as elements of the form and are objects in their own right. Following is the list of the objects that may be contained in a form.

Object	***Description***
button	A general-purpose button for implementing GUI controls.
checkbox	A clickable field that allows multiple selections from within a group.
FileUpload	A field that allows a user to specify a file to be submitted as part of the form.
hidden	A field that may contain a value but is not displayed within a form.
password	A text field in which the values that a user enters are hidden via mask characters.
radio	A clickable field that allow only a single selection from within a group.
reset	A button that is used to reset the contents of a form to its default state.
select	A list from which individual list items may be selected.
submit	A button that is used to submit the data entered into a field.
text	A single-line field for entering text.
textarea	A multiline field for entering text.

If the elements of a form are named using an HTML name attribute, then the element can be accessed using this name. In most cases, you will access the elements of a form using the elements array property of the form object. This array contains an object for each element of a form. The objects described in the above table refers to the elements of a form and have properties and methods of their own as summarized in following tables:

Properties of Form Elements

Object	*Property*	*Description*
button	name	It provides access to the button's name attribute.
	type	It identifies the object's type.
	value	It identifies the object's value.
checkbox	checked	It identifies whether the checkbox is currently checked.
	defaultChecked	It identifies whether the checkbox is checked by default.
	name	It provides access to the checkbox's HTML name attribute.
	type	It identifies the object's type.
	value	It identifies the object's value.
FileUpload	name	It provides access to the object's name attribute.
	type	It identifies the object's type attributes.
	value	It identifies the object's value.
hidden	name	It provides access to the object's name attribute.
	type	It identifies the object's type attributes.
	value	It identifies the object's value.
password	defaultValue	It identifies the object's default value
	name	It provides access to the object's name attribute.
	type	It identifies the object's type attributes.

	value	It identifies the object's value.
radio	checked	It identifies whether the radio button is currently checked.
	defaultChecked	It identifies whether the radio button is checked by default.
	name	It provides access to the object's name attribute.
	type	It identifies the object's type attributes.
	value	It identifies the object's value.
reset	name	It provides access to the object's name attribute.
	type	It identifies the object's type attributes.
	value	It identifies the object's value.
select	length	It identifies the length of the select list.
	name	It provides access to the object's name attribute.
	options	It is array that identifies the options supported by the select list
	selectedIndex	It identifies the first selected option within the select list.
	type	It identifies the object's type.
submit	name	It provides access to the object's name attribute.
	type	It identifies the object's type attributes.
	value	It identifies the object's value.
text	defaultValue	It identifies the default text to be displayed in the text field.
	name	It provides access to the object's name attribute.
	type	It identifies the object's type attributes.
	value	It identifies the object's value.
textarea	defaultValue	It identifies the default text to be displayed in the text area field.
	name	It provides access to the object's name attribute.
	type	It identifies the object's type attributes.
	value	It identifies the object's value.

Methods of Form Elements

All form elements except the hidden element provide the handlEvent() method for directly invoking the element's event handlers.

Object	*Method*	*Description*
button	blur()	It removes focus from the button.
	click()	It simulates the button being clicked.
	focus()	It gives focus to the button.
checkbox	blur()	It removes focus from the checkbox.
	click()	It simulates the checkbox being clicked.
	focus()	It gives focus to the checkbox.
FileUpload	blur()	It removes focus from the file upload field.
	focus()	It gives focus to the file upload field.
	select()	It selects the input area of the file upload field.
hidden	none	

password	blur()	It removes focus from the password field.
	focus()	It gives focus to the password field.
	select()	It highlights the text displayed in the password field.
radio	blur()	It removes focus from the radio button.
	click()	It simulates the clicking of the radio button.
	focus()	It gives focus to the radio button.
reset	blur()	It removes focus from the reset button.
	click()	It simulates the clicking of the reset button.
	focus()	It gives focus to the reset button.
select	blur()	It removes focus from the selection list.
	focus()	It gives focus to the selection list.
submit	blur()	It removes focus from the submit button.
	click()	It simulates the clicking of the submit button.
	focus()	It gives focus to the submit button.
text	blur()	It removes focus from the text field.
	focus()	It gives focus to the text field.
	select()	It highlights the text in the text field.
textarea	blur()	It removes focus from the text area.
	focus()	It gives focus to the text area.
	select()	It highlights the text in the text area.

You must be wondering why I am giving all the details here without any live example. Well, here you are. An example of multiform which has 3 forms of different nature is given here. First, the listing of the program and then the output of all the three forms.

Listing of program for MultiForm

```
<HTML>
<HEAD>
<TITLE>Learning JavaScript</TITLE>
<SCRIPT>
function displayFormData() {
    win2=open(" ","window2")
    win2.document.open("text/plain")
    win2.document.writeln("This document has "+document.forms.length+" forms.")
    for(i=0;i<document.forms.length;++i) {
    win2.document.writeln("Form "+i+" has "+ document.forms[i].elements.length+" elements.")
    for(j=0;j<document.forms[i].elements.length;++j) {
    win2.document.writeln((j+1)+" A "+ document.forms[i].elements[j].type+" element.")
    }
}
win2.document.close()
return false
}
</SCRIPT>
</HEAD>
<BODY>
<H1>Example of a Multiform</H1>
```

```
<FORM ACTION="nothing" onSubmit="return displayFormData()">
<H2>Form 1</H2>
    <P>Text field: <INPUT TYPE="TEXT" NAME="f1-1" VALUE="Write your text here"></P>
    <P>Password field:<INPUT TYPE="PASSWORD" NAME="f1-2"></P>
    <P>Text area field: <TEXTAREA ROWS="4" COLS"50" NAME="f1-3">Write details here.</
TEXTAREA></P>
    <P><INPUT TYPE="SUBMIT" NAME="f1-4" VALUE="Submit">
    <INPUT TYPE="RESET" NAME="f1-5"></P>
</FORM>
<HR>
<FORM>
<H2>Form 2</H2>
    <P><INPUT TYPE="CHECKBOX" NAME="f2-1" VALUE="1" CHECKED> Click me!</P>
    <P><INPUT TYPE="CHECKBOX" NAME="f2-1" VALUE="2"> Or me!</P>
    <P><INPUT TYPE="CHECKBOX" NAME="f2-1" VALUE="3"> At least me!</P>
    <P><INPUT TYPE="RADIO" NAME="f2-2" VALUE="1"> TRUE</P>
    <P><INPUT TYPE="RADIO" NAME="f2-2" VALUE="2" CHECKED> FALSE</P>
    <P><INPUT TYPE="RADIO" NAME="f2-2" VALUE="3"> DO NOT KNOW</P>
    <INPUT TYPE="FILE" NAME="f2-3">
</FORM>
<HR>
<FORM>
<H2>Form 3</H2>
    <INPUT TYPE="HIDDEN" NAME="f3-1">
    <SELECT NAME="f3-2" SIZE="4">
    <OPTION VALUE=" ">Item 1</OPTION>
    <OPTION VALUE=" ">Item 2</OPTION>
    <OPTION VALUE=" " SELECTED>Item 3</OPTION>
    <OPTION VALUE=" ">Item 4</OPTION>
    <OPTION VALUE=" ">Item 5</OPTION>
</SELECT>
</FORM>
</BODY>
</HTML>
```

User Actions

By the look of the above program and the listing of the above mentioned events it can be said that the most of the form events have the following:

List Selection — When a user selects an item from a selection list, even-handling code is used to verify that the selection is consistent with other inputs and to perform any processing indicated by the selection.

Clicks and checks — These are most common types of form events. A use clicks a button or checks a checkbox to provide information or to perform an action. These events are handled by event-handling functions that provide feedback to the user on the results of the actions taken in response to the click and check.

Form submission and reset — These events are generated when a user clicks a submit or reset button. Form-submission events are typically handled by validating all of the data entered by the user, performing any local processing on that data, and then forwarding the data to a CGI program or other server-side script.

Text changes — Text changes are another common type of form event. The user enters data into a text field or text area, an event is generated, and the event handler

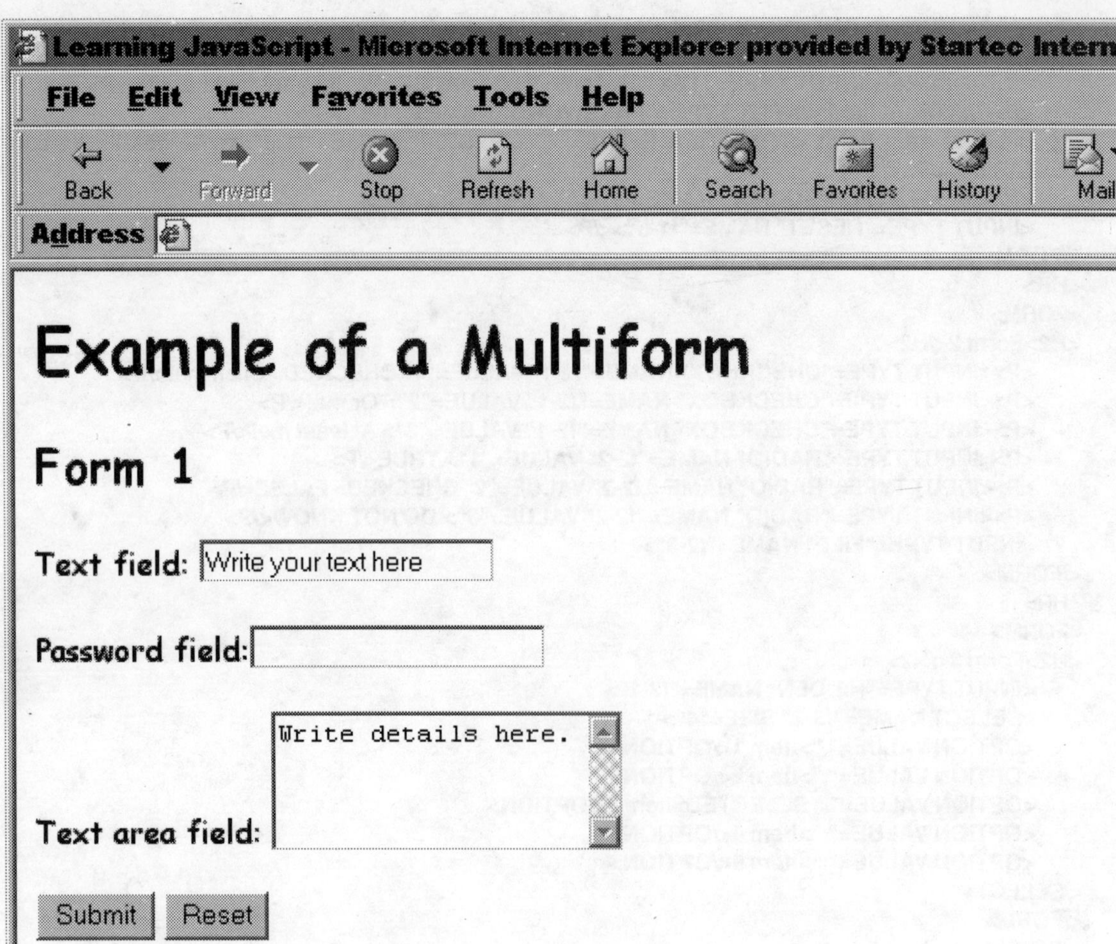

validates the user's entry and performs further processing based on the user's input.

Change of focus — Change of focus events occur when a form element, such as a text field or selection list, receives or loses the current input focus. These events usually do not require special event handling. However, JavaScript provides the capability to do so, when required.

Windows and Frames

Now, I will discuss the following 5 important objects of JavaScript. They are:

- window
- frame
- document
- navigator
- screen

Let us study them one by one.

window Object

It is a basic object of all browser scripts. In Navigator, this is listed under Window menu.

Various properties of the window object are listed below:

Property	Description
closed	It identifies whether the window is closed.
defaultStatus	It specifies the default status message that appears on the status bar on the button of the browser window.
document	It is an object that refers to the current document being displayed in a window.
frames	It is an array that consists of all frame objects contained in a window object.
history	It refers to the windows's history object, which contains a list of URLs last loaded into the window.
length	It identifies the number of frames contained in a window.
location	It is an object that identifies the URL associated with a window object.
name	It identifies the name of the window.
offscreenBuffering	It gives the Boolean value that specifies whether offscreen buffering of window information should be used. Offscreen buffering is used to load all of window's elements before displaying the window.
opener	It identifies the window object that caused a window to be created and opened.
parent	It is a synonym that identifies the window containing a particular window.
self	It is a synonym that identifies the current window being referenced.
status	It specifies a temporary message that appears on the status bar on the bottom of the browser window.
top	It is a synonym that refers to the topmost browser in a series window in a series of nested windows.
window	It is a synonym that identifies that identifies the current window being referenced.

Various methods of the window object are:

Method	Description
alert(text)	It displays an alert dialog box.
blur()	It removes focus from a window.
clearInterval(interval)	It clears a previously set interval timer.
clearTimeout(timer)	It clears a previously set timeout.
close()	It closes the specified window.
confirm(text)	It displays a confirm dialog box.
focus()	It gives focus to a window.
open(url,name,[options])	It opens a new window and creates a new window object.
prompt(text,defaultInput)	It displays a prompt dialog box.
scroll(x,y)	It scrolls a window to the specific location.
setInterval(expression, milliseconds)setInterval (function, milliseconds,	It repeatedly evaluates an expression or invokes a function after the elapse of a specified time interval. The arguments are a possible empty

File Edit Search Help

```
<HTML>
<HEAD>
<TITLE>Learning JavaScript</TITLE>
<SCRIPT>
function directions() {
open("birthday.htm","directions")
}
</SCRIPT>
<BODY>
<H1>My Birthday Party!</H1>
<P>On 28th August, 2000 at 7 P.M</P>
<P>at Hotel Radisson</P>
<P>Room No. 211</P>
<FORM>
<INPUT TYPE="BUTTON" VALUE="Click for directions"
ONLCLICK="directions()">
</FORM>
<P>RSVP: 571 5968, 572 7854<P>
</BODY>
</HEAD>
```

File Edit Search Help

```
<HTML>
<HEAD>
<TITLE>Directions</TITLE>
</HEAD>
<BODY ONLOAD="defaultStatus='It is where the arrow
is!'">
<FORM>
<INPUT TYPE="BUTTON" VALUE="Close window"
ONCLICK="windows.close()">
</FORM>
<IMG SRC="directions.gif">
</BODY>
```

[arguments])	comma-separated list of arguments to the function to be invoked. It returns an interval reference that can be cleared by clearInterval().
setTimeout(expression, milliseconds)setTimeout (function, milliseconds, [arguments])	It evaluates an expression or invokes a function after a timeout period has elapsed. The arguments are possibly an empty comma-separated list of arguments to the function to be invoked. It returns a timer reference that can be cleared by clearTimer().

In window you can either open it or close it. The open() and close() methods may be used from within JavaScript to open and close browser windows. The open() method opens a new Web browser window at a specified URL with a given set of options.

Various options of open() method are:

Option	***Values***	***Description***
directories	Y/N	The window provides directory buttons.
height	Integer	The height of the window in pixels.

My Birthday Party!

On 28th August, 2000 at 7 P.M

at Hotel Radisson

Room No. 211

Click for directions

RSVP: 571 5968, 572 7854

location	Y/N	The window displays the location field.
menubar	Y/N	The window has a menu bar.
resizeable	Y/N	The window is resizeable.
scrollbars	Y/N	The window provides scroll bars.
status	Y/N	The window has a status bar.
toolbar	Y/N	The window has a tool bar.
width	Integer	The width of the window in pixels.

As usual I have to give you an example to illustrate more about the working of the object. In the following example, you will see that I have created 2 htm programs. One for the invitation to my birthday party and the other for the directions. You have to run the first one where you will find a button which you have to press to get the directions (it will then open the second htm program). On this page see the listings and output of invitation above.

Frame Object

The frames enable you to partition a window into independent display areas and organize and control the information displayed in these areas in powerful ways. Framesets are used to organize the frames. Frameset is a HTML command and is widely used there.

Frame objects are automatically created by your browser and they enable your scripts to control how frames are used. These frame objects allow you to load, based on user-generated or other events, new documents into frames. The properties and methods of the frame object are the same as the window object with one exception—the frame object does not support the close() method. The important characteristic of frames allows multiple nested frame sets to be displayed in a single browser window.

Document Object

It allows you to update a document that is being loaded and to access the HTML elements contained in a loaded document. The following table identifies the methods of document object that are supported by both the browsers.

Properties of the Document Object

Property	*Description*
alinkColor	It identifies the value of the alink attribute of the <body> tag.
anchor	It is an object that refers to an array contained in a document.
anchors	It is an array of all the anchors contained in a document.
applet	It is an object that refers to an applet that is contained in a document.
applets	It is an array of all the applets contained in a document.
area	It is an object that refers to an image map area contained in a document.
bgColor	It identifies the value of the bgcolor attribute of the <body> tag.
cookie	It identifies the value of a cookie.
domain	It identifies the domain name of the server from which the document is loaded.
embeds	It is an array of all the plug-ins contained in a document.
fgColor	It identifies the value of the text attribute of the <body> tag.
form	It is an object that refers to a form contained in a document.
Forms[]	It is an array of all the forms contained in a document.
image	It is an object that refers to an image contained in a document.
Images[]	It is an array of all the images contained in a document.
lastModified	The date that a document was last modified.
link	It is an object that refers to a link contained in a document.
links	It is an array of all the links contained in a document.
linkColor	It identifies the value of the link attribute of the <body> tag.
plugin	It is an object that refers to a plugin contained in a document.
Plugins[]	It is an array of objects that describe the plug-ins supported by a browser.
referrer	It is the URL of the document that provided the link to a document.
title	The document's title.
URL	The URL of a document.
vlinkColor	It identifies the value of the vlink attribute of the <body> tag.

Methods of the Document Object

Method	*Description*
close()	It closes a stream used to create a document object.
open([mimeType][,"replace"])	It opens a stream used to create a document object with the optional MIME type. The "replace" parameter is used with the

	text/html MIME type to replace the current document in the history list.
write(expr1[,expr2,....,exprN])	It writes the values of the expressions to a document.
write(expr1[,expr2,...,exprN])	It writes the values of the expressions to a document followed by a new-line character.

Navigator Object

This object provides information about the type and the version of the browser that is used to run a script. The navigator properties that Navigator and Internet Explorer support differ slightly. The following table shows the properties of the navigator object, with the browser supported by them.

Properties of the Navigator Object

Property	*Browser*	*Description*
appCodeName	N2/I3	The code name of the browser.
AppMinorVersion	I4	The minor version number of the browser.
appName	N2/I3	The name of the browser.
appVersion	N2/I3	The version of the browser.
browserLanguage	I4	The language in which he browser is configured.
connectionSpeed	I4	The speed at which the browser is connected to the network.
cookieEnabled	I4	Whether the browser is configured to accept cookies.
cpuClass	I4	The type of microprocessor used to execute the browser.
onLine	I4	Whether the browser currently has an online connection.
Language	N4/I4	The language in which the browser is configured.
mimeTypes	N2/I4	An array of all MIME types currently supported by the browser.
platform	N4/I4	The operating system platform on which the browser executes.
plugins	N3/I4	An array of all plug-ins currently installed on the browser.
systemLanguage	I4	The default language of the operating system.
userAgent	N2/I3	The user-agent header sent in the HTTP protocol from the browser to the server.
userLanguage	I4	The human language used by the user.
userProfile	I4	An object that provides access to user profile information.

I: Microsoft Internet Explorer N: Netscape Navigator with versions.

The last object is screen object.

Screen Object

It provides information about the dimensions and color depth of the user's screen. It is basically used for the following:

height	current height of the user's screen (in pixels).
width	current width of the user's screen (in pixels).
colorDepth	number of bits per color currently supported by the user's screen.

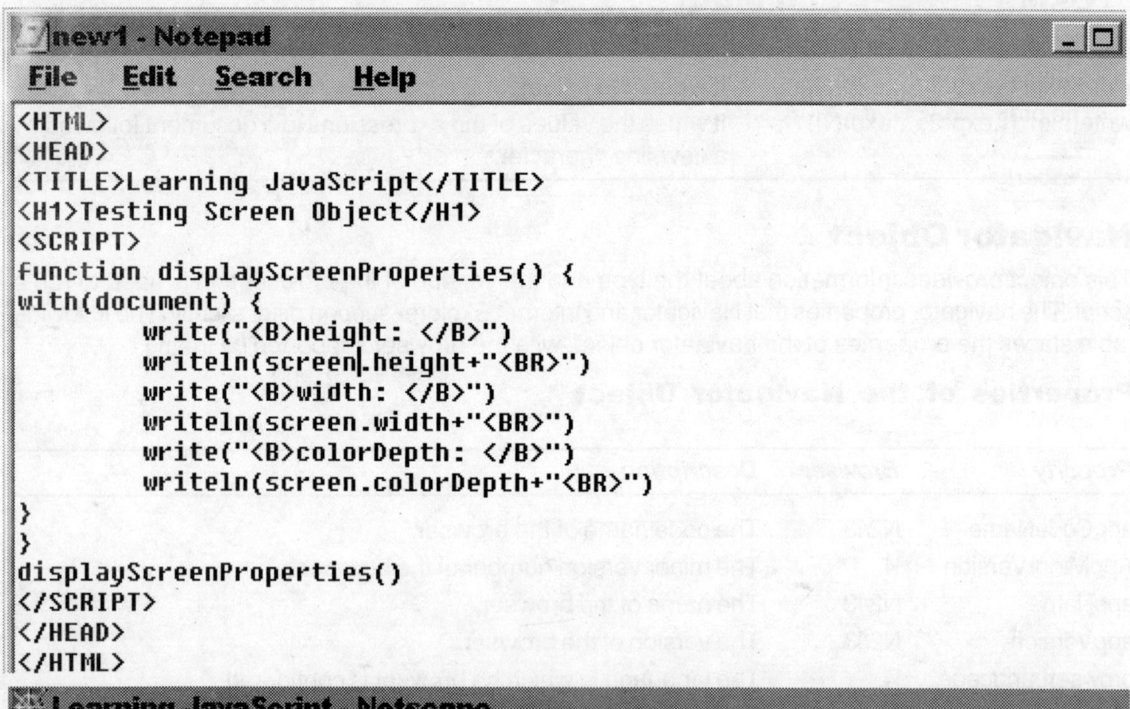


```
<HTML>
<HEAD>
<TITLE>Learning JavaScript</TITLE>
<H1>Testing Screen Object</H1>
<SCRIPT>
function displayScreenProperties() {
with(document) {
        write("<B>height: </B>")
        writeln(screen.height+"<BR>")
        write("<B>width: </B>")
        writeln(screen.width+"<BR>")
        write("<B>colorDepth: </B>")
        writeln(screen.colorDepth+"<BR>")
}
}
displayScreenProperties()
</SCRIPT>
</HEAD>
</HTML>
```

You can use this information to determine how to organize the information you display to the user. The following example would give you an idea how it is used.

As you can see that the result it has given height as 600, width as 800 and colourDepth as 24.

Using Images and Math

The image object provides access to the images that are loaded with a document. The images array is also a property of the document object.

The properties of the image object are given below:

Property	*Description*
border	The value of the <img> tag's BORDER attribute.
complete	It identifies whether an image has been completely loaded.
height	The value of the <img> tag's HEIGHT attribute.
hspace	The value of the <img> tag's HSPACE attribute.
lowsrc	The value of the <img> tag's LOWSRC attribute.
name	The value of the <img> tag's NAME attribute.
prototype	It is used to add user-specified properties to an image object.
src	The value of the <img> tag's SRC attribute.
vspace	The value of the <img> tag's VSPACE attribute.
width	The value of the <img> tag's WIDTH attribute.

The image object has no methods that are common to Internet Explorer and Navigator. Navigator defines the handleEvent() method, which is used to directly invoke an image's event handler. The image object has other properties that can be used to monitor and control image loading and display.

name — The name property is a read-only property that specifies the image's name attribute.

lowsrc — The lowsrc property can be modified to quickly load and display a low resolution image while a slower loading high-resolution image is being loaded.

src — The src property is used to load a new image in the place of a currently displayed image.

complete — The complete property is a read-only property that indicates whether an image has been completely loaded.

prototype — The prototype property allows other user-defined properties to be created for all objects of the image object type.

Images and Animation

JavaScript's dynamic image display capabilities can be used to create animation effects. Animation basically involves displaying one image after another in sequence. The quality of the animation depends on the quality of the images displayed, the delay between successive image displays, and the ability of your browser to reduce any deviations in this delay.

Event if you already know the basics of how to perform animation, there are a few tips that will improve your results.

These are:

images — Make sure that all images are loaded before you start your animation sequence. This will eliminate any additional delays associated with image loading.

schedule — You must schedule successive image displays using the onLoad event handler of the <img> tag to which the images are to be written. This will ensure that an image has been loaded and displayed before the next image is processed.

delays	Use the setTimeout() method of the window object to implement delays between images. This method will help to educe any deviations in your average delay.
test animation	Experiment with the delay value to see what delay is best for your animation. Although a lower delay value provides a smoother animation, a delay that is too small can result in images being displayed at a comically quick pace.

Area Object

This object provides a way of handling user actions related to specific image areas. These areas are defined by the <area> tag.

The properties of the area object that are common to Navigator and Internet Explorer are given below:

Property	***Description***
hash	The file offset part of an area object's HREF attribute.
host	The host name part of an area object's HREF attribute.
hostname	The host port part of an area object's HREF attribute.
href	An area object's complete HREF attribute.
pathname	The path name part of an area object's HREF attribute.
port	The port part of an area object's HREF attribute.
protocol	The protocol part of an area object's HREF attribute.
search	The query string part of an area object's HREF attribute.
target	An area object's TARGET attribute.

Math Object

It was mentioned briefly earlier. It is a built-in object that is defined by the script standard and supported by both Navigator and Internet Explorer. The Math object cannot be instantiated like Date and String to create specific object instances. Math constants which can be used as:

Constant	***Description***
E	Euler's constant. It is found everywhere in computational math and is the base for natural logarithms.
LN2	The natural logarithm of 2. This is handy constant for converting between natural logarithms and base 2 logarithms.
LN10	The natural logarithm of 10. Like LN2, it is used in logarithm conversions.
LOG2E	The base 2 logarithm of E. It is used in base 2 to base E logarithm conversions.
LOG10E	The base 10 logarithm of E. It is used in base 10 to base E logarithm conversions.
PI	Another famous mathematical constant, PI is the ratio of the circumference of a circle to its diameter.
SQRT1_2	The square root of ½ is used in many trigonometric calculations.
SQRT2	The square root of 2 is commonly used in algebraic formulas.

The Math object provides 18 mathematical functions that run the gamut from rounding to random number generation. These functions are discussed in the following categories.

Rounding Functions

Various functions under this are: round(), floor(), and ceil(). These are used to approximate floating-point numbers with integers.

round()	This function returns the closet integer to a floating-point number.

floor()	This function returns the greatest integer that is less than or equal to a floating-point number.

ceil()	This function returns the least integer that is greater than or equal to a floating-point number.

Comparison Functions

Various functions under this heading are: min() and max(). They are used to compare two numbers.

min()	This function returns the lesser of the two numbers.

max()	This function returns the greater of the two numbers.

Algebraic Functions

Various functions under this heading are: abs() and sqrt().

abs()	This function calculates the absolute value of a number. This is useful in calculating the distance between two numbers.

sqrt()	This function calculates the square root of a number.

Logarithmic and Exponential Functions

Various logarithmic functions used are: log(), LN2, LN10, LOG2, LOG10E. These are used to convert between different bases.

pow()	This function calculates a number raised to a power.

exp()	This function calculates Euler's constant raised to a power.

Trigonometric Functions

Various functions available here are: cos(), sing(), tan(), acos(), asin(), atan(), and atan2(). These functions are used to calculate the position and relationship of points on circles, ellipses, waves, and other curved objects with respect to Cartesian coordinates.

Random Number Generation

The random() function generates a pseudorandom number between 0 and 1.

Revision

1. JavaScript is a programming language, which is mainly used with Web Pages.
2. JAVA and JavaScript have only name in common and nothing else.
3. JavaScript deals with events with commands called event handlers.
4. There are various assignment operators in use in JavaScript.
5. JavaSript can run on most of the platforms available in the market.
6. You can create your text in any word processor and run in a browser.
7. You can alert the user about anything by using the alert symbol.
8. A function consists of the word function followed by the function name.
9. Statements form the integral part of any programming language.
10. As in any other programming language, assignment is used to assign value to the variables.
11. IF statement is used when you have to give more than one option.
12. The switch statement evaluates the expression and determines if any of the values match the expression's value.

13. The while statement is used where you have to perform certain statements repeatedly depending upon the variable, whose value keeps on changing.
14. Instead of FOR, you can use the option of DO WHILE too.
15. Label statement is used to assign a label to the statement.
16. Break statement is used to break an onging execution.
17. Continue statement is quite similar to the break statement. It affects the execution of the for, do while, or while statement in which it is contained.
18. Return statement is used to return a value which has been processed as a result of the processing of a function.
19. With statement helps you in eliminating retyping the name of an object that is to be referenced in a series of references and method invocations.
20. The throw statement is used to throw an exception.
21. The try statement and the catch statement work together to support exception handling.
22. An object is a collection of properties that contain data.
23. JavaScript supports both predefined and user-defined object types.
24. When a Web page is loaded by a JavaScript capable browser, the browser creates a number of JavaScript objects that provide access to the Web page and the HTML elements it contains.
25. The Date object type provides a common set of methods for working with dates and times.
26. Most of the mathematical operations like Sin, Cos; Logarithm of 2 and 10; round off; square root; are all there in Math object.
27. The string object allows strings to be accessed as objects. It provides a set of methods for manipulating strings.
28. In JavaScript you can create your own objects and use them whenever required.
29. Events are nothing but the actions which can be performed by you using a hyperlink in JavaScript.
30. Windows events produce a display; they also apply to layout documents that replace the document body with a frame set.
31. When an event simulation method is invoked, the object to which it refers acts as if the event is taking place.
32. Forms play an important part in the Web page development. It is through forms that you get the important information from the user.
33. A form may contain a wide range of fields and GUI controls.
34. The frames enable you to partition a window into independent display areas and organize and control the information displayed in these areas in powerful ways.
35. The Navigator object provides information about the type and the version of the browser that is used to run a script.
36. JavaScript's dynamic image display capabilities can be used to create animation effects.

Multiple Choice Questions

1. JavaScript is mainly used with :
   (a) Database (b) Text
   (c) Web Pages
2. Text is created by :
   (a) Word processor (b) Browser
   (c) Wizard
3. A function has name FUNCTION followed by :
   (a) Variable name (b) Constant
   (c) Function name

4. For can be replaced by :
   (a) IF..THEN (b) DO WHILE
   (c) IF..THEN..ELSE
5. For exception handling, we have a set of TRY and __________ statement:
   (a) THROW (b) END
   (c) CATCH
6. For getting information from the users, we use :
   (a) Forms (b) Tables
   (c) Macros
7. Events are nothing but actions in which a ________ is used.
   (a) Hyper (b) Hyperlink
   (c) Link
8. The String object allows strings to be accessed as :
   (a) Objects (b) Forms
   (c) Macros

Answers

1. c	2. a	3. c	4. b	5. c
6. a	7. b	8. a		

Subjective Questions

1. What can you do with JavaScript?
2. What are Events and Event Handlers?
3. What are Operators?
4. What are Assignments?
5. Which words should be avoided while writing programs in JavaScript?
6. What is the software requirement for JavaScript?
7. How would you use Alert?
8. What are Functions?
9. How would you use eval function?
10. Describe the working of Assignment Statement.
11. What are Objects?
12. What is Object Inheritance?
13. What is Classification of Objects?
14. What are Browser Objects?
15. Describe the various Date Objects.
16. Describe the various Math Objects.
17. Describe the various String Objects.
18. How are Events handled in JavaScript?
19. What are Forms?
20. What is a Frame Object?
21. What is Document Object?
22. Describe the working of Screen Object.
23. What are Math Objects?

24. Describe the following statements:

Swtich	While
If For	
Do while	Label
Break	Continue
Return	With
Throw	Try
Catch	

25. Describe the following type of functions:

Rounding	Comparision
Algebaric	Trignometric
Logarthmic and Exponential	
Random Number Generation	

6 JavaBeans

Introduction

A bean is a reusable and self-contained software component. Beans created using Java take advantages of all the security and platform independent features of Java. Some examples of pluggable software components which you can create are a spell-check utility, a calculator, a text editor, a find and replace utility, a login validator, or a balance sheet generator. These utilites can be used across multiple target applications such as MS Word and MS Excel.

Calling Your Bean from a JSP Page

This section presents a step-by-step tutorial on how to write and deploy a simple bean called CalculatorBean that is part of the com.brainysoftware pack-age. After it is deployed, you then can call the bean from a JSP page. In total, there are six steps you need to follow:

1. Write a bean class whose code is given in the following listing. Save it as CalculatorBean.Java.

```
package com.brainysoftware;
public class CalculatorBean {
    public int doubleItfint number) {
        return 2 * number;
    }
}
```

   The code has one public class called doubleIt that returns an integer as a result of multiplication of the argument by 2.
2. Compile the bean to obtain a class file called CalculatorBean.class.
3. Copy the bean class file to the classes directory under WEB-INF under your application directory. The deployment must take into account the package name. In this case, you need to create a directory called com under the classes directory. Under corn, create a directory named brainysoftware. Copy your CalculatorBean. class file to this brainysottware directory.
4. Create a JSP page that will call the bean you just created. For this example, just use the code in the following listing.

```
<jsp:useBean id="theBean" class="com.brainysoftware.CalculatorBean"/>
<HTML>
<HEAO>
</HEAD>
<BODY>
<%
    int i = 4;
    int j = theBean.doubleIt(i);
    out.print("2M=" + j);
%>
</BODY>
</HTML>
```

5. Restart Tomcat.
6. Open and direct your browser to the URL of the JSP page your wrote in Step 4.

Your browser should display something like this:

2 x 4 = 8

Components of a Java Bean

Before you can use a bean in your JSP page, you must make the bean available. using the jsp:useBean action element. This element has attributes that you can use to control the bean's behavior.The syntax for the jsp:useBean ele-ment has two forms. The first form is used when you don't need to write any initialization code, and the second form is used if you have Java code that needs to be run when the bean initializes. Code initialization is discussed in the next section.

The two forms of the jsp:useBean action element are as follows:

<jsp:useBean (attribute³"value")*/>

and

<jsp:useBean (attribute="value")⁺>
initialization code </jsp:useBean>

The (attribute="value")⁺ part of the code means that one or more attributes must be present. The five attributes that can be used in a jsp:useBean action element are as follows:

- id
- class
- type
- scope
- beanName

Either the class attribute or the type attribute must be present. The five attributes are explained in the following sections.

id

The id attribute defines a unique identifier for the bean. This identifier can be used throughout the page and can be thought of as the object reference for the bean. The value for the id attribute has the same requirements as a valid variable name in the current scripting language.

class

The class attribute specifies the fully qualified name for the JavaBean class. A fully qualified name is not required if the bean's package is imported using the page directive, however.

type

If the type attribute is present in a jsp:useBean element, it specifies the type of the JavaBean class. The type of the bean could be the type of the class itself, the type of its superclass, or an interface the bean class implements. Normally, this attribute isn't often used,

scope

The scope attribute defines the accessibility and the life span of the bean. This attribute can take one of the following values:

- page
- request
- session
- application

The default value of the scope attribute is page.

The scope is a powerful feature that lets you control how long the bean will continue to exist. Each of the attribute values is discussed in the following sections.

page

Using the page attribute value, the bean is available only in the current page after the point where the jspruseBean action element is used. A new instance of the bean will be instantiated every time the page is requested. The bean will be automatically destroyed after the JSP page loses its scope; namely, when the control moves to another page. If you use the jsp;include or jsp:forward tags, the bean will not be accessible from the included or forwarded page.

request

With the request attribute value, the accessibility of the bean is extended to the forwarded or included page referenced by a jsp:forward or jsp:include action element. The forwarded or included page can use the bean without having a jsp:useBean action element. For example, from inside a forwarded or included page, you can use the jsp:getProperty and jsp:setProperty action ele-ments that reference to the bean instantiated in the original page.

session

A bean with a session scope is placed in the user's session object. The instance of the bean will continue to exist as long as the user's session object exists. In other words, the bean's accessibility extends to other pages.

Because the bean's instance is put in the session object, you cannot use this scope if the page on which the bean is instantiated does not participate in the JSP container's session management. For example, the following code will gen-erate an error:

```
<%@ page session="false" %>
<jsp:useBean id«="theBean" scope="session"
    class="com.t)rainy3oftware.CalculatorBean°/>
```

application

A bean with an application scope lives throughout the life of the JSP container itself. It is available from any page in the application.

beanName

The beanName attribute represents a name for the bean that the instantiate method of the Java .beans .Beans class expects.

As an example, the following is a jsp:useBean that instantiates the bean called com.newriders.HomeLoanBean:

```
<jsp:useBean id="theBean" class="com.newriders.HomeLoanBean"/>
```

Alternatively, you can import the package com.newriders using a page direc-tive and refer to the class using its name, as follows;

```
<%@ page import="com.newriders" V
<jsp:useBean id="theBean" class="HQmei_oanBean"/>
```

The Generated Servlet

The jsp:useBean action element causes the JSP container to generate Java code that loads the bean's class. Let's now examine the generated servlet. Consider again the JSP page that uses a bean in Listing 10.1:

```
<jsp:useBean id="theBean" scope="page"
    class="com.brainysoftware.CalculatorBean"/>
```

```
<%
    int i = 4;
    int i = theBean.doubleIt(i);
    out.println("2*4=" + j);
%>
```

If you open the generated servlet class file, you will see that some code is added to the _jspService method that translates the jsp:useBean action element.

```
    public .void _jspService(HttpServletRequest request,
        HttpServletResponse   response)
        throws iava.io.IOException, ServletException {

        JspFactory _jspxFactory - null;
        PageContext pageContext = null;
        HttpSession session = null;
        ServletContext application = null;
        ServletConfig config = null;
        JspWriter out = null;
        Object page = this;
        String _value = null;

    try {
        if (_jspx_inited *- false) {
            synchronized (this) {
                if (_jspx_inited == false) {
                    jspx_init{);
                    _jspx_inited = true;
                }
            }
        }
    JspxFactory = JspFactory.getDefaultFactoryf) ;
    response. setContentType("text/html;charset=ISO-8859-1 ") ;
    pageContext = _jspxFactory.getPageContext(
        this, request, response, "", true, 8192, true);

    application = pageContext. getServletContext();
    config = pageContext. getServletConfig() ;
    session = pageContext. getSession();
    out = pageContext. getOut();

    // an object reference to the Bean
    com.brainysoftware.CalculatorBean theBean = null;
    boolean _jspx_specialtheBean = false;
    synchronized (pageContext) {
        theBean = (com.brainysoftware.CalculatorBean)
            pageContext . getAttribute ( " theBean " , PageContext . PAQE_SCOPE) ;
        if ( theBean == null ) {
            _jspx_specialtheBean = true;
            try {
                theBean = (com.brainysoftware.CalculatorBean)
                    Java. beans. Beans. instantiate(this.getClass().getClassLoader(),
                        "com.brainysoftware.CalculatorBean") ;
            }
            catch (Exception exc) {
```

```
                throw new ServletException (
                        " Cannot create bean of class" +
                        "com.brainysoftware.CalculatorBean" , exc) ;
            }
            pageContext. setAttributef theBean", theBean, PageContext. PAGE_SCOPE);
        }
    }
    if (_Jspx_specialtheBean == true) {
    }
    out.write("\r\n");
    int i = 4;
    int j = theBean. doubleIt(i);
    out.println("2*4=" + j);
    }
    catch (Throwable t) {
        if (out 1= null && out.getBufferSizef)  1= 0)
            out.clearBuffer();
        if (pageContext 1= null) 'pageContext. handlePageException(t) ;
        }
    }
    finally {
        if ( jspxFactory 1= null) _jspxFactory.releasePageContext(pageContext);
        }
    }
```

The most important line of code in the generated servlet is the one that instantiates the bean and casts it to the bean's class type:

```
tneBean = {com.brainysoftware.CalculatorBean)
    java. beans. Beans.instantiate(this.getClass(),getClassLoader(),
        "com.brainysoftware.CalculatorBean") ;
```

See how the jsp:useBean action element is translated into the instantiate method of java.beans.Beans? Also note that the generated servlet will be slightly different if you use a different scope for the bean. To be more precise, the getAttribute and setAttribute methods of the PageContext class will be passed a scope accord-ing to the scope used for the bean. For instance, if the bean has an application scope, APPLICATION_SCOPE will be used instead of PAGE_SCOPE. If the bean has a session scope, SESSION_SCOPE will be used.

Accessing Properties Using jsprgetProperty and jsptsetProperty

In accessing a property in a bean, you can conveniently use the jsp:getProperty and jsp:setProperty action elements. The jsp:getProperty element is used to obtain the value of an internal variable, and the bean must provide a getter method. The syntax of the jspigetProperty element is as follows:

```
<jsp:getProperty name="flean Name" property="properryName"/>
```

The name attribute must be assigned the name of the bean instance from which the property value will be obtained. The property attribute must be assigned the name of the property.

A jsprgetProperty element returns the property value converted into String. The return value is then automatically fed into an out. print method so it will be displayed in the current JSP page.

The jspisetProperty action element is used to set the value of a property. Its syntax has four forms:

```
<jsp:setP.roperty name="Bean Name" property="PropertyName" value="value"/>
<jsp:setProperty name="Bean Name" property="PropertyName"/>
<]sp:setProperty name="Bean Name" property="PropertyName" param="parameterName"/>
<jspisetProperty name="Bean Name" property="*"/>
```

In all the forms, the name attribute is assigned the name of the bean instance available in the current JSP page. In this section, however, you learn about and use only the first form. The other three forms will be explained in the next section, "Setting a Property Value from a Request."

In the first form of the syntax, the property attribute is assigned the name of the property whose value is to be set, and the value attribute is assigned the value of the property.

The following example demonstrates the use of the jsp:getProperty and jsp:setProperty action elements.

Consider the bean called CalculatorBean in the com.brainysoftware package in the following listing. It has a private integer called memory. (Note that the variable name starts with a lowercase m.) It also has a getter called getMemory and a setter named setMemory. Note that in both accessor methods, memory is spelled using an uppercase M.

```
package com.brainysoftware;
public class CalculatorBean {
    private int memory;

    public void setMemory(int number) {
        memory = number;
    }

    public int getMemory() {
        return memory;
    }
    public int doubleIt(int number) {
        return 2 * number;
    }
}
```

Using the jsp:setProperty and jsp:getProperty action elements, you can set and obtain the value of memory, as demonstrated in the JSP page in the following listing.

```
<jsp:useBean id="theBean" class="com.brainysoftware.CalculatorBean"/>
<jsp:setProperty name="theBean" property="memory" value="169"/>
The value of memory is <jsp:getProperty name="theBean" property^'memory"/^
```

Calling this page in a web browser, you should be able to see

The value of memory is 163

As shown above, the result of the jsp:getProperty is displayed on the current JSP page.

Note that you also can access the accessor methods in a conventional way, as described in next listing.

```
<jsp:useBean id="theBean" class="com.brainysoftware.CalculatorBean"/>
<%
    theBean.setMemory(987);
%>
The value of memory is <%= theBean.getMemory()%>
```

In many cases, it is also common to use an expression as the value of the value attribute. For example, the code in the following listing feeds a JSP expression to the value attribute in a jsp:setProperty action element.

```
<jsp:useBean id="theBean" class="com.brainysoftware.CalculatorBean"/>
<%
    int 1=2;
%>
```

```
<jsp:setProperty name="theBean" property="memory" value="<%- 100 * i %>"/>
The value of memory is <jsp:getProperty name="theBean" property="memory"/>
```

Setting a Property Value from a Request

You can use a more concise way of setting a property value using the jspisetProperty action element if the value happens to come from the Request object's parameter(s). In fact, there are three forms of syntax, as discussed in the previous section. The three forms are re-written here:

```
<jsp:setProperty name="Bean Name" property="PropertyName"
param="parameterName"/>

<jsp:setProperty name="Bean Name" property="PropertyName"/>

<jsp:setProperty name="Bean Name" property^"*"/*
```

The first form allows you to assign a Request object parameter value to a property. The param attribute of the jsp:setProperty action element must be assigned the name of the Request object's parameter name. The following example demonstrates the use of this form. The example has two pages. The first page is a simple HTML with a form containing a Text box into which the user can enter a number.The code for this page is shown in the following listing.

The Page Where the User Can Type a Number

```
<HTML>
<HEAD>
<TITLE>Passing a value</TITLE>
</HEAD>
<BODY>
<CENTER>
Please type in a number in the box
<BR>
<FORM METHOD=POST ACTION^SimplePage.jsp>
<INPUT TYPE=TEXT NAME=memory>
<INPUT TYPE=SUBMIT>
</FORM>
</CENTER>
</BODY>
</HTML>
```

Note that the TEXT element name is memory and the form is passed to a JSP page called SimplePage.jsp.The SimplePage.jsp is given in the next listing.

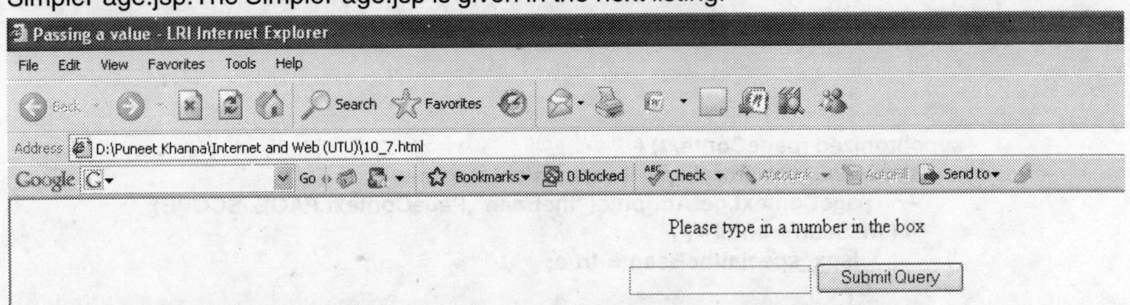


```
<jsp:useBean id="theBean" class="com.brainysoftware.CalculatorBean"/>
<jsp:setProperty name-"theBean" property="memory" param="memory"/>
The value of memory is <jsp:getProperty name="theBean" property="meinory"/>
```

If the parameter name is the same as the bean's property name, you can even omit the param attribute:

```
<jsp:setProperty name="theBean" property="memory"/>
```

This is, of course, the third form of the jsp:setProperty syntax.

The last form allows you to assign the values of Request object's parameters to multiple properties in a bean as long as the names of the properties are the same as the parameter names. The second line in the above listing can be replaced with the following:

```
<jsp:setProperty name="theBean" property="*"/>
```

JavaBeans Code Initialization

The jsp:useBean action element allows you to write code that will be executed after the bean is initialized. For a bean that has a page scope, the initialization code is run every time the JSP page is requested.

Initialization code can be as simple as writing a piece of text, such as the following:

```
<jsp:useBean id="theBean" scope="page"
    class*"com.brainysoftware.CalculatorBean">
    Initializing the Bean
</jsp:useBean>
```

Remember that any template data will be translated into an argument to an out.print method. It's not hard to guess that the initialization code will be added to the code block that instantiates the bean. Consider the following _jspService method of the servlet generated from the previous JSP page. The method has been cleaned up for clarity:

```
public void _jspService(HttpServletRequest request,
    HttpServletResponse    response)
    throws Java.i0.IOException, ServletException {

    JspFactory JspxFactory = null;
    PageContext pageContext = null;
    HttpSession session = null;
    ServletContext application = null;
    ServletConfig config = null;
    JspWriter out = null;
    Object page = this;
    String  _value = null;
    try {
        .
        .
        .
        com.brainysoftware.CalculatorBean theBean = null;
        .
        .
        .
            synchronized (pageContext) {
                theBean = (com.brainysoftware.CalculatorBean)
                    pageContext.getAttribute("theBean",PageContext.PAGE_SCOPE);
                if ( theSean == null ) {
                    _jspx_specialtheBean = true;
                    try {
                        theBean = (com.brainysoftware.CalculatorBean)
                        java.beans.Beans.instantiate(this.getClass().getClassLoader(),
                        +"com.brainysoftware.CalculatorBean");
                    }
                    catch (Exception exc) {
```

```
                throw new ServletException (" Cannot create bean of class "
                +"com.brainysoftware.CalculatorBean", exc);
            }
            pageContext.setAttribute("theBean", theBean, PageContext.PAGE_SCOPE);
        }
    }
    if(_jspx_specialtheBean == true) {
        // here is the initialization code
        out, writef "\r\nInitializing the Bean\r\n");
    }
    }
    catch (Throwable t) {
    .
    .
    .
    }
    finally {
    .
    .
    .
    }
}
```

At the time of initialization, you even can use the jsp:setProperty action ele-ment to set the value of a bean's property, as demonstrated in the code in the following listing.

```
<jsp:useBean id="theBean" class="com.brainysoftware.CalculatorBean">
<jsp:setProperty name="theBean" property="*"/>-
</jsp:useBean>
```

Enterprise JavaBeans

Now, you will learn how to develop and deploy Enterprise JavaBeans (EJB) applications. For the most part, for simple web applications you should' probably stick with servlets/JSP and avoid EJB altogether. If robustness and scalability are an issue, however, you should consider developing an EJB application. As this chapter shows, there are many benefits that make EJB appealing.

The EJB technology depends on other Java technologies to function properly. First, it uses Java Remote Method Invocation (RMI) as the communication protocol between two enterprise beans and between an enterprise bean and its client. If you are a seasoned Java programmer, you should know how RMI is used in Java-distributed computing to invoke remote methods on a remote machine.

Another technology used in EJB is Remote Method Invocation over Internet Inter-ORB Protocol (RMI-IIOP), where ORB stands for *Object Request Broker.* RMI-IIOP is a more portable version of RMI that can use the HOP from the Object Management Group (OMG). RMI-IIOP is especially used in communications between an enterprise bean and a client.

Lastly, EJB uses Java Naming and Directory Interface (JNDI) as the naming service that binds a name with an enterprise bean.

What Is an Enterprise JavaBean?

In a nutshell, an enterprise JavaBean is a server-side component. Just like any other component, an enterprise bean encapsulates business logic. Enterprise beans must conform to the EJB specifications, however, and they are deployed and can run only in an EJB container, identical to the way servlets run inside a servlet container.

A servlet container provides services for servlets, such as session manage-ment and security. Likewise, an EJB container provides system-level services for EJB applications. In fact, as you will soon find out, it's the EJB container that makes EJB so great. Despite the similarity of names, EJB has little to do with JavaBeans.

Benefits of EJB

EJB applications are much more complex and more difficult to build and administer than servlets/JSP applications. For someone new to EJB, the learning curve also is steeper. If this is the case, why is EJB so popular and why do so many organizations want to invest in it? The answer is simple—after you know the nuts and bolts of EJB, writing an application is an easy task, and, more importantly, you can enjoy some benefits provided for you by the EJB container.

The following is the list of some of the benefits of EJB:

- EJB applications are easy to develop because the application developer can concentrate on the business logic. At the same time, the developer uses the services provided by the EJB container, such as transactions and connection pooling. Again, the hardest part is the learning process.
- EJBs are components. Chances are good that there are EJB vendors who sell components that encapsulate the functionality that you need. By pur-chasing a third-party's EJBs, you can avoid needing to develop your own beans, which means your application development is more rapid. The EJB specification makes sure that beans developed by others can be used in your application.
- There is a clear separation of labor in the development, deployment, and administration of an EJB application. This makes the development and deployment process even faster. The roles are discussed in detail in the section, "The Six EJB Roles."
- The EJB container manages transactions, state management details, multi-threading, connection pooling, and other low-level APIs without you, the developer, having to understand them.
- The EJB container provides security for the applications.
- The EJB architecture is compatible with other Java APIs.

EJB Application Architecture

The EJB application architecture extends the web application architecture by adding another tier.

Clients of an enterprise bean can be a traditional Java application, an applet, a JSP page or servlet, another EJB bean, or others.

Note that a client never invokes a bean's methods directly. Communication between clients and beans is done through the EJB container. Compare this with a web application in which a client web browser has to go through the web container to use a servlet or a JSP page.

The Six EJB Roles

The six distinct roles in the EJB application development and deployment life cycle are as follows;

- Bean developer
- Application assembler
- Deployer
- System administrator
- EJB server provider
- EJB container provider

Each role can be performed by a different individual or organization. Often, however, an individual plays more than one role in a given time. For example, normally there is no clear separation between an EJB server and an EJB con-tainer; that is, they normally come in one package from a vendor. All the six roles are explained in the following sections.

Bean Developer

The bean developer is the programmer who develops enterprise beans.To be a successful bean developer, you need to understand the business logic of the application.

Application Assembler

An EJB application typically consists of more than one enterprise bean. For larger applications, several bean developers may be employed to build the beans. The application assembler is the person responsible for assembling all the beans written by the bean developer(s).The application assembler also writes the deployment descriptor.

Deployer

The deployer is the one in charge of deploying the EJB application in a particular EJB container or containers, if there is more than one EJB container used. This person takes the enterprise bean(s) and deployment descriptor from the application assembler as the input. This person must be an expert of the EJB container used.You also are introduced to this role during the development and deployment of the sample EJB application presented in this book.

System Administrator

After the EJB application is deployed successfully, the application still must be maintained on a day-to-day basis—or perhaps *minute-to-minute* is more accurate. It is the system administrators role to make sure that the application runs 24 hours a day without interruption. If a server crashes, the administrator makes sure that it is restarted immediately.

Also, the system administrator is responsible for managing security. For example, it's the system administrators job to create a new account for a new user or a group of users. In a large organization, this role is a full-time job.

EJB Container Provider

The EJB container provider is a vendor that has the resources to write an EJB container and to make sure that the software conforms to the EJB specification. The developer of the EJB container also should provide various tools for the system administrator for easily administering the EJB applications deployed. For example, there should be a tool that the system administrator can use to add a user easily. This book is not concerned with this role.

EJB Server Provider

The EJB server provider provides an EJB server which in turn hosts the EJB container. Currently, the EJB architecture does not clearly state the separation of this role from the EJB container provider role. Most EJB containers come packaged in an EJB server. This book does assume that you will write an EJB server.

Types of Enterprise Beans

You use three types of enterprise beans: session, entity, and message-driven. A session bean is a component that performs a certain task for the client. An entity bean represents an entity in the database or other persistent storage. Message-driven beans are a new addition to EJB 2.O.They serve as a listener for the Java Message Service API that enables asynchronous message process-ing. All these beans are discussed in the next few chapters.

Writing Your First Enterprise Bean

Developing and deploying an EJB application requires the following steps:

1. Writing the bean.

2. Writing the deployment descriptor.
3. Creating a deployment file.
4. Deploying the bean.
5. Writing the client application to test the bean.

The following example builds an EJB application consisting of only one enterprise bean that can perform additions of two integers. After successfully compiling the bean, you will write the deployment descriptor and deploy the bean. Lastly, you will create a servlet that serves as the client of our EJB application.

Writing and Compiling the Adder Bean

You have three Java files to write, all of which are part of the com.brainysoftware.ejb package. So the first thing you need to do is to create the appropriate directory structure.

The three Java files you need to create in the com/brainy software/ejb directory are AdderHome.java, Adder.java, and AdderBean.java. AdderHome and Adder are the home and remote interfaces of the enterprise bean.

The AdderHome Interface

```
package com.brainysoftware.ejb;

import java.rmi.RemoteException;
import javax.ejb.CreateException;
import javax.ejb.EJBHome;

public interface AdderHome extends EJBHome {
    Adder create() throws RemoteException, CreateException;
}
```

The Adder Interface

```
package com.brainysoftware.ejb;

import javax.ejb.EJBObject;
import java.rtni.RemoteException;

public interface Adder extends EJBObject {
    public int add(int a, int b) throws RemoteException;
```

The AdderBean Class

```
package com.brainysoftware.ejb;

import java.rmi.RemoteException;
import iavax.ejb.SessionBean;
import javax.ejb.SessionContext;

public class AdderBean implements SessionBean {

    public int add(int a, int b) {
        System. out. println( “from AdderBean");
        return (a + b);
    }

    public void ejbCreatef) {
    }
```

```
    public void ejbRemove() {
    }

    public void ejbActivate() {
    }

    public void ejbPassivate() {
    }

    public void setSessionContext(SessionContext sc) {
    }
}
```

Writing the Deployment Descriptor

An EJB application must have a deployment descriptor that describes each enterprise bean in that application. The deployment descriptor file is called ejb.xml. For the sample application, the deployment descriptor is given in the following listing.

The Deployment Descriptor

```
<?xml version="1.0" encoding="UTF-8"?>

<ejb-jar>
    <description>Your first EJB application </description>
    <display-name>Adder Application</display-name>
    <enterprise-beans>
        <session>
            <ejb-name>Adder</eib-name>
            <home>com.brainysoftware.ejb,AdderHome</home>
            <remote>com.brainysoftware.ejb,Adder</rermote>
            <ejb-class>com.brainysoftware.ejb.AdderBean</ejb-class>
            <session-type>Stateless</session-type>
            <transaction-type>Bean</transaction-type>
        </session>
    </enterprise-beans>
</ejb-jar>
```

Creating a Deployment File

After you finish developing your enterprise bean, you need to package all the class files into one .jar file. The ejb directory contains three files: Adder.class, AdderHome.class, and AdderBean.class. The META-INF directory contains one file: the ejb-jar.xml file, the deployment descriptor. Follow these steps to create the deployment file:

1. Change directory to the parent directory of both com and META-INF.
2. Assuniingjar.exe is already in the path, type the following:
   jar cfv adder.jar com/brainysoftware/ejb/* META-INF/ejb-jar.xml
3. This creates ajar file called adder.jar.

Deployment

Copy the adder.jar file into the deployment directory under JBoss home. Restart JBoss.JBoss should spit out some messages. In the lines of messages, look for the following:

```
.
...
[Auto deploy] Starting
```

```
[Auto deploy] Auto deploy of file:/C:/jboss/deploy/adder.Jar
...
[J2EE    Deployer]    Create application adder.jar
[J2EE    Deployer]    Installing EJB package: adder.jar
[J2EE    Deployer]    Starting module adder.jar
...
[J2EE Deployer] J2EE application: file:/C:/jboss/deploy/adder.jar is deployed.
...
```

That's it. Your EJB application has been successfully deployed.

To test your enterprise bean, however, you need a client application.Before you write one, however, you need to see how EJB works.

EJB Explained

As you have seen from the preceding section, an enterprise bean does not consist of a one class. In fact, there are always two interfaces that accompany a bean class. These two interfaces are called the *home interface* and the *remote interface.*They are there to enable communications between a client and an enterprise bean. The enterprise bean contains the implementation of the business rules. Of course, an enterprise bean can have other classes as well. The two interfaces and one implementation class serve as the minimum.

The home interface and the remote interface are crucial because they are the "door" to an enterprise bean for a client application.

The EJB specification states that an enterprise bean and its supporting com-ponents are written by implementing or extending members of the javax.ejb package.

The Home Interface

The home interface performs life cycle operations: creating, finding, and removing the enterprise bean. A home interface must extend the EJBHome interface in the javax.ejb package. The javax.ejb.EJBHome interface is given as

follows:

```
package javax.ejb;

public interface EJBHome extends Java.rmi.Remote {

    EJBMetaData getEJBMetaData()
        throws Java.rmi.RemoteException;

    HomeHandle getHomeHandle()
        throws Java.rmi.RemoteException;

    void remove(Handle handle)
        throws Java.rmi.RemoteException, javax.ejb.RemoveException;

    void removefObject primaryKey)
        throws Java,rmi.RemoteException, javax.ejb.RenoveException;
}
```

See how the EJBHome interface is derived from the Java.rmi.Remote interface? This is the first sign you've seen in this chapter that EJB is dependent on RMI. The java.rmi.Remote interface itself is used to identify interfaces whose methods may be invoked from a non-local virtual machine. Any remote object must implement this interface either directly or indirectly.

A home interface can define methods to create, find, and remove the enterprise bean.The create and find methods are added to the home interface definition; the remove methods are inherited from the javax.ejb.EJBHome interface. For instance, the home interface of the Adder EJB (the AdderHome interface), given in the following listing, defines a no-argument create method in its body, using the following code:

```
import   java.rmi.RemoteException;
import   j avax . e j b .CreateException ;
import   javax.ejb.EJBHome;

public interface AdderHome extends EJBHome {

    Adder create() throws RemoteException, CreateException;

}
```

A home interface can define multiple overloads of the create and find methods. The create and find methods also can throw additional exceptions, if required. The name for a find method always start with "find".

Both the create and find methods are optional. Not all beans need them. For example, a session bean might define one or more create methods but an entity bean might not. T

The Remote Interface

The remote interface duplicates all the business methods that are to be available to a client.This happens, again, because the client cannot access the bean directly.

A remote interface must extend the EJBObject interface in the javax.ejb package. The j avax. ejb. EJB Object is defined as follows:

```
package javax.ejb;
public interface EJBObject extends java.rmi.Remote {
    public abstract EJBHome getEJBHomef)
        throws java.rmi.RemoteException;
    public abstract Object getPrimaryKey()
        throws java.rmi.RemoteException;

    public abstract void remove()
        throws java.rmi.RemoteException,  javax.ejb.RemoveException

    public abstract Handle getHandle()
        throws java.rmi.RemoteException;

    public abstract boolean is!dentical(EJBObject obj)
        throws java.rmi.RemoteException;
}
```

Like the javax.ejb.EJBHome interface, this interface extends the java.rmi. Remote interface.

In the Adder bean example, the remote interface is. called Adder and is given in the above listing. As you can see, it clones the add method found in the bean implementation in the following listing by using the following code:

```
import javax.ejb.EJBObject;
import  java.rmi.RemoteException;

public interface Adder extends EJBObject {
    public int addfint a,  int b) throws RemoteException;
}
```

The Enterprise Bean Class

This class provides the implementations of the life cycle methods in the home interface as well as the business logic defined in the remote interface.

An enterprise bean class is either a session bean, an entity bean, or a message-driven bean. In other words, a session bean implements the javax.ejb.SessionBean interface and an entity bean implements the javax.ejb.EntityBean interface. Therefore, an enterprise bean class also must provide the implementation of the EJB container callback methods defined in either the Javax.ejb.SessionBean interface or the Javax.ejb.EntityBean interface.

The life cycle create and find methods in the home interface correspond to the ejbCreate, ejbPostCreate, and ejbFind methods in the bean class.

The Deployment Descriptor

An EJB application needs a deployment descriptor that specifies meta information for the EJB container. The deployment descriptor is an XML file called ejb.xml and must be located in the META-INF directory under the application directory.

The following is some is some of the information specified by the deployment descriptor:

- The enterprise bean name
- The type of the bean
- The fully qualified name of the home interface
- The fully qualified name of the remote interface

The deployment descriptor can define more than one bean.

Writing Client Applications

An enterprise bean is a server-side component that sits idle on the server waiting for a call from clients. Writing an EJB client can sometimes be a tricky business. Without a client, however, you can't test your enterprise beans. Understanding how the client gets access and calls your bean is therefore an important part of the EJB application.

An EJB client does not invoke methods in the enterprise bean class directly. In fact, a client can see only the home and remote interfaces of the bean.

Client applications access the enterprise bean through the Java Naming and Directory Interface (JNDI).Therefore, understanding JNDI is very helpful in writing effective and efficient client applications. A brief summary of JNDI is given next that will enable you to write client applications to access enterprise beans.

In this example, you write two types of client applications. The first is a Java application consisting of one simple class called BeanClient.The second client application is a JSP page that demonstrates how to use an enterprise bean from a servlet or a JSP page.

The first client is important because it demonstrates the basic functionality that must exist for the client to call an enterprise bean. The second client is also crucial because, as a web developer, you will need to access an EJB application from inside a servlet or a JSP page.

Let's now take a brief look at JNDI.

Java Naming and Directory Interface (JNDI)

Access to an enterprise bean is done through JNDI. Therefore, it is crucial to understand this API. JNDI comes as part ofJDK 1.3 and later. If you are currently using JDK 1.2, you need to download JNDI and install it separately. JNDI provides two services: a naming service and a directory service. A naming service is a very important feature in computing because it finds an object associated with a given name. An example of a popular naming service is the Internet Domain Name System (DNS) that maps domain names (such as newriders.com) with IP addresses. In the context of EJB, a naming ser-vice finds an enterprise bean for you if you know the name given to the bean.

If a client application wants to call an enterprise bean, it simply needs to supply the name of the bean. JNDI will find the bean for the client. As such, it is not surprising why a naming service plays an important role in EJB.

A directory service is an extension of a naming service. A directory service associates names with objects and also allows those objects to have attributes that describe the objects.These attributes enable you to search for an object without knowing its name. An example of a directory service is the Lightweight Directory Access Protocol (LDAP). Directory services are not used to access enterprise beans—therefore, they are outside the scope of this book.

The JNDI is divided into five packages. These are as follows:

- javax.naming
- javax. naming, directory
- javax.naming.event
- javax.naming.ldap
- javax. naming.spi

Knowing the javax.naming package is sufficient for accessing an enterprise bean. In writing a client application that uses the service of an enterprise bean, you need to understand well two important members of the javax.naming package: the Context interface and the InitialContext class.

The javax.naming.Context Interface

The Context interface represents a naming context consisting of a set of name-to-object association. This association in JNDI terms is called *binding.* The Context interface is important because it contains methods for examining and modifying the name-to-object bindings.

The most frequently used method in the Context interface is probably the lookup method. This method returns a reference to an object, given the name of the object. There are two overloads of this method. The signatures of both overloads are as follows:

```
public Object lookup(javax.naming.Name name)
-throws j avax.naming.NaroingException
public Object lookup(String name)  throws javax.naming.NamingException
```

You use the lookup method to obtain a reference to an enterprise bean's home object. The second overload of the lookup method is very straightfor-ward. You simply pass the string representation of the bean's name and you get a reference to the bean's home object.

The lookup method throws a javax.naming.NamingException object if the name resolution fails.

The javax.naming. InitialContext Class

Naming operations are relative to a context. The InitialContext class implements the Context interface and provides the starting context for name resolutions.

For a context, you need to define a number of properties for the environment of the context. For example, a naming resolution may be restricted to authorized users only. In this case, you need to supply a user's credentials as properties. You normally create a Java.util.Properties object and use its put method to add key/value pairs representing any necessary property names and values.

In accessing an enterprise bean, you need to supply two properties. The first is the environment property Java.naming.factory.initial. The value of this property is the fully qualified class name that will be used to create the initial context.

The second property is Java.naming.provider.url.This is an environment property whose value specifies configuration information for the service provider to use.

Both property names are conveniently contained in two static fields in the Context interface. They are'Context.INITIAL_CONTEXT_FACTORY and Context.PROVIDERJJRL, respectively.

For the purpose of accessing an enterprise bean, you need to add these two properties to a java.util.Properties method using its put method, as you see in the following code:

```
import java.util.Properties;

// Create a java.util.Properties object
Properties properties = new Properties();

// Add two properties: "Java.naming.factory.initial" and
// "java.naming.provider.url"
properties.put(Context.INITIAL JX)NTEXT_FACTORY,
      "org.jnp.interfaces.NamingContextFactory");
properties.put(Context.PROVIDERJJRL, "localhost:1099");
```

Because these properties are needed when creating an initial context, the Properties object is passed to the InitialContext class's constructor, as follows:

```
javax.naming.InitialContext jndiContext =
      new javax.naming.InitialContext(properties);
```

To sum up, obtaining a reference to an enterprise bean involves the following steps:

1. Creating a java.util.Properties object.
2. Adding necessary properties to the java.util.Properties object for constructing the initial context.
3. Constructing an Javax.naming.InitialContext object.
4. Using the javax.naming.Context interface's lookup method to get a reference to che bean's home object, passing the name of the bean.

The summary can be rewritten in the following code:

```
import java.util.Properties;
import javax.naming.*;

Properties properties = new Properties();
properties.put(Context.INITIAL JJONTEXT^FACTORY,
     "org.jnp.interfaces.NamingContextFactory");
properties.put(Context.PROVIDERJJRL, "localhost:1099");

try {
      InitialContext jndiContext = new InitialContext(properties);
      // get the reference to the home interface
      Object ref = jndiContext.lookup("AflcJer");
}
catch (NamingException e) {
}
```

Creating a Bean's Instance

Using JNDI, you managed to obtain a reference to the bean's home object. The next step is to create an instance of the bean on the server. This is done using the home interface's create method. The reference you got to the bean's home object, however, is an RMI object of type Java.lang.Object. To call the create method of the home object, you first need to downcast the object reference into the type of the home interface. For example, in the Adder bean example, the code for the downcasting looks like this:

```
Adder-Home home = (AdderHome) ref;
```

In this example, ref is the reference obtained from the JNDI initial context's name lookup.

To conform with RMI-IIOP, you must use another method for the previous downcasting. Typically, you use the static narrow method of the javax.nni.PortableRemoteObject class. This method is used to ensure that an object of a remote or abstract interface type can be cast to a desired type. The signature of this method is as follows:

```
public static Object narrow(Object narrowFrom, Class narrowTo)
    throws ClassCastException
```

Here, narrowFrom is the object to check and the narrowTo is the desired type. If narrowFrom cannot be cast to narrowTo, a ClassCastException object is thrown. If narrowFrom can be cast to narrowTo, the method returns an object of type Object that can be cast to the desired type. Your code then becomes the following:

```
AdderHome nome = (AdderHome)
    PortableRemoteObject.narrow (ref, AdderHome.class);
```

Having a home object, you then can call the create method of the home interface to create the bean, as follows:

```
Adder adder = home.create();
```

You are then free to call any exposed methods of the bean. For the Adder bean example, you can call its add method, as follows:

```
int I = adder.add(2, 5);
```

Now let's see how to create client applications based on what you've learned so far.

A Java Client Application

The following is a class called BeanClient that calls the Adder session bean you have previously created. The code is very simple and is given in the following listing.

The Client Application Class

```
import javax.naming.*;
import javax.rmi.PortableRemoteQbject;
import java.util.Properties;
import com.brainysoftware.ejb.Adder;
import com.brainysoftware.ejb.AdderHome;

public class BeanClient {

    public static void main(String[j args) {
    // preparing properties for constructing an InitialContext object
    Properties properties - new Properties!];
    properties.put(Context.INITIAL_CONTEXT_FACTORY,
    "org.jnp.interfaces.Nait)ingContextFactory");
    properties.put(Context.PROVIDERJJRL, "localhost:1099");

    try {
        // Get an initial context
        InitialContext jndiContext = new InitialContext(properties);
        System.out.println("Got context"!;

        // Get a reference to the Bean
        Object ref = jndiContext.lookup!"Adder");
        System.out.println("Got reference");
```

```
        // Get a reference from this to the Bean's Home interface
        AbderHome hone = (AdderHome)
            PortableRemoteObject.narrow (ref, AdderHome.class);

        // Create an Adder object from the Home interface
        Adder adder = home.create();
        System.out.println ("2 + 5 = " + adder.add(2, 5));
    }
    catch(Exception ej {
        System. out.println(e.toString());
    }
}
```

It is assumed that the client application will run on the same computer as the EJB container. If you run it from a different computer, you need to replace "localhost:1099" with the URL of the naming service on the server.

If you are using JBoss, when you run the client class, you need to include the following files in the classpath:

- client/ejb.jar
- client/jboss-client.jar
- client/jnp-client.jar

The lib and client directories are located under JBoss home.

Calling the Bean from a JSP Page

Calling an enterprise bean from a servlet is basically the same as calling it from another client. Copying all the necessary library files into lib directory under Tomcat is important, however. The following files are needed from the JBoss's home directory:

- client/ejb.jar
- client/jboss-client.jar
- client/jnp-client.jar

Finally, copy the Adder.jar to the lib directory under CATALINA_HOME. Alternatively, deploy the class files under WEB-INF/classes/com/b rainy soft-ware/ejb/ directory under the application directory. We need the Adder.class and AdderHome. class.

The code for the JSP page is given in the following listing.

Calling an Enterprise Bean from a JSP Page

```
<%@ page import="javax.naming.*"%>
<%@ page import="javax.rnii.PortableRenioteObject"%>
<%@ page import="java.ijtil.Properties¹^
<%@ page import="coin.brainysoftware.ejb.Adder"%>
<%@ page import="com.brainysoftware,ejb.AdderHonie"%>
<%
    // preparing a Properties ooject for constructing
    // an initial context Properties properties = new Properties();
    properties, put (Context. INITIAL JXJNTEXTJACTORY,
        "org.jnp.interfaces.NamingContextFactory");
    properties.put(Context.PROVIDEH_URL, "localhost:1099");

    try {
    // Get an initial context
    InitialContext jndiContext - new Initia.IContext(properties);
```

```
        System.out.println("Got 'context");

        // Get a reference to the Bean
        Object ref  - jndiContext.lookup("Adder");
        System.out.println("Got reference");

        // Get a reference from this to the Bean's Home interface
        Adder-Home home = (AdderHome)
            PortableRemoteObject.narrow (ref, AdderHome.class);

        // Create an Adder object from the Home interface
        Adder adder = home.create));
        out.println ("2 + 5 = " + adder.add(2, 5});
    }
    catch(Exception e) {
        System.out.println(e.toString());
    }
%>
```

Revision

1. A bean is a reusable and self-contained software component
2. The five attributes that can be used in a jsp:useBean action element are as follows: id, class, type, scope and beanName.
3. In accessing a property in a bean, you can conveniently use the jsp:getProperty and jsp:setProperty action elements.
4. You can use a more concise way of setting a property value using the jspisetProperty action element if the value happens to come from the Request object's parameter(s).
5. The jsp:useBean action element allows you to write code that will be executed after the bean is initialized.
6. The EJB technology depends on other Java technologies to function properly. First, it uses Java Remote Method Invocation (RMI) as the communication protocol between two enterprise beans and between an enterprise bean and its client.
7. EJB applications are much more complex and more difficult to build and administer than servlets/ JSP applications.
8. The six distinct roles in the EJB application development and deployment life cycle are as follows; Bean developer, Application assembler, Deployer, System administrator, EJB server provider and EJB container provider.
9. Developing and deploying an EJB application requires the following steps: Writing the bean, Writing the deployment descriptor, Creating a deployment file, Deploying the bean and Writing the client application to test the bean.
10. After you finish developing your enterprise bean, you need to package all the class files into one .jar file. The ejb directory contains three files: Adder.class, AdderHome.class, and AdderBean.class.
11. The home interface performs life cycle operations: creating, finding, and removing the enterprise bean.
12. The Enterprise Bean Class provides the implementations of the life cycle methods in the home interface as well as the business logic defined in the remote interface.
13. An EJB application needs a deployment descriptor that specifies meta information for the EJB container.
14. Writing an EJB client can sometimes be a tricky business. Without a client, however, you can't test your enterprise beans.

Multiple Choice Questions

1. The application assembly is referred to as _________:
   (a) design time (b) Runtime
   (c) libraries
2. ________ allowing beans that have been customized in an application builder to have their state saved and restored.
   (a) persistence (b) Browser
   (c) Wizard
3. Select the _______ method from the dialog to signify it as the receiver of the event action.:
   (a) get() (b) starting ()
   (c) start()
4. JAR files use a compression scheme based on the one used in ZIP files.
   (a) TRUE (b) FALSE
5. Enterprise JavaBeans (EJB) is a comprehensive technology that provides the infrastructure for building enterprise-level server-side distributed Java components.
   (a) TRUE (b) FALSE

Answers

1. a 2. a 3. c 4. a 5. a

Subjective Questions

1. What is JavaBean?
2. Desribe the various components of a JavaBean.
3. What is Enterprise JavaBean?
4. What are the six EJB roles?
5. How would you write your first enterprise bean?
6. Describe the process of writing the Deployment Descriptor.
7. Explain EJB.
8. Describe the various Interfaces.
9. How would you write Client Applications?
10. How would you call the Bean from a JSP page?

7 Java Server Pages (JSP)

Introduction

JavaServer Pages (JSP) is another java technology for developing web applications. JSP was released during the time servlet technology had gained popularity as one of the best web technologies available. JSP is not meant to replace servlets, however. In fact, JSP is an extension of the servlet technology, and it is common practice to use both servlets and JSP pages in the same web applications.

Authoring JSP pages is so easy that you can write JSP applications without much knowledge of the underlying API. If you want to be a really good Java web programmer, however, you need to know both JSP and servlets. Even if you use only JSP pages in your Java web applications, understanding servlets is still very important. In JSP you work with HTTP requests and HTTP responses, request parameters, request attributes, session management, cookies, URL-rewriting, and so on.

What's Wrong with Servlets?

The history of web server-side programming in Java started with servlets. Sun introduced servlets in 1996 as small Java-based applications for adding dynamic content to web applications. Not much later, with the increasing popularity of Java, servlets took off to become one of the most popular technologies for Internet development today.

Servlet programmers know how cumbersome it is to program with servlets, however, especially when you have to send a long HTML page that includes little code. Take the snippet in the following listing as an example. The code is a fragment from a servlet-based application that displays all parameter names and values in an HTTP request.

Display All Parameter/Value Pairs in a Request Using a Servlet

```
import javax.servlet.*;
import javax.servlet. http.*;
import java.io.*;
import java.util.*;

public class MyDearServlet extends HttpServlet {

    //Process the HTTP GET request
    public void doGetfHttpServletRequest request,
    HttpServletResponse response)
    throws ServletExceptidn, IOException {

    doPost( request, response);

    //Process the HTTP POST request
    public void doPost(HttpServletRequest request,
        HttpServletResponse response)
        throws ServletException, IOException {
```

```
response. setContentType(" text/ html");
PrintWriter out = response. getWriter() ;
out.println("<HTML>");
out.println("<HEAD><TITLE>Using Servlets</TITLE></HEAD>"
out.println("<BODY BGCOLOR=#123123>");

//Get parameter names
Enumeration parameters = request.getParameterNames();
String param = null;
while (parameters.hasMoreElementsO) {
    param = (String) parameters.nextElement();
    out.println(param + ":" + request.getParameter(param) +
        "<BR>");
}
out.println('</BODY>");
out.println("</HTML>");
out.close();
} //End of doPost method

/* other parts of the class goes here
  .
  .
  .
*/
} //End of class
```

Nearly half of the content sent from the doPost method is static HTML. However, each HTML tag must be embedded in a String and sent using the println method of the PrintWriter object. It is a tedious chore. Worse still, the HTML page may be much longer.

Another disadvantage of using servlets is that every single change will require the intervention of the servlet programmer. Even a slight graphical modification, such as changing the value of the <BODY> tag's BGCOLOR attribute from #DADADA to #FFFFFF, will need to be done by the program-mer (who in this case will work under the supervision of the more graphic-savvy web designer).

JSP solves drawbacks in the servlet technology by allowing the programmer to intersperse code with static content, for example. If the programmer has to work with an HTML page template written by a web designer, the program-mer can simply add code into the HTML page and save it as a .jsp file. If at a later stage the web designer needs to change the HTML body background color, he or she can do it without wasting the charging-by-the-hour programmer's time. He or she can just open the .jsp file and edk it accordingly.

The code in earlier listing can be rewritten in JSP as shown in the following listing.

Display All Parameter/Value Pairs in a Request Using JSP

```
<%@ page import=" Java. util. Enumeration" %>
<HTML>
<HEAD><TITLE>Using JSP</TITLE></HEAD>
<BODY BGCOLOR=#DADADA>
<%
    //Get parameter names
    Enumeration parameters = request. getParameterNames() ;
    String param = null;
    while (parameters. hasMoreElementsO) {
        param = (String) parameters. nextElement() ;
```

```
        out. println{ param + ":" + request. getParameter{param) +
            "<BR>");
    }
    out. close () ;
%>
</BODY>
</HTHL>
```

You can see that <HTML> tags stay as they are. When you need to add dynamic content, all you need to do is enclose your code in <% ... %> tags.

Again, JSP is not a replacement for servlets. Rather, JSP technology and servlets together provide an attractive solution to web scripting/programming by offering platform independence, enhanced performance, separation of logic from display, ease of administration, extensibility into the enterprise, and most importantly, ease of use.

Running Your First JSP

This section invites you to write a simple JSP page and run it. The emphasis here is not on the architecture or syntax and semantics of a JSP page; instead the section demonstrates how to configure minimally the servlet/JSP con-tainer to run JSP. Tomcat 4 is used to run JSP applications.

To make your JSP page run, all you need to do is configure your JSP container (Tomcat) and write a JSP page. Configuration is only done once, at the beginning. No compilation is necessary.

Configuring Tomcat to Run a JSP Application

The first thing you need to do before you can run your JSP application is configure Tomcat so that it recognizes your JSP application. To configure Tomcat to run a particular JSP application, follow these steps:

1. Create a directory under %CATALINA_HOME°/o/webapps called myJSPApp.
2. Add a subdirectory named WEB-INF under the myJSPApp directory.
3. Edit server.xml, the server configuration file, so Tomcat knows about this new JSP application. The server.xml file is located in the conf directory under %CATALINA_HOME%. Open the file with your text editor and look for code similar to the following:

   ```
   <Context path="/examples" docBase="examples" debug="0"
       reloadable="true">
   .
   .
   .
   </Context>
   ```

   Right after the closing tag </Context>, add the following code:

   ```
   <Context path="/myJSPApp" docBase="myJSPApp" debug="0" reloadable="true'>
   </Context>
   ```

4. Restart Tomcat.

Now you can write your JSP file and store it under the myJSPApp file. Alternatively, to make it more organized, you can create a subdirectory called jsp under myJSPApp and store your JSP files there. If you do this, you don't need to change the setting in the server.xml file.

Writing a JSP File

A JSP page consists of interwoven HTML tags and Java code. The HTML tags represent the presentation part and the code produces the contents. In its most basic form, a JSP page can include only the HTML part, like the code shown in the following listing.

The Simplest JSP Page

```
<HTML>
<HEAD>
</HEAO>
<BOOY>
JSP is easy.
</BODY>
</HTML>
```

Save this file as SimplePage.jsp in the myJSPApp directory.

Now, start your web browser, and type the following address of the file saved above to get the following result.

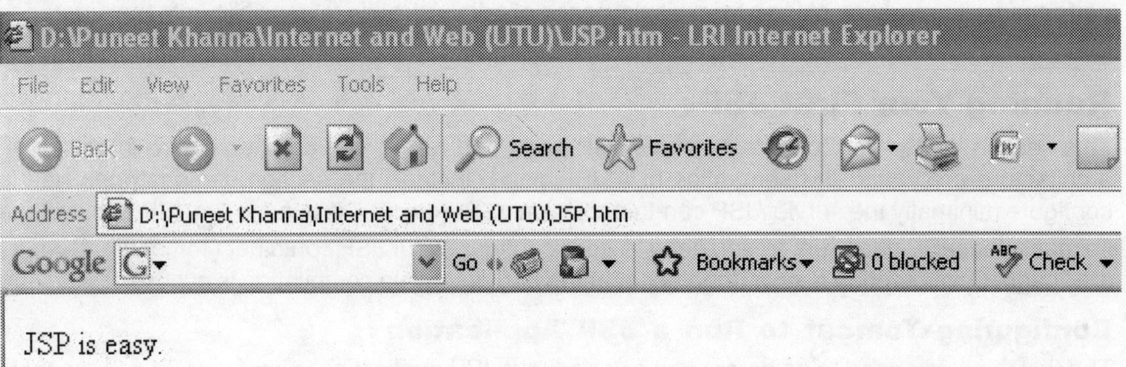

Other Examples

Of course, the code in the above listing is not a really useful page, but it illustrates the point that a JSP page does not need to have code at all. If your page is purely static, like the one in above listing, you shouldn't put it in a JSP file because JSP files are slower to process than HTML files. You might want to use a JSP file for pure HTML tags, however, if you think the code might include Java code in the future. This saves you the trouble of changing all the links to this page at the later stage.

To write Java code in your JSP file, you embed the code in <% ... %> tags. For example, the code in the following listing is an example of intertwining Java code and HTML in a JSP file.

Interweaving HTML and Code

```
<HTML>
<HEAD>
</HEAD>
<BODY>
<%
    out.println("JSP is easy"; );
%>
</BODY>
</HTML>
```

The code in above listing produces the same output as the one in the previous listing. Notice, however, the use of the Java code to send the text. If you don't under-stand what out.println does, bear with me for a moment—it is discussed in detail in the next section. For now, knowing that out.println is. used to send a String to the web browser is sufficient.

Notice also that the output of a JSP page is plain text consisting of HTML tags. No code section of the page will be sent to the browser.

Another example is given in the next listing. This snippet displays the string "Welcome. The server time is now" followed by the server time.

Displaying the Server Time

```
<HTML>
<HEAD>
    <TITLE>Displaying the server time</TITLE>
</HEAD>
<BODY>
Welcome. The server time is now
<%
    java.util. Calendar now = Java. util. Calendar. getInstance() ;
    int hour = now. get(java. util. Calendar. HOUR_OF_DAY) ;
    int minute = now. get(java. util. Calendar. MINUTE) ;

    if (hour<10)
        out.println("0" + hour);
    else
        out.println(hour);

    out.println( " : ") ;

    if (minute<10)
        out.println("0" + minute);
    else
        out.println(minute);

%>
</BODY>
</HTML>
```

The code in above listing displays the time in the hh:mm format. Therefore, if the hour is less than 10, a "0" precedes, which means that nine will be displayed as 09 instead of 9.

How JSP Works?

Inside the JSP container is a special servlet called the page compt'/er. The servlet container is configured to forward to this page compiler all HTTP requests with URLs that match the jsp file extension. This page compiler turns a servlet container into a JSP container. When a jsp page is first called, the page compiler parses and compiles the jsp page into a servlet class. If the compilation is successful, the jsp servlet class is loaded into memory. On subsequent calls, the servlet class for that jsp page is already in memory; however, it could have been updated. Therefore, the page compiler servlet will always compare the timestamp of the jsp servlet with the jsp page. If the jsp page is more cur-rent, recompilation is necessary. With this process, once deployed, JSP pages only go through the time-consuming compilation process once.

You may be thinking that after the deployment, the first user requests for a jsp page will experience unusually slow response due to the time spent for compiling the jsp file into a jsp servlet. To avoid this unpleasant situation, a mechanism in JSP allows the jsp pages to be pre-compiled before any user request for them is received. Alternatively, you deploy your JSP application as a web archive file in the form of a compiled servlet.

The JSP Servlet Generated Code

When the JSP is invoked, Tomcat creates two files in the C:\%CATALINA_HOME%\work\localhost\

examples\jsp directory.Those two files are SimplePage_ jsp Java and SimplePage_ jsp. class. When you open the SimplePage_ jsp Java file, you will see the following:

```
package org.apache.jsp;

import javax.servlet.*;
import javax.servlet.http.*;
import javax.servlet.Jsp.*;
import javax.servlet.jsp.tagext.*;
import org.apache.j asper.runtime.*;

public class SimplePagej'sp extends HttpJspBase {
    static {
    continues
    }

    public SimplePage_jsp( ) {
    }

    private static boolean _jspx_inited = false;

    public final void _jspx_init() throws org.apache.jasper.JasperException {
    }

    public void JspService(HttpServletRequest request,
        HttpServletResponse response)
        throws java.io.IOException, ServletException {

        JspFactory JspxFactory = null;
        PageContext pageContext = null;
        HttpSession session = null;
        ServletContext application = null;
        ServletConfig config = null;
        JspWriter out = null;
        Object page = this;
        String _value = null;
        try {
            if (_jspx_inited == false) {
                synchronized (this) {
                    if (_jspx_inited == false) {
                        _jspx_init() ;
                        _jspx_inited = true;
                    }
                }
            }
        _jspxFactory - JspFactory. getOefaultFactory( );
        response. setContentType("text/ntnil;charset=ISO-8859-1");
        pageContext = jspxFactory.getPageContext{this,
            request, response, "", true, 8192, true);
        application = pageContext. getServletContext();
        config = pageContext. getServletConfig() ;
        session = pageContext. getSession() ;
        out = pageContext. getOutf);
```

```
// begin
[file="C: \\tomcat4\\bin\\. .\\webapps\\examples\\jsp\\SimplePage. jsp" ;from=(0, 2);to=(2,0)]
out.println("JSP is easy");
// end
// HTML // begin
[file="C: \\tomcat4\\bin\\. . \\webapps\\examples\\ jsp\ \SimplePage. jsp" ;from=(2, 2);to=(3,0)]
out.write("\r\n");
// end
catch (Tnrowable t) {
    if (out != null && out.getBufferSize()  1= 0)
        out.clearBuffer();
    if (pageContext 1= null)
        pageContext.handlePageException(t);
}
finally {
    if (_jspxFactory != null)
    _jspxFactory.releasePageContext( pageContext) ;
    }
}
```

The JSP API

The JSP technology is based on the JSP API that consists of two packages: javax.servlet.jsp and javax.servletjsp.tagext. When you study the javax.servletjsp package, you will know why we say that JSP is an extension of servlet technology and understand why it is important that a JSP application programmer understands the servlet technology well.

The javax.servletjsp package has two interfaces and four classes. The interfaces are as follows:

- JspPage
- HttpJspPage

The four classes are as follows:

- JspEngineInfo
- JspFactory
- JspWriter
- PageContext

In addition, there are also two exception classes :JspException and JspError.

The JspPage Interface

The JspPage is the interface that must be implemented by all JSP servlet classes. This may remind you of the javax. servlet. TheJSPPage interface has two methods, Jsp I nit and JspDestroy, whose signatures are as follows:

```
public void jspInit()
public void jspDestroyt)
```

jspInit, which is similar to the init method in the javax. servlet. Servlet interface, is called when the JspPage object is created and can be used to run some initialization. This method is called only once during the life cycle of the JSP page: the first time the JSP page is invoked. The JspDestroy method is analogous with the destroy method of the javax. servlet. Servlet interface. This method is called before the JSP servlet object is destroyed.You can use this method to do some clean-up, if you want. Most of the time, however, JSP authors rarely make full use of these two methods. The following example illustrates how you can implement these two methods in your JSP page:

```
<%!
    public void jspInit() {
        System. out. println("Init");
    }
    public void jspDestroy() {
        System . out . print ln ( "Destroy ");
    }
%>
<%
    out.println("JSP is easy");
%>
```

Notice that the first line of the code starts with <%!.

The HttpJspPage Interface

This interface directly extends the JspPage interface. There is only one method: __JspService. This method is called by the JSP container to generate the content of the JSP page. The _jspService has the following signature:

```
public void _jspService(HttpServletRequest request,
HttpServletRfcsponse response) throws ServletException, IOException.
```

You can't include this method in aJSP page, such as in the following code:

```
<%!
    public void jspInit() {
        System, out, println("Init");
    }
    public void jspOestroyf) {
        System. out.println( "Destroy");
    }
    public void _jspService(HttpServletRequest request,
        HttpServletResponse response) throws ServletException, IQException {
        System. out. println( "Service");
    }
%>
```

This is because the page content itself represents this method.

The JspFactory Class

The JspFactory class is an abstract class that provides methods for obtaining other objects needed for the JSP page processing. The class has the static method getDefaultFactory that returns a JspFactory object. From the JspFactory object, a PageContext and a JspEngineInfo object can be obtained that are useful for the JSP page processing. These objects are obtained using the JspFactory class's getEngineInfo method and the getPageContext method, whose signatures are given here:

```
public abstract JspEngineInfo getEngineInfo()
public abstract PageContext getPageContext {
Servlet requestingServlet, ServletRequest request,
ServletResponse response, String errorPageURL,
boolean needsSession, int buffer, boolean autoFlush)
```

The following code is part of the _jspService method that is generated by the JSP container:

```
JspFactory JspxFactory = null;
PageContext pageContext = null;
jspxFactory = JspFactory. getDefaultFactory( );
.
.
```

```
pageContext = JspxFactory.getPageContext(this, request,
    response, "", true, 8192, true);
```

The JspEngineInfo Class

The JspEngineInfo class is an abstract class that provides information on the JSP container. Only one method, getSpecificationVersion, returns the JSP con-tainer's version number. Because this is the only method currently available, this class does not have much use.

You can obtain a JspEngineInfo object using the getEngineInfo method of the JspFactory class.

The PageContext Class

PageContext represents a class that provides methods that are implementation-dependent. The PageContext class itself is abstract, so in the _jspService method of a JSP servlet class, a PageContext object is obtained by calling the getPageContext method of the JspFactory class.

The PageContext class provides methods that are used to create other objects. For example, its getOut method returns a JspWriter object that is used to send strings to the web browser. Other methods that return servlet-related objects include the following:

- getRequest, returns a ServletRequest object
- getResponse, returns a ServletResponse object
- getServletConfig, returns a ServletConfig object
- getServletContext, returns a ServletContext object
- getSession, returns an HttpSession object

The JspWriter Class

The JspWriter class is derived from the java.io. Writer class and represents a Writer that you can use to write to the client browser. Of its many methods, the most important are the print and println methods. Both provide enough overloads that ensure you can write any type of data. The difference between print and println is that println always adds the new line character to the printed data.

Additional methods allow you to manipulate the buffer. For instance, the clear method clears the buffer. It throws an exception if some of the buffer's content has already been flushed. Similar to clear is the clearBuffer method, which clears the buffer but never throws any exception if any of the buffer's contents have been flushed.

The Generated Servlet Revisited

Now that you understand the various interfaces and classes in the javax.servlet.jsp package, analyzing the generated JSP servlet will make more sense.

When you study the generated code, the first thing you'll see is the following:

```
public class SimplePageJsp extends HttpJspBase {
```

First, remember that the JSP specification defines only standards for writing JSP pages. A JSP page itself will be translated into a Java file, which in turn will be compiled into a servlet class. The two processes are implementation depen-dent and do not affect the way a JSP page author codes. Therefore, a JSP container has the freedom and flexibility to do the page translation in its own way. The JSP servlet-generated code presented in this chapter is taken from Tomcat. You can therefore expect a different Java file to be generated by other JSP containers.

In Tomcat, HttpJspBase is a class whose source can be found under the src\jasper\src\share\org\apache\jasper\runtime directory under the %CATALINA_HOME% directory. Most importandy, the signature of this class is as follows:

```
public abstract class HttpJspBase extends HttpServlet
implements HttpJspPage
```

Now you can see that HttpJspBase is an HttpServlet and it does implement the javax.servlet.jsp.HttpJspPage interface. The HttpJspBase is more like a wrapper class so that its derived class does not have to provide implementation for the interfaces methods if the JSP page author does not override them. The complete listing of the class follows:

```
/*
*The Apache Software License, Version 1.1
*Copyright (c)  1999 The Apache Software Foundation. All rights reserved.
*Redistribution and use in source and binary forms, with or without modification, are permitted
*provided that the following conditions are met:
* 1. Redistributions of source code must retain the above copyright
*    notice, this list of conditions and the following disclaimer.
*
* 2.  Redistributions in binary form must reproduce the above copyright
*     notice, this list of conditions and the following disclaimer in
*     the documentation and/or other materials provided with the
*     distribution,
*
* 3.  The end-user documentation included with the redistribution, if
*     any, must include the following acknowlegement:
*     "This product includes software developed by the
*     Apache Software Foundation (http://www.apache.org/)."
*     Alternately, this acknowlegement may appear in the software itself,
*     if and wherever such third-party acknowlegements normally appear.
* *
* 4.  The names "The Jakarta Project", "Tomcat", and "Apache Software
*     Foundation" must not be used to endorse or promote products derived
*     from this software without prior written permission. For written
*     permission, please contact apache@apache.org.
*
* 5.  Products derived from this software may not be called "Apache"
*     nor may "Apache" appear in their names without prior written
*     permission of the Apache Group.
*
* THIS SOFTWARE IS PROVIDED "AS IS" AND ANY EXPRESSED OR IMPLIED
* WARRANTIES, INCLUDING, BUT NOT LIMITED TO, THE IMPLIED WARRANTIES
* OF MERCHANTABILITY AND FITNESS FOR A PARTICULAR PURPOSE ARE
* DISCLAIMED.  IN NO EVENT SHALL THE APACHE SOFTWARE FOUNDATION OR
* ITS CONTRIBUTORS BE LIABLE FOR ANY DIRECT, INDIRECT, INCIDENTAL,
* SPECIAL, EXEMPLARY, OR CONSEQUENTIAL DAMAGES (INCLUDING, BUT NOT
* LIMITED TO, PROCUREMENT OF SUBSTITUTE GOODS OR SERVICES; LOSS OF
* USE, DATA, OR PROFITS; OR BUSINESS INTERRUPTION) HOWEVER CAUSED AND
* ON ANY THEORY OF LIABILITY, WHETHER IN CONTRACT, STRICT LIABILITY,
* OR TORT (INCLUDING NEGLIGENCE OR OTHERWISE) ARISING IN ANY WAY OUT
* OF THE USE OF THIS SOFTWARE, EVEN IF ADVISED OF THE POSSIBILITY OF
* SUCH DAMAGE.
*
*     This software consists of voluntary contributions made by many
*     individuals on behalf of the Apache Software Foundation. For more
*     information on the Apache Software Foundation, please see
*     <http://www.apache.org/>.
*
```

```
*/
package org.apache.jasper.runtime;

import java.io.IOException;
import java.io.FileInputStream;
import java.io.InputstreamReader;

import java.net.URL;
import java. net. MalformedURLException;
import javax.servlet.*; import javax. sen/let. http,*; import javax. servlet.jsp.*;
import org .apache. jasper. JasperExceptionj import org. apache. jasper. Constants;
/**
* This is the subclass of all JSP- generated servlets.
*
*/
public abstract class HttpJspBase extends HttpServlet
    implements HttpJspPage {

protected PageContext pageContext;

protected HttpJspBasef)
{

public final void init(ServletConfig config)
    throws ServletException {
    super. init(config);
    jspInit();
}

public String getServletInfo() {
    return Constants. getString ("jsp. engine. info");
}

public final void destroy() {
    jspDestroy();
}

/**
* Entry point into service.
*/
public final void service {HttpServletRequest request,
    HttpServletResponse response)
    throws ServletException, IOException {
    _jspService( request, response);
}

public void jspInit() {
}

public void jspDestroyf) {
}

public abstract void _jspService(HttpServletRequest request,
```

```
        HttpServletResponse  response)
        throws ServletException,  IOException;
    }
```

Back to the generated JSP servlet class, remember that the JSP page source code is simply the following three lines of code:

```
    <%
    Out.println("JSP is easy");
    %>
```

There are no jspInit and jspDestroy methods in the source code, so there are no implementations for these two methods in the resulting servlet code. The three lines of code, however, translate into the _jspService method. For read-ing convenience, the method is reprinted here. Notice that, for clarity, the comments have been removed.

```
    public void _jspService(HttpServletRequest request,
        HttpServletResponse   response)
        throws Java. io. IOException, ServletException {

        JspFactory JspxFactory = null;
        PageContext pageContext = null;
        HttpSession session = null;
        ServletContext application = null;
        ServletConfig config = null;
        JspWriter out = null;
        Object page = this;
        String  _value = null;
        try { ,
            if  (_jspx_inited == false)  {
                synchronized  (this)  {
                    if  (_jspx_inited == false)  {
                        _jspx_init();
                        _jspx_inited = true;
                    }
                }
            }
        _jspxFactory = JspFactory. getDefaultFactory() ;
        response. setContentType(" text /ntml;charset=ISO- 8859-1 ") ;
        pageContext = jspxFactory.getPageContext(thiSj
            request, response, "", true, 8192, true);
        application = pageContext. getServletContext() ;
        config = pageContext, getServletConfigO ;
        session - pageContext. getSession() ;
        out = pageContext. getOutf);

        Out.printlnf "JSP is easy");
        out,write( "\r\n" ) ;
        }
        catch (Throwable t)  {
            if  (out  1= null && out.getBufferSize()   1= 0}
                out.clearBuffer();
            if (pageContext 1= null)
                pageContext.handlePageException(t);
        }
```

```
        finally {
            if (_jspxFactory != null)
            _jspxFactory,releasePageContext(pageContext);
        }
    }
```

Implicit Objects

Having examined the generated JSP servlet source code, you know that the code contains several object declarations in its _jspService method. Recall this part from the code in the preceding section:

```
    public void _jspService<HttpServletRequest request,
        HttpServletResponse response)
        throws java.io.IOException, ServletException {

        JspFactory _jspxFactory = null;
        PageContext pageContext = null;
        HttpSession session = null;
        ServletContext application = null;
        ServletConfig config = null;
        JspWriter out - null;
        Object page = this;
        String _value = null;
        try {
            .
            .
            .
        _JspxFactory = JspFactory.getDefaultFactory();
        response.setContentType("text/ntml;charset=ISO -8859-1"};
        pageContext = jspxFactory.getPageContext(this,
            request, response, "", true, 8192, true);
        application = pageContext.getServletContextf);
        config = pageContext.getServletConfig();
        session = pageContext.getSession();
        out = pageContext.getOut();
        .
        .
        .
    }
```

You see that there are object references, such as pageContext, session, applica-tion, config, out, and so on. These object references are created whether they are used from inside the page. They are automatically available for the JSP page author to use! These objects are called implicit objects and are summarized in the following Table.

JSP Implicit Objects

Object	Type
request	javax, sen/let.http.HttpServletRequest
response	javax. servlet.http.HttpServletResponse
out	javax.ser vie t.jsp.Jsp Writer
session	javax. servlet:.http.HttpSession
application	javax.servlet.ServletContext
config	javax. sen/let. ServletConfig
pageContext	javax.servlet.jsp.PageContext

page javax.servlet.jsp.HttpJspPage
exception java.lang.Throwable

By looking at the table, you now know why you can send a String of text by simply writing:

```
<%
    out.println("JSP is easy.");
%>
```

In the case of the code above, you use the implicit object out that represents a javax.servlet.jspjspWriter. All the implicit objects are discussed briefly in the following subsections.

request and response Implicit Objects

In servlets, both objects are passed in by the servlet container to the service method of the javax.servlet.http.HttpServlet class. Remember its signature?

```
protected void service(HttpServletRequest request,
    HttpServletResponse response)
    throws ServletException, IOException
```

In a servlet, before you send any output, you are required to call the setContentType of the HttpServletResponse to tell the browser the type of the content, as in the following code:

```
response.SetContentType("text/html");
```

In JSP, this is done automatically for you in the generated JSP servlet class, as follows:

```
response.setContentType("text/html;charset=ISO-8859-1");
```

Having an HttpServletRequest and an HttpServletResponse, you can do any-thing you like as you would in a servlet. For example, the following JSP page retrieves the value of a parameter called firstName and displays it in the browser:

```
<HTML>
<HEAD>
<TITLE>Simple Page</TITLE>
</HEAD>
<BODY>
<%
    String firstName = request.getParameterf"firstName");
    out.println("First name: " + firstName);
%>
</BODY>
</HTML>
```

The following example uses the sendRedirect method of the javax.servlet.http.HttpServletResponse to redirect the user to a different URL:

```
<%
response.sendRedirect("http://www.newriders.com");
%>
```

out Implicit Object

out is probably the most frequently used implicit object.You call either its print method or its println method to send text or other data to the client browser. In a servlet, you always need to call the getWriter method of the javax.servlet.http.HttpServletResponse interface to obtain a PrintWriter before you can output anything to the browser, as follows:

```
PrintWriter out = response.getWriter();
```

In JSP, you don't need to do this because you already have an out that represents a javax.servlet.jsp.JspWriter object.

session Implicit Object

The session implicit object represents the HttpSession object that you can retrieve in a servlet by calling the getSession method of the javax.servlet.http.HttpServletRequest interface, as in the following code: request.getSession();

The following code is a JSP page that uses a session object to implement a counter:

```
<HTML>
<HEAD>
<TITLE>Counter</TITLE>
</HEAD>
<BODY>
<%
    String counterAttribute = (String) session.getAttribute("counter");
    int count = 0;
    try {
        count = Integer.parseInt(counterAttribute);
    }
    catch (Exception e) {
    }
    count++;
    session.setAttribute("counter", Integer.toString(count));
    out.println("This is the " +• count + "th time you visited this page in this session.");
%>
</BODY>
</HTML>
```

See how session is readily available without efforts from the programmer's side?

application

The application implicit object represents the javax.servlet.ServlctContext object. In an HttpServlet, you can retrieve the ServletContext method by using the getServletContext method.

config

The config implicit object represents a javax.servlet.ServletConfig object that in a servlet can be retrieved by using the getServletConfig method.

pageContext

The pageContext implicit object represents the javax.servlet.jsp.

page

The page implicit object represents the javax.servlet.jsp.

exception

The exception object is available only on pages that have been defined as error pages.

Revision

1. JavaServer Pages (JSP) is another java technology for developing web applications.
2. Sun introduced servlets in 1996 as small Java-based applications for adding dynamic content to web applications.
3. Nearly half of the content sent from the doPost method is static HTML. However, each HTML tag must be embedded in a String and sent using the println method of the PrintWriter object.
4. A JSP page consists of interwoven HTML tags and Java code.
5. The servlet container is configured to forward to this page compiler all HTTP requests with URLs that match the jsp file extension.
6. The JSP technology is based on the JSP API that consists of two packages: javax.servlet.jsp and javax.servletjsp.tagext.

7. The JspPage is the interface that must be implemented by all JSP servlet classes.
8. The JspFactory class is an abstract class that provides methods for obtaining other objects needed for the JSP page processing.
9. The JspEngineInfo class is an abstract class that provides information on the JSP container.
10. PageContext represents a class that provides methods that are implementation-dependent.
11. There are no jspInit and jspDestroy methods in the source code, so there are no implementations for these two methods in the resulting servlet code.

Multiple Choice Questions

1. The JSP request vaiarble is used to obtain information from the request as sent by the browser.
   (a) TRUE (b) FALSE
2. RESPONSE can be used to affect the response being sent to the browser.
   (a) TRUE (b) FALSE
3. JSP simply puts Java inside ____________.:
   (a) HTML (b) XML
   (c) Java Script
4. Predefined tags start with jsp _______ characters.
   (a) % (b) :
   (c) #
5. Java Server Pages (JSP) is not a technology that lets you mix regular, static HTML with dynamically generated HTML.
   (a) TRUE (b) FALSE

Answers

1. a 2. a 3. c 4. a 5. a

Subjective Questions

1. What is JSP?
2. How would you run your first JSP?
3. Describe the process of writing a JSP file.
4. What are JSP Servlet Generated Codes?
5. Describe the following:
   The JspPage interface
   The HttpJspPage interface
   The JspFactory Class
   The JspEngineInfo Class
   The JspWriter Class
   The PageContext Class

8 JSP Directives and Standard Actions

Introduction

JSP page has elements and template data. The elements, which also are called *JSP tags,* make up the syntax and semantics of JSP. Template data is everything else. Template data includes parts that the JSP container does not understand, such as HTML tags. There are three types of elements:

- Directive elements
- Scripting elements
- Action elements

To write an effective JSP page, you need to understand all these elements well. Elements have two forms: the XML form and the <% ... %> alternative form. The conversion between the XML form and the alternative form is presented at the end of the chapter.

Directives

Directives are messages to the JSP container containing information on how the JSP container must translate a JSP page into a corresponding servlet. Directives have the following syntax:

<%@ directive (attribute="value")* %>

The asterisk (*) means that what is enclosed in the brackets can be repeated zero or more times. The syntax can be re-written in a more informal way as follows:

<%@ directive attribute1="value1" attribute2="value2" ... %>

White spaces after the opening <%@ and before the closing %> are optional, but are recommended to enhance readability.

The three types of directives are as follows:

- Page directives
- Include directives
- Tag library directives

The Page Directive

The Page directive has the following syntax:

<%@ page (attribute="value")* %>

Or, if you want to use the more informal syntax:

<%@ page attribute1="value1" attribute2="value2" ... %>

The Page directive supports 11 attributes. These attributes are summarized in the following table.

The Page Directive's Attributes

Attribute	*Value Type*	*Default Value*
language	Scripting language name	"Java"
info	String	Depends on the JSP container

contentType	MIME type, character set	"text/html;charset=ISO-8859-1"
extends	Class name	None
import	Fully qualified class name or package name	None
buffer	Buffer size or false	8192
autoFlush	Boolean	"true"
session	Boolean	"true"
isThreadSafe	Boolean	"true"
errorPage	URL	None
isErrorPage	Boolean	"false"

An example of the use of the Page directive is as follows:

```
<%@ page buffer="16384" session="false" %>
```

With JSP, you can specify multiple page directives in your JSP page, such as the following:

```
<%@ page buffer="16384" %> <%@ page session="false" %>
```

Except for the import attribute, however, JSP does not allow you to repeat the same attribute within a page directive or in multiple directives on the same page.

The following is illegal because the info attribute appears more than once in a single page:

```
<%@ page   info="Example Page" %>
<%@ page   buffer="16384" %>
<%@ page   info="Unrestricted Access" %>
```

The following also is illegal because the buffer attribute appears more than once, in the same page directive:

```
<%@ page buffer="16384" info="Example Page" buffer="8192" %>
```

The following is legal, however, because the import attribute can appear multiple times:

```
<%@ page   import="java.io.*" info="Example Page" %>
<%@ page   import="java.util.Enumeration" %>
<%@ page   import="com.brainysoftware.web.FileUpload" %>
```

Alternatively, imported libraries can appear in a single import attribute, separated by commas, as shown in the following code:

```
<%@ page import="java.io.*, Java.util.Enumeration" %>
```

The following section discusses each attribute in more detail and examines the effect on the generated servlet.

The language Attribute

The language attribute specifies the scripting language used in the JSP page. By default, the value is "Java" and all JSP containers must support Java as the scripting language. With Tomcat, this is the only accepted language. Other JSP containers might accept different languages, however.

Specifying this attribute, as in the following code, does not have any effect on the generated servlet:

```
<%@ page language="java" %>
<%
    out.println("JSP is easy");
%>
```

In fact, this attribute is useful only in a JSP container that supports a language other than Java as the scripting language.

The info Attribute

The info attribute allows you to insert any string that later can be retrieved using the getServletInfo method. For example, you can add the following page directive with an info attribute:

```
<%@ page info="Written by Bulbul" %>
```

The JSP container then will create a public method getServletInfo in the resulting servlet. This method returns the value of the info attribute specified in the page. For the previous page directive, the getServletInfo method will b written as follows:

```
public String getServletInfo;) {
return "Written by Bulbul";
}
```

In the same JSP page, you can retrieve the info attribute's value by calling the getServletInfo method. For instance, the following JSP page will return "Written by Bulbul" on the client browser.

```
<%@ page info="Written by Bulbul" %>
<%
    out.println(getServlet!nfo());
%>
```

The default value for this attribute depends on the JSP container.

The contentType Attribute

The contentType attribute defines the Multipurpose Internet Mail Extension (MIME) type of the HTTP response sent to the client. The default value is "text/htrnl;charset=ISO-8859-1" and this is reflected in jspService method in the generated servlet. For example, this is the cleaned up version of the servlet that is generated from a JSP page that does not specify the contentType attribute in its page directive; that is, the default value is used:

```
public void _jspService(HttpServletRequest request,
    HttpServletResponse    response)
    throws java.io.IOException, ServletException {
        .
        .
        .
    try {
        .
        .
        .
        _jspxFactory = JspFactory.getDefaultFactory();
        response.setConterrtType("text/html;charset=ISO-8859-1");
        .
        .
        .
    }
    catch (Throwable t) {
        .
        .
        .
    } finally {
        .
        .
        .
    }
}
```

You will want to use this attribute when working with pages that need to send characters in other encoding schemes. For example, the following directive tells the JSP container that the output should be sent using simplified Chinese characters:

```
<%@ page contentType="text/html;charset=GB2312" %>
```

The extends Attribute

The extends attribute defines the parent class that will be inherited by the generated servlet. You should use this attribute with extra care. In most cases, you should not use this attribute at all. In Tomcat, the parent class that will be subclassed by the resulting servlet is HttpJspBase.

The import Attribute

The import attribute is similar to the import keyword in a Java class or interface. The attribute is used to import a class or an interface or all members of a package. You will definitely use this attribute often. Whatever you specify in the import attribute of a page directive will be translated into an import statement in the generated servlet class. By default, Tomcat specifies the following import statements in every generated servlet class. You don't need to import what has been imported by default:

```
import   javax.servlet.*;
import   javax.servlet.http.*;
import   javax.servlet.jsp,*;
import   javax.servlet Jsp.tagext. *;
import   org.apache.j asper.runtime.*;
```

As an example, consider the following JSP page that imports the Java.io package and the Java.util.Enumeration interface:

```
<%@ page import="java.io.*" %>
<%@ page import="Java.util.Enumeration" %>
```

The two will be added before the default import statements in the generated servlet class,,as described in the following code fragment:

```
import java.io.*;
import Java.util.Enumeration;
import javax.servlet.*;
import javax.servlet.http.*;
import javax.servlet.jsp.*;
import javax.servlet.jsp.tagext.*;
import org.apache.j asper.runtime.*;
```

The buffer Attribute

By default, a JSP page's content is buffered to increase performance. The default size of the buffer is 8Kb or 8192 characters.

Consider an example that specifies the buffer attribute with a size of 16Kb:

```
<%@ page buffer="16kb" %>
```

This attribute is reflected in the _jspService method in the generated servlet class, as in the following cleaned up code. The buffer size is passed to the getPageContext method of the javax.servlet.jsp.JspFactory class when creating a PageContext object.

```
public void _jspService(HttpServletRequest request,
    HttpServletResponse    response)
    throws java.io.IOException, ServletException {
    .
    .
    .
    try {
    .
    .
    .
    _jspxFactory = JspFactory.getDefaultFactory();
    response.setContentType("text/html;charset=ISO-8859-1");
```

```
        // 8192 in the following line represents.the buffer size
        pageContext = jspxFactory.getPageContextfthis, request,
            response, "", true, 16384, true);
        .
        .
        .
    }
    catch (Throwable t) {
        .
        .
    } finally {
        .
        .
        .
    }
    }
```

With the buffer attribute in a page directive, you can do two things:

- Eliminate the buffer by specifying the value "none", as in the following code:

  <%@ page buffer="none" %>

  If you decide not to use a buffer, the getPageContext method will be invoked passing 0 as the buffer size, as in the following code:

  ```
  pageContext = _jspxFactory.getPageContext(
  this, request,  response, "", true, 0, true);
  ```

- Change the size of the buffer by assigning a number to the attribute. The attribute value represents the number in kilobytes. Therefore, "16" means 16Kb alias 16384 characters. You can either write the number only, or the number plus "Kb". For example, "16" is the same as "16Kb".

The following example shows two page directives in two different JSP pages. The first cancels the use of the buffer, and the second changes the size of the buffer to 12Kb.

```
<%@ page import="java.io.*" buffer="none" %>
<%@ page import="java.io.*" buffer="12" %>
```

The autoFlush Attribute

The autoFlush attribute is related to the page buffer. When the value is "true", the JSP container will automatically flush the buffer when the buffer is full. If the value is "false", however, the JSP author needs to flush the buffer manually using the flush method of the JspWriter object, such as the following:

```
out.flush();
```

For example, the following code specifies a false value for the autoFlush attribute:

```
<%@ page autoFlush="false" %>
```

The session Attribute

By default, a JSP page participates in the JSP container's session management. This is indicated by the declaration of a javax.servlet.http.HttpSession object reference and the creation of it through the getSession method of the javax.servlet.jsp.PageContext, as illustrated by the following code:

```
public void JspService(HttpServletRequest request,
    HttpServletResponse    response)
    throws java.io.IOException, ServletException {

    JspFactory _jspxFactory = null;
```

```
PageContext pageContext = null;
// declare an HttpSession object reference
HttpSession session = null;

try {
    .
    .
    .
jspxFactory = JspFactory.getDefaultFactory();
response.setContentType("text/html;charset=ISO-8859-1");
pageContext = _jspxFactory.getPageContext(
    this, request, response, "", true, 0, true);
application = pageContext.getServletContext();
config = pageContext.getServletConfig();
// create a HttpSession object
session = pageContext.getSession();
    .
    .
    .
catch (Throwable t) {
    .
    .
    .
}
finally {
    .
    .
    .
}
}
```

Whether your JSP page participates in the session management is determined by the value of the session attribute in a page directive. By default this value is true; that is, the JSP page is part of the session management. I want to make an important point, however: You should not let your JSP page participate in the session management unnecessarily because this consumes resources. Because of this, you should always use the session attribute and assign it the value of "false" unless you are specifically needing session management.

When the value of this attribute is "false", no javax.servlet.http.HttpSession object reference is declared and no HttpSession object is created. As a result, the session implicit object is not available in the JSP page.

The following shows an example of the use of the session attribute:

```
<%@ page session="false" %>
```

The isThreadSafe Attribute

As you know that you can make a servlet thread-safe by inheriting the javax.servlet.SingleThreadModel.You can control this behavior in the JSP page by using the isThreadSafe attribute, which by default has the value of "true". When the value of this attribute is true, you guarantee that simultaneous access to this page is safe. However, you can assign "false" to this attribute as in the following code:

```
<%@ page isThreadSafe="false" %>
```

By setting the isThreadSafe attribute to false, you are telling the JSP translator that you (the programmer of this page) cannot guarantee that simultaneous accesses to this page will be safe.

Therefore, you are asking the JSP translator to make this page thread-safe. This will make the generated servlet implement the javax.servlet.SingleThreadModel interface, as in the following signature of the JSP servlet class:

```
public class SimplePage_jsp extends HttpJspBase
implements SingleThreadModel
```

In other words, when the value of this attribute is false, the JSP container will serialize multiple requests for this page; that is, the JSP container will wait until the JSP servlet finishes responding to a request before passing another request to it.

The errorPage Attribute

The errorPage attribute specifies the URL of an error page that will be displayed if an uncaught exception occurs in this current JSP page. Referring to the URL of the error page is sometimes tricky. The easiest solution is to store the error page in the same directory as the current JSP page. In this case, you need only to mention the name of the error page, as in the following example:

```
<%@ page errorPage="ErrorPage.jsp" %>
<%
    String name = request. getParameterf"otherName");
    // this will throw an exception because the parameter "otherName"
    // does not exist,  so name will be null.
    // this will cause the ErrorPage.jsp to be displayed
    name.substring(1, 1);
%>
```

When the page is called, the error page is displayed. An error page must specify the isErrorPage attribute in its page directive and the value of this attribute must be "true".

The following is an errorPage attribute that is assigned an error page in a URL:

```
<%@ page errorPage="/myJspApp/ErTorPage.jsp" %>
```

The isErrorPage Attribute

The isErrorPage attribute can accept the value of "true" or "false", and the default value is "false". It indicates whether the current JSP page is an error page; that is, the page that will be displayed when an uncaught exception occurs in the other JSP page. If the current page is an error page, it has access to the exception implicit object, as explained in Chapter 8.

The include Directive

The include directive is the second type of the JSP directive elements. This directive enables JSP page authors to include the contents of other files in the current JSP page.

The include directive is useful if you have a common source that will be used by more than one JSP page. Instead of repeating the same code in every JSP page, thus creating a maintenance problem, you can place the common code in a separate file and use an include directive from each JSP page.

The included page itself can be static, such as an HTML file, or dynamic, such as another JSP page. If you are including aJSP page, the included JSP page itself can include another file. Therefore, nesting include directives is permitted in JSP.

The syntax for the include directive is as follows:

```
<%@ include -file-"relativeURL" %>
```

The following is an example of how to include HTML files:

```
<%@ include file="includes/header.html" %>
<%
    out.println("<BODY>");
    // other content;
%>
<%@ include file="includes/footer.htnl" %>
```

Now, we'll look at how included files are translated into the JSP servlet file. The following example is a simple JSP page that includes two HTML files: Header.html and Footer.html. Both included files are located under the includes subdirectory, which itself is under the directory that hosts the current JSP file. The JSP page is called SimplePage.jsp and is given in the following listing. The page simply displays the current server time. The Header.html is given in second listing and the Footer.html in the third listing. What you are interested in here is the generated servlet code.

A Simple JSP Page that Includes Two Files

```
<%@ page session="false" %>
<%@ page import="java.util.Calendar" %>
<%@ include file="includes/Header.html" %>
<%
    out.println("Current time:  " + Calendar.getInstance().getTimef));
%>
<%@ include file="includes/Footer.html" %>
```

The Header.html File

```
<HTML>
<HEAD>
<TITLE>Welcome</TITLE>
<BODY>
```

The Footer.html File

```
</BODY>
</HTML>
```

The generated servlet is as follows: package org. apache. jsp;

```
import  java. util. Calendar;
import  javax. servlet.*;
import  j avax. servlet .http.*;
import  javax. servlet. jsp.*;
import  j avax . servlet .jsp. tagext . * ;
import  org . apache . j asper . runtime . * ;

public class SimplePage_jsp extends HttpJspBase {

    static {
    }

    public SimplePage_jsp( ) {
    }

    private static boolean _jspx_inited = false;

    public final void _Jspx_init() throws org. apache. jasper. JasperException {
    }

    public void _jspService(HttpServletRequest request,
        HttpServletResponse response)
        throws java.io.IOException, ServletException {

        JspFactory _jspxFactory = null;
        PageContext pageContext = null;
        ServletContext application = null;
        ServletConfig config = null;
```

```
JspWriter out - null;
Object page = this;
String _value = null;
try {
    if (_jspx_inited == false) {
        synchronized (this) {
            if (_jspx_inited == false) {
                _jspx_init();
                jspx_inited = true;
            }
        }
    }

    _jspxFactory = JspFactory. getDefaultFactory();
    response. setContentType(" text /html;charset=ISO- 8859-1 "
    pageContext = _JspxFactory.getPageContext(
        this, request, response, "", false, 8192, true);
    application = pageContext.getServletContext();
    config = pageContext.getServletConfig();
    out = pageContext.getOut();

    // HTML // begin
[file="C:\\tomcat4\\bin\\..\\webapps\\myJSPApp\\SimplePage.jsp";from=(0,27); to=(1,0)]
    out.write("\r\n");

    // end
    // HTML // begin
[file="C:\\tomcat4\\bin\\..\\webapps\\myJSPApp\\SimplePage.jsp";from=(1,39); to=(2,0)]
    out.write("\r\n");

    // end
    // HTML
    // begin
[file="C:\\tomcat4\\bin\\..\\webapps\\myJSPApp\\includes\\Header.html"; from=(0,0);to=(4,0)]

    out. write ("<HTML>\r\n<HEAD>\r\n<TITLE>Welcome</TITLE>\r\n<BODY>\r\n");

    // end
    // HTML // begin
[file="C:\\tomcat4\\bin\\..\\webapps\\myJSPApp\\SimplePage.jsp";from=(2,42); to=(3,0)]
    out.write("\r\n");

    // end
    // begin
[file="C:\\tomcat4\\bin\\..\\webapps\\myJSPApp\\SimplePage.jsp";from=(3,2);to =(5,0)]

    out.println("Current time: " + Calendar.getInstance().getTime());
    //end
    // HTML // begin
[file="C:\\tomcat4\\bin\\..\\webapps\\myJSPApp\\SimplePage.jsp";from=(5,2);to =(6,0)]
    out.write("\r\n");

    // end
```

```
        // HTML // begin
    [file="C: \\tomcat4\\bin\ \..\\webapps\ \myJSPApp\\includes\ \Footer.html"; from=(0,0);to=(2,0)]
        out.write("</BODY>\r\n</HTML>\r\n");

        // end
        // HTML
        // begin
[file="C: \\tomcat4\\bin\\ . . \\webapps\\myJSPApp\\SimplePage.jsp" ;from=(6,42) ; to=(7,0)]
        out .write( " \r\n" ) ;
        // end

    }
    catch  (Throwable t) {
        if (out  != null && out.getBufferSize()  != 0)
            out.clearBuffer( ) ;
        if (pageContext != null) pageContext.handlePageException(t) ;
    }
    finally {
        if (_jspxFactory  != null) JspxFactory. releasePageContext(pageContext) ;
        }
    }
}
```

The lines in bold denote the beginning of the included file. Because all the included files are placed inline before being compiled, including files does not have any performance effect on the application.

The taglib Directive

The taglib, or tag library, directive can be used to extend JSP functionality. This is a broad topic and has been given a chapter of its own.

Scripting Elements

Scripting elements allow you to insert Java code in your JSP pages. There are three types of scripting elements:

- Scriptlets
- Declarations
- Expressions

The three elements are discussed in the following sections.

Scriptlets

So far you have seen scriptlets in the examples. Scriptlets are the code blocks of a JSP page. Scriptlets start with an opening <% tag and end with a closing %> tag.

The following JSP page is an example of using scriptlets. In the JSP page, you try to connect to a database and retrieve all records from a table called Users. Among the fields in the Users table are FirstName, LastName, UserName, and Password. Upon obtaining a ResultSet object, you display all the records in an HTML table. The JSP page is given in the following listing.

Displaying Database Records

```
<%@ page session="false" %>
<%@ page import="java.sql.*" %>
<%
    try {
        Class. forName(" sun. jdbc.odbc.JdbcOdbcDriver") ;
```

```
            System. out. println("JDBC driver loaded");
        }
        catch (ClaSsNotFoundException e) {
            System . out . println (e . toString ( ) ) ;
        }
    %>
    <HTML>
    <HEAD>
    <TITLE>Display All Users</TITLE>
    </HEAD>
    <BODY>
    <CENTER>
    <BR><H2>Displaying All Users</H2>
    <BR>
    <BR>
    <TABLE>
    <TR>
    <TH>First Name</TH>
    <TH>Last Name</TH>
    <TH>User Name</TH>
    <TH>Password</TH>
    </TR>
    <%
    String sql = "SELECT FirstName, LastName, UserName, Password" + "FROM Users";
        try {
            Connection con = Driver-Manager. getConnection("jdbc:odbc:JavaWeb") ;

            Statement s = con.createStatement() ;
            ResultSet rs = s.executeduery(sql);

        while  (rs.next())  {
        out.println("<TR>");
        out.println("<TD>" + rs.getString(1) + "<TD>");
        out.println("<TD>" + rs.getString(2) + "<TD>");
        out.println("<TD>" + rs.getString(3) + "<TD>");
        out.println("<TD>" + rs.getString(4) + "<TD>");
        out.println("</TR>");
        }
        rs.close();
        s.close();
        con.close();
        }
        catch (SQLException e) {
        }
        catch (Exception e) {
        }
    %>
    </TABLE>
    </CENTER>
    </BODY>
    </HTML>
```

The result of this JSP page in a browser is shown in the next figure.

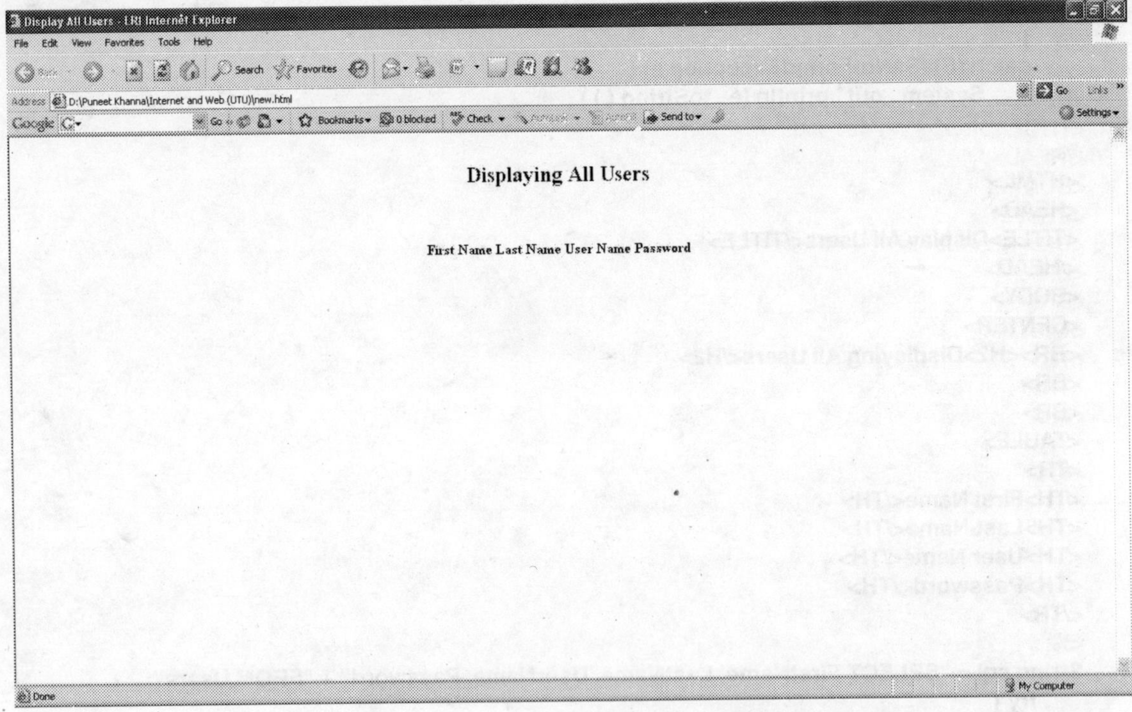

Note that you don't do any HTML encoding to the data returned by the getString method of the ResultSet because you don't know yet how to write and use a method in a JSP page. If, for example, the data contains something like <BR>, this will ruin the display in the web browser.

When writing scriptlets, it is commonplace to interweave code with HTML tags, for writing convenience. For instance, you can modify the while statement block of the code in the above listing to the following:

```
while (rs.next()) {
%>
<TR>
<TD><% out.print(rs.getString(1)) ; %></TD>
<TD><% out.print(rs.getString(2)) ; %></TD>
<TD><% out.print(rs.getString(3)) ; %></TD>
<TD><% out.print(rs.getString(4)) ; %></TD>
</TR>
<%
}
```

This is perfectly legal syntax.

JSP Page that Interweaves HTML Tags with Scriptlets

```
<%@ page session="false" %>
<%@ page import="java.sql.*" %>
<%
    try {
        Class. forName(" sun. jdbc.odbc.JdbcOdbcDriver") ;
        System. out. println("JDBC driver loaded");
    }
    catch (ClassNotFoundException e) {
        System.out.println(e.toString()) ;
```

```
}
%>
<HTML>
<HEAD>
<TITLE>Display All Users</TITLE>
</HEAD>
<BODY>
<CENTER>
<BR><H2>Displaying All Users</H2>
<BR>
<BR>
<TABLE>
<TR>
<TH>First Name</TH>
<m>Last Name</TH>
<TH>User Name</TH>
<TH>Password</TH>
</TR>
<%
    String sql = "SELECT FirstName, LastName, UserName, Password" +
        "FROM Users";
    try {
        Connection con = DriverManager.getConnection("jdbc:odbc:JavaWeb");

        Statement s = con.createStatement();
        ResultSet rs = s.executeQuery(sql);

        while (rs.next()) {
    %>
    <TR>
    <TD><% out.print(rs.getString(1)); %></TD>
    <TD><% out.print(rs.getString(2)); %></TD>
    <TD><% out.print(rs.getString(3)); %></TD>
    <TD><% out.print(rs.getString(4)); %></TD>
    </TR>
    <%
        }
        rs.close();
        s. close () ;
        con. close () ;
    }
    catch (SQLException e) {
    }
    catch (Exception e) {
    }
%>
</TABLE>
</CENTER>
</BODY>
</HTML>
```

Scriptlets alone are not sufficient to write efficient and effective JSP pages. As mentioned earlier, you can't declare a method with scriptlets. Additionally, in the code in above listings, the JSP page will try to load the JDBC driver every time the page is requested. This is unnecessary because the page needs to load it only once. How do you run an initialization code as you do in servlets? The answer is explained in the following section: declarations.

Declarations

Declarations allow you to declare methods and variables that can be used from any point in the JSP page. Declarations also provide a way to create initialization and clean-up code by utilizing the jspInit and jspDestroy methods.

A declaration starts with a <%! and ends with a %> and can appear anywhere throughout the page. For example, a method declaration can appear above a page directive that imports a class, even though the class is used in the method.

The code in the following listing shows the use of declarations to declare a method called getSystemTime and an integer i.

Using Declarations to Declare a Method and a Variable

```
<%!
    String getSystemTime () {
        return Calendar. getInstance() .getTimef) .toString() ;
    }
%>
<%@ page import="java.util. Calendar" %>
<%@ page session="false" %>
<%
    out. printlnf "Current Time:  " + getSystemTime()) ;
%>
<%.' int i; %>
```

The resulting servlet follows. For clarity, some comments have been added to the code.

```
package org. apache. jsp;

import Java. util. Calendar;
import j avax. servlet. *;
import javax. servlet. http.*;
import j avax. servlet. jsp.*;
import javax. servlet. jsp.tagext.*;
import org. apache. jasper. runtime.*;

public class SimplePage_jsp extends HttpJspBase {   ;
    // method declaration       1
    String getSystemTime () {
        return Calendar. getInstance() .getTimef) .toString() ;
    }
    // variable declaration
    int i;

    static {
    }
    public SimplePage_jsp( ) {
    }

    private static boolean _jspx_inited = false;
    public final void _jspx_init() throws org.apache.jasper.JasperException {
    }
```

```
    public void _jspService(HttpServletRequest request;
        HttpServletResponse    response)
        throws java.io.IOException, ServletException {
        // the content has been removed
        .
        .
        .
    }
}
```

See how the variable and method are added as a class-level variable and method in the servlet class? Having the capability to declare a method, we will present the example in the following listings to also apply HTML encoding to the data from the database.The modified code is presented in the next listing.

JSP Page with a Method Declaration

```
<%
    String encodeHtmlTag (String tag) {
        if (tag==null)
            return null;
        int length = tag. length) );
        StringBuffer encodedTag = new StringBuffer(2 * length);
        for (int i=0; i<length; i++) {
            char c = tag.charAt(i) ;
            if (c=='<')
                encodedTag. append ("&lt; ") ;
            else if (c=='>' )
                encodedTag. append("&gt; ") ;
            else if (c=='&')
                encodedTag . append ( "&amp ; " ) ;
            else if (c==' " ' )
                encodedTag . append ( " Squot ; " ) ;
                //when trying to output text as tag's value as in
                // values="???".
            else if (c=='  ' )
                encodedTag. append( "  " ) ;
            else
                encodedTag. append(c);

        }
        return encodedTag. toString( );
    }
%>

<%@ page session=" false" %>
<%@ page import="java.sql.*" %>
<%
    try {
        Class . forName ( " sun . j dbc . odbc . JdbcQdbcDriver " ) ;
        System. out. println("JDBC driver loaded");
    }
    catch (ClassNotFoundException e) {
        System, out. println(e.toString());
    }
%>
```

```
<HTML>
<HEAD>
<TITLE>Display All Users</TITLE>
</HEAD>
<BODY>
<CENTER>
<BR><H2>Displaying All Users</H2>
<BR>
<BR>
<TABLE>
<TR>
<TH>First Name</™>
<TH>Last Name</TH>
<TH>User Name</TH>
<TH>Password</TH>
</TR>
<%
    String sql = "SELECT FirstName, LastName, UserNarae, Password" + " FROM Users";
    try {
        Connection con = 0riverManager.getConnection("jdbc:odbc:JavaWeb");

        Statement s = con.createStateitient();
        ResultSet rs = s.executeduery(sql);

        while (rs.next()) {
%>
<TR>
<TD><% out.print(encodeHtmlTag(rs.getString(1))); %></TD>
<TD><% out.print(encodeHtmlTag(rs.getString(2))); %></TD>
<TD><% out.print(encodeHtmlTag(rs.getString(3))); %></TD>
<TD><% out.print (encodeHtmlTag(rs.getstring (4))); %></TD>
</TR>
%>
    }
    rs.close) );
    s.close () ;
    con.close();
}
catch (SQLException e) {
}
catch (Exception e) {
}
%>
</TABLE>
</CENTER>
</BODY>
</HTML>
```

Writing Initialization and Clean-up Code

The code in above listings suffer from unnecessary repetition:The JSP page tries to load the JDBC driver every time the page is called. This is just a small example. In real life, you may have similar cases where you need to do initialization and clean-up; that is, you want a piece of code to be run only when the JSP servlet is first initialized or when it is destroyed.

Earlier you learned that every servlet generated from a JSP page must directly or indirectly implement the javax.servlet.jsp.JspPage interface. This interface has two methods: jspInit and jspDestroy. The JSP container calls the jspInit the first time the JSP servlet is initialized and calls the jspDestroy when the JSP servlet is about to be removed. These two methods provide a way for initialization and clean-up code. With declarations, you can override these two methods.

You can now modify the code in listing by moving the part that loads the JDBC driver to the jspInit method. The code is given in the next listing.

Utilizing the jspInit Method

```
>%!
    public void jspInit() {
        try {
            Class. forName(" sun. jdbc.odbc.JdbcOdbcDriver") ;
            System. out. println("JDBC driver loaded");
        }
        catch (ClassNotFoundException e) {
        System . out . println (e . toString ());
        }
    }

    String encodeHtmlTag (String tag) {
        if (tag==null)
            return null;
        int length = tag. length () ;
        StringBuffer encodedTag = new StringBuffer(2 * length);
        for (int i=0; i<length;
            char c = tag.charAt(i) ;
            if (c=='<')
                encodedTag. append ("&lt; ") ;
            else if (c=='>' )
                encodedTag . append ( "&gt ; " ) ;
            else if (c=='&' )
                encodedTag . append (" &amp ;");
            else if (c==' " ' )
                encodedTag . append ( "Squot ;");
                //when trying to output text as tag's value as in
                // values="???".
            else if (c==' ')
                encodedTag. append ("  ") ;
            else
                encodedTag. append(c);
        }
        return encodedTag. toString( );
    }

<%
<%@ page session="false" %>
<%@ page import="java.sql.*" %>
<HTML>
<HEAD> •
<TITLE>Display All Users</TITLE>
</HEAD>
<BODY>
```

```
<CENTER>
<BR><H2>Displaying All Users</H2>
<BR>
<BR>
<TABLE>
<TR>
<TH>First Name</TH>
<TH>Last Name</TH>
<TH>User Name</TH>
<TH>Password</TH>
</TR>
<%
String sql = "SELECT FirstName, LastName, UserName, Password" + " FROM Users";
    try {
        Connection con = DriverManager.getConnection("jdbc:odbc:JavaWeb");

        Statement s = con.createStatement();
        ResultSet rs = s.executeduery(sql);

        while (rs.next()) {
%>
<TR>
<TD><% out.print(encodeHtmlTag(rs.getString(1))); %></TD>
<TD><% out.print(eneodeHtmlTag(rs.getString(2) )); %></TD>
<TD><% out.print(encodeHtmlTag(rs.getString(3))); %></TD>
<TD><% out.print(encodeHtm!Tag(rs.getString(4))); %></TD>
</TR>
<%
        }
        rs.close() ;
        s.close();
        con.closef);
    }
    catch (SCILException e) {
    }
    catch (Exception e) {
    }
%>
</TABLE>
</CENTER>
</BODY>
</HTML>
```

The generated servlet is presented here. Note that some parts of the servlet code have been removed for brevity.

```
public class SimplePage_jsp extends HttpJspBase {
    public void ]'spInit() {
        try {
            Class .forName ("sun. jdbc. odbc.JdbcOdbcDriver " ) ;
            System.out. println("JDBC driver loaded");
        }
        catch (ClassNotFoundException e) {
            System. out. println(e.toString());
        }
```

```
    }
    String encodeHtmlTag(String tag) {
        .
        .
        .
    }

    static {
    }

    public Simp]ePage_jsp( ) {
    }

    private static boolean _jspx_inited = false;
    public final void _jspx_init() throws org. apache. jasper. JasperException {
    }

    public void _jspService(HttpServletRequest request,
        HttpServletResponse response)
        throws java.io.IOException, ServletException {
        :
        .
        .
    }
}
```

Expressions

Expressions are the last type of JSP scripting elements. Expressions get evaluated when the JSP page is requested and their results are converted into a String and fed to the print method of the out implicit object. If the result cannot be converted into a String, an error will be raised at translation time. If this is not detected at translation time, at request-processing time, a ClassCastException will be raised.

An expression starts with a <%— and ends with a %>.You don't add a semicolon at the end of an expression.

An example of an expression is given in the following listing.

Using an Expression

```
Current Time: <%= Java.util.Calendar.getInstance().getTime() %>
```

This expression will be translated into the following statement in the _jspService method of the generated servlet:

```
out.print( Java.util.Calendar.getInstance().getTime() );
```

The expression in the following listing is equivalent to the following scriptlet:

```
Current Time: <% out.print(java.util.Calendar.getInstancef).getTime()); %>
```

As you can see, an expression is shorter because the return value is automatically fed into the out.print.

As another example, the while block shown in above listing can be re-written as follows using expressions:

```
while (rs.nextf)) {
%>
<TR>
<TD><%= encodeHtmlTag(rs.getString(1)) %></TD>
```

```
<TD><%= encodeHtmlTag(rs.getString(2)) %></TD>
<TD><%= encodeHtmlTag(rs.getString(3)) %></TD>
<TD><%= encodeHtmlTag(rs.getString(4)) %></TD>
</TR>
<%
}
```

Standard Action Elements

Standard action elements basically are tags that can be embedded into a JSP page. At compile time, they also are replaced by Java code that corresponds to the predefined task.

The following is the list of JSP standard action elements:

- jsp:useBean
- jsp:setProperty
- jsp:getProperty
- jsprparam
- jsp:include
- jsp: forward
- jsp:plugin
- jsp:params
- jsp:fallback

The jsp:useBean,jsp:setProperty, and jsp:getProperty elements are related to Bean. The jsp:param element is used in the jsp:include, jsp:forward, and jsp:plugin elements to pro-vide information in the name/value format, and therefore will be discussed with the three other elements.

jsp:include

The jsp:include action element is used to incorporate static or dynamic resources into the current page. This action element is similar to the include directive, but jsp:include provides greater flexibility because you can pass information to the included resource.

The syntax for the jsp:include action element has two forms. For the jsp:include element that does not have a parameter name/value pair, the syntax is as follows:

```
<jsp:include paQe="relativeURL" flush="true"/>
```

If you want to pass information to the included resource, use the second syntax:

```
<jsp:include page="relativeURL" flush="true">
    ( <jsp:param ..../>)*
</jsp:include>
```

The page attribute represents the URL of the included resource in the local server. The flush attribute indicates whether the buffer is flushed. In JSP 1.2, the value of the flush attribute must be true.

In the second form, the * indicates that there can be zero or more elements in the brackets. This means that you also can use this form even though you are not passing any information to the included resource.

jsp:forward

The jsp:forward action element is used to terminate the execution of the current JSP page and switch control to another resource. You can forward control either to a static resource or a dynamic resource.

The syntax for the jspiforward action element has two forms. For the jsp:forward element that does not have a parameter name/value pair, the syntax is as follows:

```
<jsp:forward page="relativel/fIL"/>
```

If you want to pass information to the included resource, use the second syntax:

```
<jsp:forward page="relativeURL">
    ( <jsp:param ,../>)*
</jsp:include>_
```

The page attribute represents the URL of the included resource in the local server.

jsp:plugin

The jsp:plugin action element is used to generate HTML <OBJECT> or <EMBED> tags containing appropriate construct to instruct the browser to download the Java Plugin software, if required, and initiates the execution of a Java applet or aJavaBeans component specified.

jsp:params

The jsp:params action element can occur only as part of the <jsp:plugin> action.

jsp:fallback

The jsp: fallback action element can occur only as part of the <jsp:plugin> action.

Comments

Commenting is part of good programming practice. You can write two types of comments inside a JSP page:

- Comments that are to be displayed in the resulting HTML page at the client browser
- Comments used in the JSP page itself

For comments that are meant to be displayed in the HTML page, you use the comments tags in HTML. This kind of comment must be sent as normal text in a JSP page. For example, the following code sends an HTML comment to the browser:

```
<%
    out. println ("<!— Here is a comment —>");
%>
```

This is equivalent to the following:

```
<!— Here is a comment —>
```

For comments in the JSP page itself, you use the <%— . . . —%> tag pair. For example, here is a JSP comment:

```
<%—
    Here is a comment
—%>
```

A JSP comment can contain anything except the closing tag. For example, this is an illegal comment:

```
<%— Here is a comment —%>
```

Converting into XML Syntax

XML is getting more and more important in the computing world. JSP pages can be represented using XML, and representing a JSP page as an XML document presents the following benefits:

- The content of the JSP page can be validated against a set of descriptions.
- The JSP page can be manipulated using an XML tool.
- The JSP page can be generated from a textual representation by applying an XML transformation.

Except for the standard action elements, the other JSP programming elements have been presented using the alternative syntax throughout this chapter.

Directives

The non-XML syntax for a JSP directive takes the following form:

<%@ directive (attribute="value")* %>

The XML syntax for a directive is as follows:

<jsp:directiveidirectiveName attribute_list />

Scripting Elements

A scripting element can be one of the following: declaration, scriptlet, or expression. The XML syntax for each of the three is given in the following sections.

Declarations

The alternative syntax for a declaration is as follows:

<%1 declaration code %>

This is equivalent to the following XML syntax:

<jsp:declaration> declaration code </jsp:declaration>

Scriptlets

The alternative syntax for a scriptlet is as follows:

<% scriptlet code %>

This is equivalent to the following XML syntax:

<jsp:scriptlet> scriptlet code </jsp:scriptlet>

Expressions

The alternative syntax for an expression is as follows:

<%= expression %>

This is equivalent to the following XML syntax:

<jsp:expression> *expression* </jsp:expression>

Template Data

The <jsp:text> XML tag is used to enclose template data in JSP. The syntax is as follows:

<jsp:text> text </jsp:text>

Revision

1. JSP page has elements and template data. The elements, which also are called JSP tags, make up the syntax and semantics of JSP.
2. Directives are messages to the JSP container containing information on how the JSP container must translate a JSP page into a corresponding servlet.
3. The language attribute specifies the scripting language used in the JSP page.
4. The info attribute allows you to insert any string that later can be retrieved using the getServletInfo method.
5. The contentType attribute defines the Multipurpose Internet Mail Extension (MIME) type of the HTTP response sent to the client.
6. The extends attribute defines the parent class that will be inherited by the generated servlet.
7. The import attribute is similar to the import keyword in a Java class or interface.
8. By default, a JSP page's content is buffered to increase performance.
9. The errorPage attribute specifies the URL of an error page that will be displayed if an uncaught exception occurs in this current JSP page.
10. The include directive is the second type of the JSP directive elements. This directive enables JSP page authors to include the contents of other files in the current JSP page.
11. The taglib, or tag library, directive can be used to extend JSP functionality.
12. Scriptlets are the code blocks of a JSP page.

13. Declarations allow you to declare methods and variables that can be used from any point in the JSP page.
14. The JSP page tries to load the JDBC driver every time the page is called.
15. Expressions get evaluated when the JSP page is requested and their results are converted into a String and fed to the print method of the out implicit object.
16. The jsp:include action element is used to incorporate static or dynamic resources into the current page.
17. The jsp:forward action element is used to terminate the execution of the current JSP page and switch control to another resource.
18. The jsp:plugin action element is used to generate HTML <OBJECT> or <EMBED> tags containing appropriate construct to instruct the browser to download the Java Plugin software, if required, and initiates the execution of a Java applet or aJavaBeans component specified.
19. Commenting is part of good programming practice. You can write two types of comments inside a JSP page: Comments that are to be displayed in the resulting HTML page at the client browser and Comments used in the JSP page itself.
20. JSP pages can be represented using XML, and representing a JSP page as an XML document.
21. A scripting element can be one of the following: declaration, scriptlet, or expression.

Multiple Choice Questions

1. JSP directives are ________________.
   (a) Page directive (b) Include directive
   (c) Taglib directive
2. ________ is a JSP tag that will be used in almost every JSP file.
   (a) Page directive (b) Include directive
   (c) Taglib directive
3. An error page can also be displayed using the ________ directi ve.
   (a) Page directive (b) Include directive
   (c) Taglib directive
4. The <%@Page%> directive applies to an entire jsp file and any of its static include files, which together are called a __________
   (a) Session unit (b) Translation unit
   (c) Application unit
5. <jsp: include> action allows a ________ resource to be included in the JSP at request time.
   (a) static (b) dynamic
   (c) static or dynamic
6. The include directive includes ________ file in a JSP file.
   (a) static (b) dynamic
   (c) HTML
7. The JSP model revolves around _______.
   (a) ASP (b) EJB
   (c) beans

Answers

1. a	2. a	3. a	4. b	5. c
6. a	7. c			

Subjective Questions

1. Describe directives.
2. What are scripting elements?
3. What are Declarations?
4. What are expressions?
5. How would you convert a JSP file to XML syntax file?
6. Write short notes on the following:
   language attribute
   info attribute
   contentType attribute
   extends attribute
   import attribute
   errorPage atribute
   jsp.include
   jsp.forward
   jsp.plugin

9 Java Servlets

Introduction

The servlet technology is the foundation of web application development using the Java programming language. It is one of the most important Java technologies, and it is the underlying technology for another popular Java technology for web application development: JavaServer Pages (JSP).Therefore, understanding the servlet technology and its architecture is important if you want to be a servlet developer. Even if you plan to develop your Java web application using JSP pages alone, understanding the servlet technology helps you build a more efficient and effective JSP application.

The Benefits of Servlets

When it first emerged, this great thing we call the Internet consisted of only static contents written using Hypertext Markup Language (HTML). At that time, anyone who could author HTML pages was considered an Internet expert.

This did not last long, however.

Soon dynamic web contents were made possible through the Common Gateway Interface (CGI) technology. CGI enables the web server to call an external program and pass HTTP request information to that external program to process the request. The response from the external program is then passed back to the web server, which forwards it to the client browser. CGI programs can be written in any language that can be called by the web server. Over the course of time, Perl became the most popular language to write CGI programs.

As the Internet became more and more popular, however, the number of users visiting a popular web site increased exponentially, and it became apparent that CGI had failed to deliver scalable Internet applications. The flaw in CGI is that each client request makes the web server spawn a new process of the requested CGI program. As we all know, process creation is an expensive operation that consumes a lot of CPU cycles and computer memory.

Gradually, new and better technologies will replace CGI as the main technology for web application development. The world has witnessed the follow-ing technologies trying to dominate web development:

- **ColdFusion.** Allaire's ColdFusion provides HTML-like custom tags that can be used to perform a number of operations, especially querying a data-base. This technology had its glamorous time in the history of the World Wide Web as the main technology for web application programming. Its glorious time has since gone with the invention of other technologies.
- **Server-side JavaScript (SSJS).** SSJS is an extension of the JavaScript language, the scripting language that still rules client-side web programming. SSJS can access Java classes deployed at the server side using the Live Wire technology from Netscape.
- **PHP.** PHP is an exciting open-source technology that has matured in recent years. The technology provides easy web application development with its session management and includes some built-in functionality, such as file upload. The number of programmers embracing PHP as their technology of choice has risen sharply in recent years.

- **Servlet.** The servlet technology was introduced by Sun Microsystems in 1996.
- **JavaServer Pages (JSP).**JSP is an extension of the servlet technology. This, too, is the center of attention in this book.
- **Active Server Pages (ASP).** Microsoft s ASP employs scripting technologies that work in Windows platforms, even though there have been efforts to port this technology to other operating systems. Windows ASP works with the Internet Information Server web server. This technology will soon be replaced by Active Server Pages.NET.
- **Active Server Pages.NET (ASP.NET).** This technology is part of Microsoft's .NET initiative. Interestingly, the .NET Framework employs a runtime called the Common Language Runtime that is very similar to Java Virtual Machine and provides a vast class library available to all .NET languages and from ASP.NET pages. ASP.NET is an exciting technology. It introduced several new technologies including state management that does not depend on cookies or URL rewriting.

In the past, ASP and servlet/JSP have been the main technologies used in web application development. With the release of ASP.NET, it is not hard to predict that this technology will become the servlet/JSP s main competitor. ASP (and ASP.NET) and servlet/JSP each have their own fans, and it is not easy to pre-dict which one will come out the winner. The most likely outcome is that neither will be an absolute winner that corners the market; instead the tech-nologies will probably run head-to-head in the coming years.

Servlet (and JSP) offers the following benefits that are not necessarily avail-able in other technologies:

- **Performance.** The performance of servlets is superior to CGI because there is no process creation for each client request. Instead, each request is handled by the servlet container process. After a servlet is finished pro-cessing a request, it stays resident in memory, waiting for another request.
- **Portability.** Similar to other Java technologies, servlet applications are portable.You can move them to other operating systems without serious hassles.
- **Rapid development cycle. As** a Java technology, servlets have access to the rich Java library, which helps speed up the development process.
- **Robustness.** Servlets are managed by the Java Virtual Machine. As such, you don't need to worry about memory leak or garbage collection, which helps you write robust applications.
- **Widespread acceptance.** Java is a widely accepted technology. This means that numerous vendors work on Java-based technologies. One of the advantages of this widespread acceptance is that you can easily find and purchase components that suit your needs, which saves precious development time.

Servlet Application Architecture

A servlet is a Java class that can be loaded dynamically into and run by a special web server. This servlet-aware web server is called a *servlet container,* which also was called a *servlet engine* in the early days of the servlet technology.

Servlets interact with clients via a request-response model based on HTTP. Because servlet technology works on top of HTTP, a servlet container must support HTTP as the protocol for client requests and server responses. However, a servlet container also can support similar protocols, such as HTTPS (HTTP over SSL) for secure transactions.

In a JSP application, the servlet container is replaced by a JSP container. Both the servlet container and the JSP container often are referred to as the *web con-tainer* or *servlet /JSP container,* especially if a web application consists of both servlets and JSP pages.

A servlet application also can include static content, such as HTML pages and image files. Allowing the servlet container to serve static content is not preferable because the content is faster if served by a more robust HTTP server, such as the Apache web server or Microsoft Internet Information Server. As such, it is common practice to put a web server at the front to handle all client requests. The web server serves static content and passes to the servlet containers all client requests for servlets.

How a Servlet Works?

A servlet is loaded by the servlet container the first time the servlet is requested. The servlet then is forwarded the user request, processes it, and returns the response to the servlet container, which in turn sends the response back to the user. After that, the servlet stays in memory waiting for other requests—it will not be unloaded from the memory unless the servlet container sees a shortage of memory. Each time the servlet is requested, however, the servlet container compares the timestamp of the loaded servlet with the servlet class file. If the class file timestamp is more recent, the servlet is reloaded into memory. This way, you don't need to restart the servlet container every time you update your servlet.

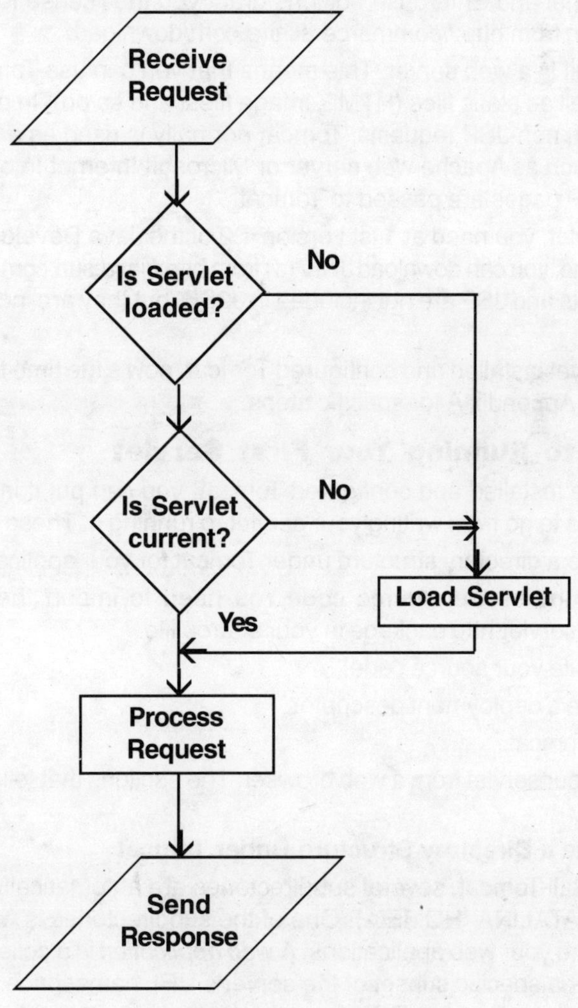

The Tomcat Servlet Container

A number of servlet containers are available today. The most popular one—and the one recognized as the official servlet/JSP container—is Tomcat. Originally d'esigned by Sun Microsystems, Tomcat source code was handed over to the Apache Software Foundation in October 1999. In this new home, Tomcat was included as part of the Jakarta Project, one of the projects of the Apache Software Foundation. Working through the Apache process, Apache, Sun, and other companies—with the help of volunteer programmers worldwide— turned Tomcat into a world-class servlet reference implementation. Two months after the handover, Tomcat version 3.0 was released. Tomcat went through several *3.x* releases until version 3.3 was introduced.

The successor of version 3.3 is the current version, version 4.0. The 4.0 servlet container (Catalina) is based on a completely new architecture and has been developed from the ground up for flexibility and performance. Version 4.0 implements the Servlet 2.3 and JSP 1.2 specifications, and it is this version you will be using in this book.

Another popular servlet container is JRun from Allaire Corporation. JRun is available in three editions: Developer, Professional, and Enterprise. The Developer edition is free but not licensed for deployment. The Professional and Enterprise editions grant you the license for deployment with a fee. You can download JRun from http://commerce.allaire.com/download.

Tomcat by itself is a web server. This means that you can use Tomcat to service HTTP requests for servlets, as well as static files (HTML, image files, and so on). In practice, however, since it is faster for non-servlet, non-JSP requests, Tomcat normally is used as a module with another more robust web server, such as Apache web server or Microsoft Internet Information Server. Only requests for servlets or JSP pages are passed to Tomcat.

To write a servlet, you need at 1 ist version 1.2 of the Java Development Kit. If you have not already downloaded one, you can download JDK 1.2 from http://java.sun.com/j2se.The reference implementation for both servlets and JSP are not included in J2SE, but they are included in Tomcat. Tomcat is written purely in Java.

If you haven't yet installed and configured Tomcat, now's the time to do it. If you need help with these tasks, refer to Appendix A for specific steps.

Six Steps to Running Your First Servlet

After you have installed and configured Tomcat, you can put it into service. Basically, you need to follow six steps to go from writing your servlet to running it. These steps are summarized as follows:

1. Create a directory structure under Tomcat for your application.
2. Write the servlet source code.You need to import the javax.servlet package and the javax.servlet.http package in your source file.
3. Compile your source code.
4. Create a deployment descriptor.
5. Run Tomcat.
6. Call your servlet from a web browser. The sections that follow walk you through each of these steps.

Step 1: Create a Directory Structure Under Tomcat

When you install Tomcat, several subdirectories are automatically created under the Tomcat home directory (%CATALINA_HOME%). One of the subdirectories is webapps.The webapps directory is where you store your web applications. A *web application* is a collection of servlets and other content installed under a specific subset of the server's URL namespace. A separate directory is dedicated for each servlet application. Therefore, the first thing to do when you build a servlet application is create an application directory. To create a directory structure for an application called myApp, follow these steps:

1. Create a directory called myApp under the webapps directory. The directory name is important because this also appears in the URL to your servlet.
2. Create the and WEB-INF directories under myApp, and create a directory named classes under WEB-INF. The directory classes under WEB-INF is for your Java classes. If you have HTML files, put them directly under the myApp directory. You may also want to create a directory called images under myApp for all your image files.

   Note that the examples, manager, ROOT, tomcat-doc, and webdav directories are for applications that are created automatically when you install Tomcat.

Step 2: Write the Servlet Source Code

In this step, you prepare your source code. You can write the source code yourself using your favorite text editor.

The code in the following listing shows a simple servlet called TestingServlet. The file is named TestingServlet.java. The servlet sends a few HTML tags and some text to the browser. For now, don't worry if you haven't got a clue about how it works.

TestingServlet.java

```
import  javax.servlet.*;
import  javax.servlet.http.*;
import  java.io.*;
import  java.util.*;

public class TestingServlet extends HttpServlet {

    public void doGet(HttpServletRequest request,
        HttpServletResponse response)
        throws ServletException, IOException {

        PrintWriter out = response.getWriter();
        out.println("<HTML>");
        out.println("<HEAD>");
        out.println("<TITLE>Servlet Testing</TITLE>");
        out.println("</HEAD>");
        out.println("<BODY>");
        out.printlnf"Welcome to the Servlet Testing Center");
        out.println("</BODY>");
        out.println("</HTML>");
    }
}
```

Now, save your TestingServlet.java file to the WEB-INF/classes directory under myApp. Placing your source code here will make it inaccessible from a web browser. Static files, such as HTML files and image files, should be placed directly under the myApp directory or a directory under it.

Step 3: Compile Your Source Code

For your servlet source code to compile, you need to include the path to the servlet.jar file in your CLASSPATH environment variable. The servlet.jar is located in the common\lib\ subdirectory under %CATALINA_HOME%.

For example, if you installed Tomcat under the C:\drive on Windows and you named the install directory tomcat, type the following command from the directory where TestingServlet.java resides.

```
javac classpath C:\tomcat\common\lib\servlet.jar TestingServlet.java
```

Alternatively, to save you typing the class path every time you compile your source code, you can add the complete path to the servlet.jar file to your CLASSPATH environment variable. Again, if you have installed Tomcat under C:\and named the install directory tomcat, you must add C:\tomcat\common\lib\servlet.jar to the CLASSPATH environment variable. Afterward, you can compile your source by simply typing the following.

```
javac TestingServlet.java
```

If you are using Windows, remember that the new environment variable takes effect only for new console windows. In other words,after changing a new environment variable, open a new console window for typing in your com-mand lines.

Step 4: Create the Deployment Descriptor

A deployment descriptor is an optional component in a servlet application. The descriptor takes the form of an XML document called web.xml and must be located in the WEB-INF directory of the servlet application. When present, the deployment descriptor contains configuration settings specific to that application.

To create the deployment descriptor, you now need to create a web.xml file and place it under the WEB-INF directory under myApp.

The web.xml for this example application must have the following content.

```
<?xml version="1.0" encoding="ISO-8859-1"?>

<!DOCTYPE web-app
    PUBLIC "-//Sun Microsystems, Inc.//DTD Web Application 2.3//EN"
    "http://java.sun.com/dtd/web -app_2_3.dtd">

<web-app>
    <servlet>
    <servlet-name>Testing</servlet-name>
    <servlet-class>TestingServlet</servlet-class>
</servlet>
</web-app>
```

The web.xml file has one element—web-app.You should write all your servlets under <web-app>. For each servlet, you have a <servlet> element and you need the <servlet-name> and <servlet-class> elements. The <servlet-name> is the name for your servlet, by which it is known Tomcat. The <servlet-class> is the compiled file of your servlet without the .class extension.

Having more than one servlet in an application is very common. For every servlet, you need a <servlet> element in the web.xml file. For example, the following shows you how web.xml looks if you add another servlet called Login:

```
<?xml version="1.0" encoding="ISO-8859-1"?>

<!DOCTYPE web-app
    PUBLIC "-//Sun Microsystems, Inc.//DTD Web Application 2.3//EN"
    "http: / / j ava. sun. com/dtd/web - app_2_3. dtd" >

<web-app>
    <servlet>
        <servlet-name>Testing</servlet-name>
        <servlet-class>TestingServlet</servlet-class>
    </servlet>
    <servlet>
        <servlet-name>Login</servlet-name>
        <servlet-class>LoginServlet</servlet-class>
```

</servlet>
</web-app>

Step 5: Run Tomcat

If Tomcat is not already running, you need to start it.

Step 6: Call Your Servlet from a Web Browser

Now, you can call your servlet from a web browser. By default, Tomcat runs on port 8080 in the myApp virtual directory under the servlet subdirectory. The servlet that you wrote in the preceding steps is named Testing. The URL for that servlet has the following format:

http: / /domain - nam&/ victual- directory/servlet/servlet-name

Any static file can be accessed using the following URL:

http: //domain-name/virtual-directory/staticFile.html

For example, a Logo.gif file under the myApp /images/ directory can be accessed using the following URL.

http: //domain -name /virtual -directory /images /Logo.gif

If you run the web browser from the same computer as Tomcat, you can replace the *domain-name* part with "localhost". In that case, the URL for your servlet is

http:/ /localhost :8080/myApp/servlet/Testing

In the deployment descriptor you wrote in Step 4, you actually mapped the servlet class file called TestingServlet with the name "Testing," so that your servlet can be called by specifying its class file (TestingServlet) or its name (Testing) . Without a deployment descriptor, you must call the servlet by speci-fying its class name; that is, TestingServlet. This means that if you did not write a deployment descriptor in Step 4, you need to use the following URL to call your servlet:

http:/ /localhost :8080/myApp/ servlet /TestingServlet

Typing the URL in the Address or Location box of your web browser will give you the string "Welcome to the Servlet Testing Center."

Congratulations. You have just written your first servlet.

If you don't want to type the port number each time, you can change the default port of Tomcat so that it runs on port 80, the default port for a web server.

Inside Servlets

Two packages are available for servlet programmers: javax.servlet and javax.servlet.http.The first one contains basic classes and interfaces that you can use to write servlets from the scratch. The second package, javax.servlet.http, offers more advanced classes and interfaces that extend classes and interfaces from the first package. It is much more convenient to program using the second package.

When you learn something, it is best to start with the basics and build a strong foundation. For example, understanding the javax.servlet.Servlet inter-face is very important because it encapsulates the life cycle methods of a servlet and it is the interface that all servlets must implement. You also need to know the servlets context, which represents a servlets environment, and the servlets configuration. Because of the importance of these items, this chapter introduces you to members of the javax.servlet package. In this chapter, you also will see that oftentimes several ways exist to do the same thing.

After a few examples, I will introduce you to the GenericServlet class, a member of the javax.servlet package that acts as a wrapper for the javax.servlet.Servlet interface. Extending this class makes your code simpler because you need to provide implementations only for methods that you need to use.

To run each example in this chapter, you need to compile the source code and copy the resulting class file into the classes directory under the WEB-INF directory of your application.

The rest of this chapter explains and uses the interfaces and classes of the javax.servlet package. The chapters that follow focus more on the second package.

The javax.servlet Package

The javax.servlet package contains seven interfaces, three classes, and two exceptions. Instead of using the conventional approach by explaining each interface and class in alphabetical order we present the discussions based on functions and offer examples that demonstrate each function.

The seven interfaces are as follows:

- RequestDispatcher
- Servlet
- ServletConfig
- ServletContext
- ServletRequest
- ServletResponse
- SingleThreadModel

The three classes are as follows:

- GenericServlet
- ServletInputStream
- ServletOutputStream

And, finally, the exception classes are these:

- ServletException
- UnavailableException

A Servlet's Life Cycle

Let there be servlet.This interface in the javax.servlet package is the source of all activities in servlet programming. Servlet is the central abstraction of the Java servlet technology. Every servlet you write must implement this javax.servlet.Servlet interface, either directly or indirectly. The life cycle of a servlet is determined by three of its methods: init, service, and destroy.

The init( ) Method

The init method is called by the servlet container after the servlet class has been instantiated. The servlet container calls this method exactly once to indicate to the servlet that the servlet is being placed into service. The init method must complete successfully before the servlet can receive any requests.

You can override this method to write initialization code that needs to run only once, such as loading a database driver, initializing values, and so on. In other cases, you normally leave this method blank.

The signature of this method is as follows:

```
public void init(ServletConfig config) throws ServletException
```

The init method is important also because the servlet container passes a ServletConfig object, which contains the configuration values stated in the web.xml file for this application.

This method also can throw a ServletException. The servlet container cannot place the servlet into service if the init method throws a ServletException or the method, does not return within a time period defined by the web server.

The service( ) Method

The service method is called by the servlet container after the servlet's init method to allow the servlet to respond to a request. Servlets typically run inside multithreaded servlet containers that can handle multiple requests concurrently. Therefore, you must be aware to synchronize access to any shared resources, such as files, network connections, and the servlets class and instance variables. For example, if you open a file and write to that file from a servlet, you need to remember that a different thread of the same servlet also can open the same file.

This method has the following signature:

```
public void service(ServletRequest request, ServletResponse response)
    throws ServletException, java.io.IOException
```

The servlet container passes a ServletRequest object and the ServletResponse object.The ServletRequest object contains the client's request and the ServletResponse contains the servlet's response. These two objects are important because they enable you to write custom code that determines how the servlet services the client request.

The service method throws a ServletException if an exception occurs that interferes with the servlet's normal operation. The service method also can throw a java.io.IOException if an input or output exception occurs during the execution of this method.

As the name implies, the service method exists so that you can write code that makes the servlet function the way it is supposed to.

The destroy( ) Method

The servlet container calls the destroy method before removing a servlet instance from service. This normally happens when the servlet container is shut down or the servlet container needs some free memory.

This method is called only after all threads within the servlet's service method have exited or after a timeout period has passed. After the servlet container calls this method, it will not call the service method again on this servlet.

The destroy method gives the servlet an opportunity to clean up any resources that are being held (for example, memory, file handles, and threads) and make sure that any persistent state is synchronized with the servlet's current state in memory.

The signature of this method is as follows:

```
public void destroy()
```

Demonstrating the Life Cycle of a Servlet

The following listing contains the code for a servlet named PrimitiveServlet, a very simple servlet that exists to demonstrate the life cycle of a servlet. The PrimitiveServlet class implements javax.servlet.Servlet (as all servlets must) and provides implementations for all the five methods of servlet. What it does is very simple. Each time any of the init, service, or destroy methods is called, the servlet writes the method's name to the console.

Primitive Servlet .Java

```
import javax. servlet.*;
import java.io.IOException;

public class PrimitiveServlet implements Servlet {

    public void init(ServletConfig config) throws ServletException {
        System . out . print In ( " init " ) ;
    }
```

```
    public void service(ServletRequest request, ServletResponse response)
        throws ServletException, IOException {
        System . out . print ln ( " service " ) ;
    }

    public void destroy () {
        System. out. println( "destroy" ) ;
    }

    public String getServlet!nfo() {
        return null;
    }

    public ServletConfig getServletConfig() {
        return null;
    }
}
```

After you compile the source code into the myApp\WEB-INF\classes directory, add the servlet to the web.xml under the name Primitive, as shown in the following listing.

The web.xml File for PrimitiveServlet

```
<?xml version="1.0" encoding="ISO-8859-1"?>

<!DOCTYPE web-app
    PUBLIC "-//Sun Microsystems, Inc.//DID Web Application 2.3//EN"
    "http://java.sun.com/dtd/web-app_2_3.dtd">

<web-app>
    <servlet>
        <servlet-name>Primitive</servlet-name>
        <servlet-class>PrimitiveServlet</servlet-class>
    </servlet>
</web-app>
```

You should then be able to call this servlet from your browser by typing the following URL:

http://localhost:8080/myApp/servlet/Primitive

The first time the servlet is called, the console displays these two lines:

```
init
service
```

This tells you that the init method is called, followed by the service method. However, on subsequent requests, only the service method is called. The servlet adds the following line to the console:

```
service
```

This proves that the init method is called only once.

What are the getServletInfo and getServletConfig doing in the above listing? Nothing. They can be useful, but in the PrimitiveServlet class, they are just there to meet the specification that a class must provide implementations for all methods in the interface it implements.

You can return any string in the getServletInfo method, such as your company name or the author name or other information deemed necessary. Other people might extend your servlet class and might want to know what useful information the designer of the servlet has provided.

The getServletConfig is more important. We will see how it can be of use next.

Obtaining Configuration Information

The servlet specification allows you to configure your application. We present an example that demonstrates how you can retrieve configuration information from the application web.xml file.

For each servlet registered in the web.xml file, you have the option of specifying a set of initial parameter name/value pairs that you can retrieve from inside the servlet. The following web.xml file contains a servlet called ConfigDemo whose class is named ConfigDemoServlet.class. The servlet has two initial parameter name/value pairs. The first parameter is named adminEmail and its value is admin@brainysoftware.com.The second parameter is named adminContactNumber and the value for this parameter is 04298371237.

```
<?xml version="1.0" encoding="ISO-8859-1"?>
<!DOCTYPE web-app
    PUBLIC "-//Sun Microsystems, Inc.//DID Web Application 2.2//EN"
    "http://Java.sun.com/j2ee/dtds/web-app_2.2.dtd">

<web-app>
<servlet>
    <servlet-name>ConfigDemo</servlet-name>
    <servlet-class>ConfigDemoServlet</servlet-class>
    <init-param>
        <param-name>adminEmail</param-name>
        <param-value>admin@brainysoftware.com</param-value>
    </init-param>
    <init-param>
        <param-name>adminContactNumber</param-name>
        <param-value>04298371237</param-value>
    </init-param>
</servlet>
</web-app>
```

Why would you want to use an initial parameter? For practicality. Hardcoding information in the servlet code means that you have to recompile the servlet if the information changes. A web.xml file is plain text.You can edit its content easily using a text editor.

The code that retrieves the initial parameter name and values is given in the following listing.

To retrieve initial parameters, you need the ServletConfig object passed by the servlet container to the servlet. After you get the ServletConfig object, you then can use its getInitParameterNames and getInitParameter methods. The getInitParameterNames does not take an argument and returns an Enumeration containing all the parameter names in the ServletConfig object. The getInitParameter takes a String containing the parameter name and returns a String containing the value of the parameter. Because the servlet container passes a ServletConfig object to the init method, it is easiest to write the code in the init method. The code in the following listing loops through the Enumeration object called parameters that is returned from the getInitParameterNames method. For each parameter, it outputs the parameter name and value. The parameter value is retrieved using the getInitParameter method.

Retrieving Initial Parameters

```
import javax. servlet.*;
import Java. util. Enumeration;
import java.io.IOException;

public class ConfigDemoServlet implements Servlet {
```

```
public void init (ServletConfig config) throws ServletException {
    Enumeration parameters = config. getInitParameterNames() ;
    while (parameters. hasMoreElements() ) {
        String parameter = (String) parameters. nextElement( );
        System. out. println(" Parameter name : " + parameter);
        System. out. println( "Parameter value :  " +
            config. getInitParameter(parameter) ) ;
    }
}

public void destroy () {
}

public void service(ServletRequest request, ServletResponse response)
    throws ServletException,  IOException {
}

public String getServletInfo() {
    return null;
}

public ServletConfig getServletConfig () {
    return null;
}
}
```

The output of the code in the console is as follows:

```
Parameter name  : adminContactNumber
Parameter value  : 04298371237
Parameter name  : adminEmail
Parameter value  : admin@brainysoftware.com
```

Preserving the ServletConfig

The code in the last listing shows how you can use the ServletConfig object in the init method. Sometimes, however, you may want more flexibility. For example, you may want to have access to the ServletConfig object from the service method, when you are servicing the user. In this case, you need to pre-serve the ServletConfig object to a class level variable. This task is not difficult. You need to create a ServletConfig object variable and set it to the ServletConfig object returned by the servlet container in the init method.

The next listing gives the code that preserves the ServletConfig object for later use. First, you need a variable for the ServletConfig object.

```
ServletConfig ServletConfig;
```

Then, in the init method, you write the following code:

```
ServletConfig = config;
```

Now the ServletConfig variable references the ServletConfig object returned by the servlet container. The getServletConfig method is provided to do just that: return the ServletConfig object.

```
public ServletConfig getServletConfig() {
    return ServletConfig;
}
```

If you extend the ReserveConfigServlet class, you can still retrieve the ServlerConfig object by calling the getServletConfig method.

Preserving the ServletConfig Object

```
import javax. servlet.*;
import java.io.IOException;

public class ReserveConfigServlet implements Servlet {
    ServletConfig servletConfig;

public void init(ServletConfig config) throws ServletException {
    servletConfig = config;
}

public void destroy () {
}

public void service(ServletRequest request, ServletResponse response)
    throws ServletException, IOException {
}

public String getServletInfo() {
    return null;
}

public ServletConfig getServletConfig() {
    return servletConfig;
}
}
```

The Servlet Context

In servlet programming, the servlet context is the environment where the servlet runs. The servlet container creates a ServletContext object that you can use to access information about the servlet's environment.

A servlet also can bind an object attribute into the context by name. Any object bound into a context is available to any other servlet that is part of the same web application.

How do you obtain the ServletContext object? Indirectly, from the ServletConfig object passed by the servlet container to the servlet's init method. The ServletConfig interface has a method called getServletContext that returns the ServletContext object.You then can use the ServletContext interface's various methods to get the information you need. These methods include the following:

- **getMajorVersion.** This method returns an integer representing the major version for the servlet API that the servlet container supports. If the servlet container supports the servlet API version 2.3, this method will return 2.
- **getMinor Version.** This method returns an integer representing the minor version of the servlet API that the servlet container supports. For the servlet API version 2.3, this method will return 3.
- **getAttributeNames.** This method returns an enumeration of strings representing the names of the attributes currently stored in the ServletContext.
- **getAttribute.** This method accepts a String containing the attribute name and returns the object bound to that name.
- **set Attribute.** This method stores an object in the ServletContext and binds the object to the given name. If the name already exists in the ServletContext, the old bound object will be replaced by the object passed to this method.

- **removeAttribute.** This method removes from the ServletContext the object bound to a name. The removeAttribute method accepts one argument: the name of the attribute to be removed.

The code in the next listing shows a servlet named ContextDemoServlet that retrieves some of the servlet context information, including attribute names and values, minor and major versions of the servlet container, and the server info.

Retrieving Servlet Context Information

```
import  javax. servlet.*;
import  Java. util. Enumeration;
import  java.io.IOException;

public class ContextDemoServlet implements Servlet {
    ServletConfig servletConfig;

    public void init (ServletConfig config) throws ServletException {
        servletConfig = config; }
    }

    public void destroy () {
    }

    public void service(ServletRequest request, ServletResponse response)
        throws ServletException,  IOException {

    ServletContext servletContext = servletConfig.getServletContext() ;
    Enumeration attributes = servletContext. getAttributeNames() ;
    while (attributes. hasMoreElements()) {
        String attribute = (String) attributes. nextElement( );
        System. out. println( “Attribute name : “ + attribute);
        System. out. println(“ Attribute value : “ + servletContext. getAttribute(attribute)) ;
    }
        System. out. println( “Major version : “ + servletContext. getMajorVersion() ) ;
        System. out. println( “Minor version : “ + servletContext. getMinorVersion() ) ;
        System. out. println( “Server info : “ + servletContext. getServerInfo() );
    }
    public String getServletInfof) {
    return null;
    }

    public ServletConfig getServletConfig() {
    return null;
    }
}
```

The output of the code is as follows. This output may be different on your computer, depending on the version of Tomcat you are using, the operating system, and so on.

```
Attribute     name : javax.servlet.context.tempdirAttribute value :
..\work\localhost\myApp
Attribute     name : org.apache.catalina.resources
Attribute     value : org.apache.naming.resources.ProxyDirContext@24e2e3
Attribute     name : org.apache.catalina.WELCOME_FILES
Attribute     value : [Ljava.lang.String;@2l j?e0
Attribute     name : org.apache.catalina.jsp_classpath
```

```
Attribute      value : C:\tomcat4\webapps\myApp\WEB-INF\classes;
.
.
Major version : 2Minor version : 3
Server info : Apache Tomcat/4.0-b5
```

Sharing Information Among Servlets

For some applications, you want to make certain types of information available to all the servlets. You can share this information — such as a database connection string or a page count — among the servlets by using attributes in the ServletContext object.

The following example uses two servlets: AttributeSetterServlet and Display AttributesServlet. The AttributeSetterServlet servlet, shown in the following listing, binds the name password to a String object containing the word "ding-dong". The servlet does this by first obtaining the ServletContext object from the ServletConfig object passed by the servlet container to the init method. Then the servlet uses the setAttribute method to bind "password" with "ding-dong".

The AttributeSetterServlet

```
import javax. servlet.*;
import java.io.IOException;

public class AttributeSetterServlet implements Servlet {

public void init (ServletConfig config) throws ServletException {
    // bind an object that is to be shared among other servlets
    ServletContext ServletContext = config. getServletContext() ;
    ServletContext. setAttribute ("password" ,  "dingdong") ;
}

public void service(ServletRequest request, ServletResponse response)
    throws ServletException, IOException {
    }

public void destroy () {
    }

public String getServletInfof) {
    return null;
    }

public ServletConfig getServletConfigf) {
    return null;
    }
}
```

The code in the following listing is the servlet that retrieves all attribute name/value pairs in the ServletContext object. The init method of this servlet preserves the ServletConfig object into ServletConfig. The service method then uses the ServletConfig interface s getServletContext method to obtain the ServletContext object. After you get the ServletContext object, you can then use its getAttributeNames method to get an Enumeration of all attribute names and loop through it to obtain each attribute s value, which it outputs to the console along with the attribute name.

DisplayAttributesServlet

```
import   javax.servlet.*;
import   java.io.IOException;
```

```
import   java.util. Enumeration;

public class ResponseDemoServlet implements Servlet {
    ServletConfig servletConfig;

    public void init (ServletConfig config) throws ServletException {
        servletConfig = config;
    }

    public void destroy () {
    }

    public void service(ServletRequest request, ServletResponse response)
        throws ServletException, IOException {

        ServletContext ServletContext = servletConfig. getServletContext ( );
        Enumeration attributes = ServletContext. getAttributeNames () ;
        while (attributes. hasMoreElementsO) {
            String attribute = (String) attributes.nextElement();
            System. out. printlnf "Attribute name :  " + attribute);
            System. out. println( "Attribute value :  " + ServletContext. getAttribute(attribute) ) ;
        }
    }

    public String getServletInfof) {
        return null;
    }

    public ServletConfig getServletConfig() {
        return null;
    }
}

Enumeration attributes = ServletContext.getAttributeNames();
    while (attributes.hasMoreElementsO) {
        String attribute = (String) attributes.nextElement();
        System.out.println("Attribute name : " + attribute);
        System.out.printlnf"Attribute value : " + ServletContext.getAttribute(attribute));
    }
```

To see the servlets work, first you need to call the AttributeSetterServlet servlet to set the attribute "password".You then call the DisplayAttributesServlet to get an Enumeration of the names of all attributes and display the values.

The output is given here:

```
Attribute    name :  javax.servlet.context.tempdir
Attribute    value :  C:\123data\JavaProjects\JavaWebBook\work\localhost_8080
Attribute    name : password
Attribute    value : dingdong
Attribute    name  :  sun.servlet.workdir
Attribute    value :  C:\123data\JavaProjects\JavaWebBook\work\localhost_8080
```

Requests and Responses

Requests and responses are what a web application is all about. In a servlet application, a user using a web browser sends a request to the servlet container, and the servlet container passes the request to the servlet.

In a servlet paradigm, the user request is represented by the ServletRequest object passed by the servlet container as the first argument to the service method. The service method's second argument is a ServletResponse object, which represents the response to the user.

The ServletRequest Interface

The ServletRequest interface defines an object used to encapsulate information about the user's request, including parameter name/value pairs, attributes, and an input stream.

The ServletRequest interface provides important methods that enable you to access information about the user. For example, the getParameterNames method returns an Enumeration containing the parameter names for the current request. To get the value of each parameter, the ServletRequest interface provides the getParameter method.

The getRemoteAddress and getRemoteHost methods are two methods that you can use to retrieve the user's computer identity. The first returns a string representing the IP address of the computer the client is using, and the second method returns a string representing the qualified host name of the computer.

The following example, shown in listings, shows a ServletRequest object in action. The example consists of an HTML form in a file named index.html that you need to put in the application directory—that is, under myApp—and a serVlet called RequestDemoServlet.

index.htrnl

```
<HTML>
<HEAD>
<TITLE>Sending a request</TITLE>
</HEAD>
<BODY>
<FORM ACTION=servlet/RequestDemoServlet METHOD="POST">
<BR><BR>
Author: <INPUT TYPE="TEXT" NAME="Author">
<INPUT TYPE="SUBMIT" NAME="Submit">
<INPUT TYPE="RESET" VALUE="Reset">
</FORM>
</BODY>
</HTML>
```

RequestDemoServlet

```
import  javax.servlet.*;
import  Java. util. Enumeration;
import  java.io.IOException;

public class RequestDemoServlet implements Servlet {

    public void init(ServletConfig config) throws ServletException {
    }

    public void destroy () {
    }

    public void service(ServletRequest request, ServletResponse response)
        throws ServletException, IOException {
```

```
        System. out. println( "Server Port: " + request. getServerPort() );
        System. out. println( "Server Name: " + request. getServerName() );
        System. out. println( "Protocol: " + request. getProtocol() );
        System. out. println( "Character Encoding: " +
            request .getCharacterEncoding( ) ) ;
        System. out. println( "Content Type: " + request. getContentType() );
        System. out. println( "Content Length: " + request. getContentLength() );
        System. out. println( "Remote Address: " + request. getRemoteAddr() );
        System. out. println( "Remote Host: " + request .getRemoteHost() );
        System. out. println( "Scheme: " + request. getScheme() );
        Enumeration parameters = request. getParameterNames( );
        while (parameters. hasMoreElements( )) {
            String parameterName = (String) parameters. nextElement() ;
            System. out. println( "Parameter Name: " + parameterName);
            System. out. println( "Parameter Value: " +
                request. getParameter (parameterName) ) ;
        }

        Enumeration attributes = request.getAttributeNames();
        while (attributes.hasMoreElementsO) {
            String attribute = (String) attributes.nextElement();
            System.out.printlnf"Attribute name: " + attribute);
            System.out.println("Attribute value: " +
                request.getAttribute(attribute));
        }
    }

    public String getServlet!nfo() {
        return null;
    }

    public ServletConfig getServletConfigf) {
        return null;
    }
}
```

To run the example, first request the index.html file by using the following URL:

http://localhost:8080/myApp/index.html

When you submit the form, you should see the list of attribute names and values in your console.

The ServletResponse Interface

The ServletResponse interface represents the response to the user. The most important method of this interface is getWriter, from which you can obtain a java.io.PrintWriter object that you can use to write HTML tags and other text to the user.

The code in the following listings offer an HTML file named index2.html and a servlet whose service method is overridden with code that outputs some HTML tags to the user.This servlet modifies the example in earlier listings that retrieves various information about the user. Instead of sending the information to the console, the service method sends it back to the user.

index2.html

```
<HTML>
<HEAD>
<TITLE>Sending a request</TITLE>
</HEAD>
```

```
<BODY>
<FORM ACTION=servlet/ResponseDemoServlet METHOD="POST">
<BR><BR>
Author: <INPUT TYPE="TEXT" NAME="Author">
<INPUT TYPE="SUBMIT" NAME="Submit">
<INPUT TYPE="RESET" VALUE="Reset">
</FORM>
</BODY>
</HTML>
```

The ResponseDemoServlet

```
import javax.servlet.*;
import java.io.PrintWriter;
import java.io.IOException;
import java.util.Enumeration;

public class ResponseDemoServlet implements Servlet {

    public void init(ServletConfig config) throws ServletException {
    }

    public void destroy () {
    }

    public void service(ServletRequest request, ServletResponse response)
        throws ServletException, IOException {

        PrintWriter out = response. getWriter( );
        out.println("<HTML>");
        out . println ( " <HEAD> " ) ;
        out. println ("<TITLE>");
        out. printlnf "ServletResponse") ;
        out.println("</TITLE>");
        out. printing "</HEAD>");
        out . println ( " <BODY> " ) ;
        out.println("<B>Demonstrating the ServletResponse object</B>") ;
        out.println("<BR>") ;

        out.println("<BR>Server Port: " + request. getServerPortf) );
        out.println("<BR>Server Name: " + request. getServerNamef) );
        out.println("<BR>Protocol: " + request. getProtocol() );
        out.println("<BR>Character Encoding: " + request. getCharacterEncoding() );
        out.println("<BR>Content Type: " + request. getContentType() );
        out.println("<BR>Content Length: " + request. getContentl_ength() );
        out. println ("<BR>Remote Address: " + request. getRemoteAddr() ) ;
        out. println ("<BR>Remote Host: " + request. getRemoteHost() );
        out. println ("<BR>Scheme: " + request. getScheme()) ;
        Enumeration parameters = request. getParameterNames( );
        while (parameters. hasMoreElementsf )) {
            String parameterName = (String) parameters. nextElement('
            out. println ("<br>Parameter Name: " + parameterName);
            out. println ("<br>Parameter Value: " + request. getParameter(parameterName) ) ;
        }
```

```
        Enumeration attributes = request. getAttributeNamesf );
        while (attributes. hasMoreElements()) {
            String attribute = (String) attributes. nextElementf );
            out. println ("<BR>Attribute name: " + attribute);
            out. println ("<BR>Attribute value: " + request. getAttribute(attribute) );
        }
        out.println("</BODY>");
        out.println("</HTML>");
    }

    public String getServlet!nfo() {
        return null;
    }

    public ServletConfig getServletConfig() {
        return null;
    }
}
```

To run the example, first request the index2.html file by using the following URL:

http://localhost:8080/myApp/index2.html

The GenericServlet Wrapper Class

Throughout this chapter, you have been creating servlet classes that implement the javax.servlet.Servlet interface. Everything works fine, but there are two annoying things that you've probably noticed:

1. You have to provide implementations for all five methods of the Servlet interface, even though most of the time you only need one. This makes your code look unnecessarily complicated.
2. The ServletConfig object is passed to the init method.You need to pre-serve this object to use it from other methods. This is not difficult, but it means extra work.

The javax.servlet package provides a wrapper class called GenericServlet that implements two important interfaces from the javax.servlet package: Servlet and ServletConfig, as well as the Java.io.Serializable interface. The GenericServlet class provides implementations for all methods, most of which are blank. You can extend GenericServlet and override only methods that you need to use. Clearly, this looks like a better solution.

The code in the next listing is a servlet called SimpleServlet that extends GenericServlet. The code provides the implementation of the service method that sends some output to the browser. Because the service method is the only method you need, only this method needs to appear in the class. Compared to all servlet classes that implement the javax.servlet.Servlet interface directly, SimpleServlet looks much cleaner and clearer.

Extending GenericServlet

```
import javax.servlet.*;
import java.io.IOException;
import java.io.PrintWriter;

public class SimpleServlet extends GenericServlet {

    public void service(ServletRequest request, ServletResponse response)
        throws ServletException, IOException {

        PrintWriter out = response.getWriter();
        out.println("<HTML>");
```

```
            out.println("<HEAD>");
            out.println("<TITLE>");
            out.println("Extending GenericServlet");
            out.println("</TITLE>");
            out.println("</HEAD>");
            out.println("<BODY>");
            out.println("Extending GenericServlet makes your code simpler.");
            out,println("</BODY>");
            out.println("</HTML>");
        }
}
```

Revision

1. The servlet technology is the foundation of web application development using the Java programming language
2. CGI enables the web server to call an external program and pass HTTP request information to that external program to process the request.
3. PHP technology provides easy web application development with its session management and includes some built-in functionality, such as file upload.
4. Server-side JavaScript (SSJS) is an extension of the JavaScript language, the scripting language that still rules client-side web programming. SSJS can access Java classes deployed at the server side using the Live Wire technology from Netscape.
5. Microsoft s ASP employs scripting technologies that work in Windows platforms, even though there have been efforts to port this technology to other operating systems.
6. A servlet is a Java class that can be loaded dynamically into and run by a special web server.
7. A number of servlet containers are available today. The most popular one—and the one recognized as the official servlet/JSP container—is Tomcat.
8. Two packages are available for servlet programmers: javax.servlet and javax.servlet.http.
9. The init method is called by the servlet container after the servlet class has been instantiated. The servlet container calls this method exactly once to indicate to the servlet that the servlet is being placed into service
10. The service method is called by the servlet container after the servlet's init method to allow the servlet to respond to a request. Servlets typically run inside multithreaded servlet containers that can handle multiple requests concurrently.
11. The servlet container calls the destroy method before removing a servlet instance from service. This normally happens when the servlet container is shut down or the servlet container needs some free memory.
12. The servlet specification allows you to configure your application.
13. In servlet programming, the servlet context is the environment where the servlet runs. The servlet container creates a ServletContext object that you can use to access information about the servlet's environment.
14. You can share the information — such as a database connection string or a page count — among the servlets by using attributes in the ServletContext object.
15. Requests and responses are what a web application is all about. In a servlet application, a user using a web browser sends a request to the servlet container, and the servlet container passes the request to the servlet.
16. The ServletRequest interface defines an object used to encapsulate information about the user's request, including parameter name/value pairs, attributes, and an input stream.
17. The javax.servlet package provides a wrapper class called GenericServlet that implements two

important interfaces from the javax.servlet package: Servlet and ServletConfig, as well as the Java.io.Serializable interface.

18. The ServletResponse interface represents the response to the user. The most important method of this interface is getWriter, from which you can obtain a java.io.PrintWriter object that you can use to write HTML tags and other text to the user.

Multiple Choice Questions

1. doGet and doPost are called by the service method, and sometime you may want to override service directly, e.g., for a servlet that handles both get and post request.
   (a) TRUE (b) FALSE.
2. Use __________ to read incoming HTTP headers (e.g., cookies).
   (a) cokkies (b) request
   (c) response
3. __________ is used to send output back to the client.
   (a) PrintWriter (b) ScriptWriter
   (c) Printin
4. To be a servlet, a class should extend :
   (a) HttpServlet (b) HttpRequest
   (c) HttpResponse
5. Use ______________ to specify the HTTP response line and headers.
   (a) feedback (b) write
   (c) response

Answers

1. a 2. b 3. a 4. a 5. c

Subjective Questions

1. What are Java Servlets?
2. Desribe the Java Servlet Architecture.
3. Describe the various steps in running servlet.
4. Write a note on A Servlet's Life Cycle.
5. How would you obtain the configuration information?
6. How would you preserve the ServletConfig?
7. Describe the method of sharing information among servlets.

10 JDPC (Java Database Connectivity)

Introduction

Most web applications use a database. Database accessing and programming therefore play a significant role in web development. In this chapter, we examine how data can be stored, retrieved, and manipulated. In Java, the technology that enables database access and manipulation is called Java Database Connectivity (JDBC).

JDBC has two parts: the JDBC Core Application Programming Interface (API) and the JDBC Optional Package API. The JDBC Core API is the main part of JDBC and when people talk about JDBC, they often are referring to this part.

The JDBC Core API is part of the Java 2, Standard Edition (J2SE) and takes the form of the classes and interfaces in the java.sql package. At the time of writing, the most recent version is 3.0. You use the members of the java.sql package primarily for basic database programming.

The JDBC Optional Package API is specified in the javax.sql package and is currently in version 2.0. The javax.sql package supports connection pooling, distributed transactions, rowsets, and so on. Of these, connection pooling is the most important in servlet programming; it is therefore explored in this chapter. You can download this package from http://java.sun.com/products/jdbc/ download.html.

JDBC is a broad topic, and a complete discussion of its object model requires a book of its own. This chapter starts with a look at the object model in the Java.sql package, but does not attempt to explain JDBC thoroughly. Some of the features in JDBC do not even suit our needs in web program-ming. For example, JDBC supports cursors that can scroll back, but using this type of cursor in a servlet is not tolerable for efficiency reasons. If you want to know more about JDBC, refer to the JDBC web site, http://java.sun.com/ products/jdbc/index.html.

The Java.sql Package

The Java.sql package provides the API for accessing and processing data in a data source. The most important members of the Java.sql package are as follows:

- The DriverManager class
- The Driver interface
- The Connection interface
- The Statement interface
- The ResultSet interface
- The PreparedStatement interface
- The ResultSetMetaData interface

Each of the types is briefly discussed in the following sections.

The DriverManager Class

The DriverManager class provides static methods for managing JDBC drivers. Each JDBC driver you want to use must be registered with the DriverManager. The JDBC driver of the database to which you want to con-nect is supplied either by the database vendor or a third party. You use different JDBC drivers for different database servers. For example, the JDBC driver for Microsoft SQL Server is different from the one used to access Oracle data-bases. JDBC drivers are discussed in more detail in the section, "Four Steps to Getting to the Database."

To load a JDBC driver from a servlet or JSP page, you copy the JDBC driver file for a particular database server (normally a jar file) to the WEB-INF\lib directory under your application directory. Create the lib directory, if it doesn't exist, and then use the following code:

```
try {
    Class.forName("JDBC.driver");
}
catch (ClassNotFoundException e) {
    // driver not found }
```

In this *case, JDBC. driver* is the fully qualified name of the JDBC driver class. This name can be found in the documentation accompanying the JDBC driver.

The DriverManager class's most important method is getConnection that returns a java.sql.Connection object. This method has three overloads whose signatures are as follows:

```
public static Connection getConnection(String url)
public static Connection getConnection(String url, Properties info)
public static Connection getConnection(String url, String user, String
password)
```

The Driver Interface

The Driver interface is implemented by every JDBC driver class. The driver class itself is loaded and registered with the DriverManager, and the DriverManager can manage multiple drivers for any given connection request. In the case where there are multiple drivers registered, the DriverManager will ask each driver in turn to try to connect to the target URL.

The Connection Interface

The Connection interface represents a connection to the database. An instance of the Connection interface is obtained from the getConnection method of the DriverManager class.

close

The close method immediately closes and releases a Connection object instead of waiting for it to be released automatically. Its signature is as follows:

```
public void close() throws SQLException
```

isClosed

You use this method to test whether the Connection object is closed. The sig-nature of this method is as follows:

```
public boolean isClosed() throws SQLException
```

createStatement

The createStatement method is used to create a Statement object for sending SQL statements to the database. If the same SQL statement is executed many times, it is more efficient to use a PreparedStatement object. This method has two overloads with the following signatures:

```
public Statement createStatement() throws SQLException
public Statement createStatement
(int resultSetType,
int resultSetConcurrency) throws SQLException
```

prepareStatement

You use the prepareStatement method to create a PreparedStatement object. Its signature is as follows:

public PreparedStatement prepareStatement()throws SQLException

getAutoCommit

The getAutoCommit method returns a boolean specifying the current auto-commit state. The signature of this method is as follows:

public boolean getAutoCommit() throws SQLException

This method returns true if auto-commit is enabled and false if auto-commit is not enabled. By default, auto-commit is enabled.

setAutoCommit

The setAutoCommit method sets the auto-commit state of the Connection object. Its signature is as follows:

public void setAutoCommit(boolean autocommit) throws SQLException

commit

You use the commit method to commit a transaction. The signature of this method is as follows:

public void commit() throws SQLException

rollback

You use the rollback method to roll back a, transaction. Its signature is as follows:

public void rollback() throws SQLException

The Statement Interface

You use the statement interface method to execute an SQL statement and obtain the results that are produced. The two most important methods of this interface are executeQuery and executeUpdate.

executeQuery

The executeQuery method executes an SQL statement that returns a single ResultSet object. Its signature is as follows:

public ResultSet executeQueryfString sql) throws SQLException

executeUpdate

The executeUpdate method executes an insert, update, and delete SQL statement. The method returns the number of records affected by the SQL statement execution, and its signature is as follows:

public int executeUpdate(String sql)

The ResultSet Interface

The ResultSet interface represents a table-like database result set. A ResultSet object maintains a cursor pointing to its current row of data. Initially, the cursor is positioned before the first row. The following are some important meth-ods of the ResultSet interface.

isFirst

The isFirst method indicates whether the cursor points to the first record in the ResultSet. Its signature is as follows:

public boolean isFirst() throws SQLException

isLast

The isLast method indicates whether the cursor points to the last record in the ResultSet. Its signature is as follows:

public boolean isLast() throws SQLException

next

The next method moves the cursor to the next record, returning true if the current row is valid and false if there are no more records in the ResultSet object. The method's signature is as follows:

public boolean next() throws SQLException

getMetaData

The getMetaData method returns the ResultSetMetaData object representing the meta data of the ResultSet. The signature of the method is as follows:

public ResultSetMetaData getMetaDate() throws SQLException

In addition to the previous methods, you can use several getXXX" methods to obtain the value of the specified column in the row pointed by the cursor. In this case, XXX represents the data type returned by the method at the speci-fied index, and each getXXX method accepts the index position of the col-umn in the ResultSet. The column index 1 indicates the first column. The signature of this method is as follows:

public XXX getXXX(int columnIndex) throws SQLException

For example, the getString method has the following signature and returns the specified cell as String:

public String getString(int columnIndex) throws SQLException

The PreparedStatement Interface

The PreparedStatement interface extends the Statement interface and represents a precompiled SQL statement. You use an instance of this interface to execute efficiently an SQL statement multiple times.

The ResultSetMetaData Interface

The ResultSetMetaData interface represents the meta data of a ResultSet object. The following sections introduce the most important methods.

getColumnCount

The getColumnCount method returns the number of columns in the ResultSet whose meta data is represented by the ResultSetMetaData object. Its signature is as follows:

public int getColumnCount() throws SQLException

getColumnName

The getColumnName method returns the column name as the specified col-umn index. Its signature is as follows:

public String getColumnName(int columnIndex) throws SQLException

The first column is indicated by index number 1.

Four Steps to Getting to the Database

This section explains what it takes to access a database and manipulate data in it.

Before you can manipulate data in a database, you need to connect to that database server. After you get the connection, you can communicate with the database server.You can send an SQL query to retrieve data from a table, update records, insert new records, or delete data you no longer need. You also can do more than manipulate data: You can invoke a stored procedure, create a table, and more.

Accessing a database with JDBC can be summarized in the following four steps:

1. Load the JDBC database driver.
2. Create a connection.
3. Create a statement.
4. Create a resultset, if you expect the database server to send back some data.

The discussion of each of these steps can be found in the following sections.

Step 1: Load a JDBC Database Driver

Database servers have their own proprietary "languages" for communication. This means that if you want to communicate with a database server, you need to understand its language. Fortunately, there are "translators" that can interface Java code with these database servers. These translators take the form of JDBC drivers. In other words, if you want to access a particular database, you need to get the JDBC driver for that database.

JDBC Drivers are available for most popular databases today. Oracle, Sybase, DB2, Microsoft SQL Server, MySQL—you name it. Drivers are even available for Open Database Connectivity (ODBC), which means that if you can't find a JDBC driver for a certain'database but that database can be connected through ODBC, you still can access the database through ODBC.

These JDBC drivers come at various prices and quality. Some are free and some are pricey. Some are slow, and some are lightning fast. The list of JDBC drivers can be found at `http: //industry, java.sun.com/products/ jdbc/drivers`.

Driver Types

Not all JDBC drivers are created equal. As mentioned previously, some are fast and some are slow. Slower drivers are not necessarily inefficient code. Some drivers are slower than others due to their architectural limitation. Depending on the architecture, a JDBC driver can fall into one of four types: Type 1, Type 2, Type 3, and Type 4.

JDBC started with Type 1 drivers. Today this type is normally the slowest, used only when you have no other alternative. Type 1 drivers are drivers that provide access to ODBC drivers and also are called *theJDBC-ODBC Bridge* driver. This driver is included in the J2SE and J2EE distribution and originally played an important role to help JDBC gain market acceptance. Why the bridge? At the moment JDBC was conceived, ODBC had already ruled the database connectivity world. Sun played a good strategy by enabling Java programmers to access any database server that could be accessed through ODBC.

As previously mentioned, you don't use Type 1 drivers in production, unless you can't find other drivers for that database. Fpr rapid prototyping, however, these drivers are acceptable. For example, if you need to prove a concept and a Microsoft Access database is all you have, the JDBC-ODBC Bridge can allow access to it in no time.

Note that you need to set up an ODBC Data Source Name (DSN) on the computer you use to connect to the database. Some ODBC native code and, in many cases, native database client code must be loaded on each machine that uses this type of driver. Hence, this kind of driver is generally most appropriate when automatic installation and downloading of a Java technology application is not important.

Setting Up an ODBC Data Source Name

To set up an ODBC data source name, follow these steps:

1. Open the ODBC Data Source Administration dialog box from the Control Panel.
2. Click the System DSN tab.
3. Click the Add button.
4. Select a driver name from the list of ODBC drivers.
5. Enter the needed information in the Setup dialog box that appears.
6. Click OK to close the dialog box.

Type 2 drivers are drivers written in part Java and part native API to convert JDBC calls into calls on the client API for Oracle, Sybase, Informix, DB2, or other DBMS. Note that, like the bridge driver, this style of driver requires that some binary code be loaded on each client machine.

Type 3 drivers are drivers written to use the network protocol to translate JDBC API calls into a DBMS-independent net protocol, which is then trans-lated to a DBMS protocol by a server. This net server middleware is able to connect all of its Java technology-based clients to many different databases. The specific protocol used depends on the vendor. In general, this is the most flexible JDBC API alternative—and most likely, all vendors of this solution will provide products suitable for Intranet use. In order for these products also to support Internet access, they must handle the additional requirements for secu-rity, access through firewalls, and so forth, that the web imposes. Several ven-dors are adding JDBC technology-based drivers to their existing database middleware products.

Type 4 drivers are pure Java. They are written to use the native protocol to convert JDBC technology calls into the network protocol directly used by DBMSs.This allows a direct call from the client machine to the DBMS server and is a practical solution for Intranet access. Because many of these protocols are proprietary, the database v< ndors themselves will be the primary source for this style of driver.

Installing JDBC Drivers

Because JDBC drivers normally come in a .jar file, the first thing you need to do after you get the driver is tell Tomcat where to find it by copying the jar file into the WEB-INF\lib directory under your application directory. Create the lib directory if it doesn't exist.

Now that the driver is within reach, you can create an instance of it using the static forName method of the Class class, passing the driver's fully qualified class name. The following is the code:

```
Class. forName(QfriverA/a/fle)
```

For example, for a MySQL database, the most popular driver is the Type 4 driver developed by Mark Matthews of Purdue University and downloadable from http://www.worldserver.eom/mm.mysql/.This driver is available under the GNU General Public License. Assuming that you have downloaded it and made it available to your Java code, you can load it using the following code:

```
Class.forName("org.gjt.mm.mysql.Driver");
```

Of course, the *driverName* argument will be different if you are using a different driver for MySQL.

For an ODBC database, the code that loads the driver is the following:

```
Class.forName("sun.j dbc.odbc.JdbcOdbcDriver");
```

Or, if you are using the FreeTdsType 4 JDBC driver to connect to a Microsoft SQL Server, you use this code to load the driver:

```
Class.forName("com.internetcds.jdbc.tds.Driver");
```

Your JDBC driver should come with documentation that tells you the driver class name to use.

If you plan to connect to different database servers in your code, you need to load all JDBC drivers for every database. For example, if you need to connect to an ODBC database as well as a MySQL database, you load both drivers by using the following code:

```
Class.forName("sun.jdbc.odbc.JdbcOdbcDriver");
Class.forName("org.gjt.mm.mysql.Driver");
```

Step 2: Creating a Connection

After you register a JDBC driver with the DriverManager, you can use it to get a connection to the database. In JDBC, a database connection is represented by the Java.sql.Connection interface. You use the DriverManager class's getConnection method to obtain a Connection object.

As mentioned in the previous section, the getConnection method has three overloads, two of which are normally used. The first overload takes three argu-ments: url, userName, and password. Its signature is as follows:

```
public static Connection getConnection(String url,
    String user, String password)
    throws SQLException
```

The url part is the trickiest element in this method. The url is of this syntax:

jdbc:subprotocol:subname

The subprotocol and subname parts depend on the database server you use. The documentation of the JDBC driver should tell you the subprotocol and subname to use.

If you are using a JDBC-ODBC bridge driver, the subprotocol is "odbc" and the subname is the Data Source Name (DSN) for that database. For instance, for a DSN called MarketingData, your URL will be:

jdbc:odbc:MarketingData

If you want to connect to a MySQL database, the subprotocol part is "mysql" and the subname part should be given the name of the machine and the database. For example, for a database named Fred, use the following:

jdbc:mysql///Fred

If one of the drivers you loaded recognizes the JDBC URL passed to the getConnection method, that driver will try to establish a connection to the database specified in the URL. For example, the following code shows how you obtain a Connection object to a MySQL database named Fred in a server called robusta. You also pass the user "admin" and the password "secret" to the getConnection method.

```
Connection connection =
    DriverManager.getConnection("jdbc:mysql///robusta/Fred", "admin", "secret");
```

You use the following code to get access to an ODBC database whose DSN is MarketingData, using the user name "sa" and password "1945":

```
Connection connection =
    DriverManager.getConnection("jdbc:odbc:MarketingData", "sa", "1945");
```

If the connection does not require you to login, you can pass null as arguments for the user and password parts.

The other overload of the getConnection method allows you to pass the user and password information in the URL. Its signature is as follows:

```
public static Connection'getConnection(String url)
    throws SQLException
```

For example, you can use the following URL to connect to a MySQL database named Fred in robustathat requires the user name "me" and password "pwd":

```
DriverManager.getConnection("jdbc:mysql://robusta/Fred?user=me&password=pwd");
```

Or, to connect to an ODBC database named MarketingData that is accessible without a login name and password, you use the following:

```
Connection connection =
    DriverManager.getConnection("jdbc:odbc:MarketingData");
```

If one of the drivers you loaded recognizes the JDBC URL supplied to the method DriverManager.getConnection, that driver will establish a connection to the DBMS specified in the JDBC URL. The getConnection method is probably the first and the last method of the DriverManager class you need to know to write Java programs that access databases. Other methods are needed only if you are writing a JDBC driver yourself.

The Connection object obtained through the DriverManager class's getConnection method is an open connection you can use to pass your SQL statements to the database server. As such, you do not need to have a special method to open a Connection.

Step 3: Creating a Statement

After you have a Connection object, your SQL skill takes over. Basically, you can pass any SQL statement that your database server understands. The level of understanding is different from one server to another. For example, an Oracle database is comfortable with subqueries, whereas a MySQL server is not. Also, whether your SQL statement will be executed successfully on the database server depends on the permission level of the user who passed the query to the database. If the user has the permission to view all the tables in the database, you can pass an SQL SELECT statement to it and expect some return. If the user has the permission to update records but not to delete retords, you can only update, not delete, records. You should understand these factors before you program your JDBC code.

After you have a Connection object from the DriverManage.getConnection method, you are ready to pass a SQL statement. To do this, you need to create another JDBC object called Statement. You can do this using the createStatement method of the Connection interface. Although there are two overloads of this method, the one that you almost always use is the no-argument overload. Its signature is as follows:

```
public Statement createStatement() throws SQLException
```

Therefore, to create a Statement object from an open Connection object, you write the following:

```
// connection is an open Connection object
Statement statement = connection.createStatement();
```

Next, use the methods in the Statement class interface to manipulate your data or data structure. You will use two important methods: executeUpdate and executeQuery. The signatures for both methods are as follows:

```
public int executeUpdate(String sql) throws SQLException
public ResultSet executeQuery(String sql) throws SQLException
```

For example, to create a table named Addresses with two fields, you can use the following code:

```
String sql = "CREATE TABLE Addresses " +
    "(FirstName VARCHAR(32), LastName VARCHAR(32)";
statement.executeUpdate(sql);
```

And, to insert a record in the Addresses table, you use the following code:

```
String sql = "INSERT INTO Addresses " +
"VALUES ('Don1, 'Bradman1)";
statement.executeUpdate(sql);
```

You use executeQuery when you expect a ResultSet object. You learn about the ResultSet interface in the next section.

Step 4: Creating a ResultSet

A ResultSet is the representation of a database table that is returned from a Statement object. A ResultSet object maintains a cursor pointing to its current row of data. When the cursor is first returned, it is positioned before the first row. To access the first row of the ResultSet, you need to call the nextQ method of the ResultSet interface.

The nextQ method moves the cursor to the next row and can return either a true or false value. It returns true if the new current row is valid; it returns false if there are no more rows. Normally, you use this method in a while loop to iterate through the ResultSet object.

To get the data from the ResultSet, you can use one of many the getXXX methods of ResultSet, such as getInt, getLong, and so on. getShortYou use the getInt method, for example, to obtain the value of the designated column in the current row of this ResultSet object as an int in the Java programming language. The getLong gets the cell data as a long, etc. The most commonly used method is getString, which returns the cell data as a String. Using getString is preferable in many cases because you don't need to worry about the data type of the table field in the database.

The getString method, similar to other getXXX methods, has two overloads that allow you to retrieve a cell's data by passing either the column index or the column name. The signatures of the two overloads of getString are as follows:

```
public String getString(int columnIndex) throws SQLException
public String getString(String columnName) throws SQLException
```

The following example illustrates the use of the nextQ method as well as the getString method. The code is used to retrieve the FirstName and LastName columns from a table called Users. It then iterates the returned ResultSet and prints to the console all the first names and last names in the format "first name:last name". As you consider the following code, assume that you already have a Statement object called statement:

```
String sql = "SELECT FirstName,  LastName FROM Users";
ResultSet resultSet = statement.executeQuery(sql);
while (resultSet.next()) {
      System.out.printlnfresultSet.getString(1) + ":" +
resultSet.getString("LastName")  );
}
```

The first column is retrieved by passing its column index, which is 1 for the first column. The second column is obtained by passing its column name.

Another important method is the close method that closes the ResultSet object when it is no longer used. The close method releases this ResultSet object's database and JDBC resources immediately instead of waiting for this to happen when it is automatically closed. A ResultSet object will be closed automatically when it is garbage collected or when the Statement object that generated it is closed, re-executed, or used to retrieve the next result from a sequence of multiple results. Always call the close method to explicitly close the ResultSet object. Completing your example code, you should put the fol-lowing code right after you finish accessing the ResultSet object:

```
resultSet.close();
```

Note that if an SQL query results in zero row, the Statement object will return a ResultSet object containing no row, not a null.

Putting It All Together

This section summarizes the four steps just discussed by presenting an example that uses the FreeTds Type 4 JDBC driver to access the Users table in a Microsoft SQL Server database named Registration. The database server is called Lampoon and you need to pass the user name "sa" and password "s3mlcOnductOr" to login to the database server. The SQL statement queries two columns: FirstName and LastName. Upon retrieving the ResultSet, the statement will loop through it to print all the first names and last names in the ResultSet. The code is given in the following listing.

Accessing a Database

```
try {
      Class.forName("com.internetcds.j dbc.tds.Driver");
      Connection con = DriverManager.getConnectionf
            "jdbc:freetds:sqlserver://Lampoon/Registration",
            "sa", "s3m1 conductor");
      System.out.println("got connection");

      Statement s = con.createStatement();
      String sql =
            "INSERT INTO UserReg VALUES ('a', 'b', '12/12/2001', 'f')",
      s.executeUpdate(sql);
      sql = "SELECT FirstName,  LastName FROM Users";
      ResultSet rs = s.executeQuery(sql);
      while (rs.nextO)  { System.out.println(rs.getString(1) + "  " + rs,getString(2));
      }
      rs.close();
```

```
            s.close();
            con.close();
        }
        catch (ClassNotFoundException e1) {
            System.out.println(e1.toString());
        }
        catch (SQLException e2) {
            System.out.println(e2.toString());
        }
        catch (Exception e3) {
            System.out.println(e3.toString());
        }
```

On my machine, the code returns the following at the console. The result displayed depends on the data stored in your table, of course.

```
Jimmy Barnes
Richard Myers
George Lucas
```

A Database-Based Login Servlet

As your first example of connecting a database from a servlet, we would like to present the type of servlet that is probably built most often: a Login servlet. Unlike the Login servlet in which you hard-coded the user name and password that log in a user, this servlet matches the entered user name and password against the values of the UserName and Password columns in a table called Users. This table may have other columns, but those other columns are not relevant here.

When the user types the URL in the Location or Address box and first calls the servlet, its doGet method is invoked. The servlet then does what a login servlet should do—challenges the user to type in a user name and password. What the doGet does is call the private method sendLoginForm, which sends the HTML page for the user to login.

```
sendLoginForm(response, false);
```

The sendLoginForm method accepts two arguments: a HttpServletResponse object that the method can use to send output to the browser, and a boolean. This boolean is a flag to indicate whether an error message should be sent along with the form. The error message tells the user that the previous login has failed. When the servlet is first called; of course, no error message is sent; thus the false value is given for the second argument of the sendLoginForm method called from doGet.The sendLoginForm method is given in the following listing.

The LoginServlet's sendLoginForm Method

```
private void sendLoginForm(HttpServletResponse response,
    boolean withErrorMessage)
    throws ServletException, IOException {

    response.setContentType("text/html");
    PrintWriter out = response.getWriter();
    out.println("<HTML>");
    out.println("<HEAD>");
    out.println("<TITLE>Login</TITLE>");
    out.println("</HEAD>");
    out.println("<BODY>");
    out.println("<CENTER>");

    if (withErrorMessage)
```

```
            out.printing Login failed. Please try again.<BR>");

        out.println("<BR>");
        out.println("<BR><H2>Login Page</H2>");
        out.println("<BR>");
        out.println("<BR>Please enter your user name and password.");
        out.println("<BR>");
        out.println("<BR><FORM METHOD=POST>");
        out.println("<TABLE>");
        out.println("<TR>");
        out.println("<TD>User Name:</TD>");
        out.println("<TD><INPUT TYPE=TEXT NAME=userName></TD>");
        out.println("</TR>");
        out.println("<TR>");
        out.println("<TD>Password:</TD>");
        out.println("<TD><INPUT TYPE=PASSWORD NAME=password></TD>");
        out.println("</TR>");
        out.println("<TR>");
        out.println("<TD ALIGN=RIGHT COLSPAN=2>");
        out.println("<iNPUT TYPE=SUBMIT VALUE=Login></TD>");
        out.println("</TR>");
        out.println("</TABLE>");
        out.println("</FORM>");
        out.println("</CENTER>");
        out.println("</BODY>");
        out.println("</HTML>");
    }
```

The user can then type the user name and password to login and click the Submit button. Upon submit, the doPost method is invoked. The first thing it does is obtain the userName and password from the HttpServletRequest object, as follows:

```
    String userName = request.getParameter("userName");
    String password = request.getParameter("password");
```

Then, the method passes the userName and password to the login method. This method returns true if the user name and password are correct; other-wise, it returns false. On successful login, the request is dispatched to another servlet, as follows:

```
    RequestDispatcher rd =
        request.getRequestDispatcher("AnotherServlet");
    rd.forward(request, response);
```

If the login failed, the doPost calls the sendLoginForm method again, this time with an error message, as follows:

```
    sendLoginForm(response, true);
```

The doPost method is given in the following listing.

The LoginServlet's doPost Method

```
    public void doPost(HttpServletRequest request,
        HttpServletResponse response)
        throws ServletException, IOException {   .

        String userName = request. getParameter( "userName" );
        String password = request. getParameter( "password" );
        if (login(userName, password)) {
            RequestDispatcher rd =
```

```
            request. getRequestDi3patcher("AnotherServlet") ;
        rd. forward (request, response);
    }
    else {
        sendLoginForm(response, true);
    }
}
```

Here is the part that does the authentication: the login method. The method tries to authenticate the user by trying to retrieve a record whose UserName field is userName and Password field is the same as password. If a record is found, the login succeeds; otherwise, the login fails.

The login method first loads the JDBC driver, as follows:

```
Class . forName ( " sun . j dbc . odbc . JdbcOdbcDriver " ) ;
```

This example uses a JDBC-ODBC driver; however, as you learned in the previous section, you can use any other driver without changing the rest of your code.

With a JDBC driver, you can create a Connection object and call its createStatement method to create a Statement object, as follows:

```
Connection con =
    DriverManager . getConnection ( " j dbc : odbc : JavaWeb " ) ;
Statement s = con. createStatement () ;
```

Note that you use the getConnection method overload that accepts an argument — the URL.

When a Statement object is available, you can call its executeQuery method, passing an SQL statement of the following syntax:

```
SELECT UserName FROM Users
WHERE UserName=userName
AND Password=password
```

To determine whether there is any record in the ResultSet object, you can call its next() method. This method returns true if there is a record; otherwise, it returns false. On true, the login method returns a true, indicating a successful login. A false return value indicates otherwise. The login method is given in the following listing, and the complete code for LoginServlet is given in the next listing.

The LoginServlet's login Method

```
boolean login(String userName, String password) {
    try {
        Class.forName("sun.jdbc.odbc.JdbcOdbcDriver");
        Connection con =
            DriverManager.getConnection("jdbc:odbc:JavaWeb");
        System, out. printing got connection");

        Statement s = con.createStatement();
        String sql = "SELECT UserName FROM Users" +
            "WHERE UserName='" + userName + "''" +
            "AND Password='" + password + "''";
        ResultSet rs = s.executeQuery(sql);
        if (rs.nextO) {
            rs.close();
            s.close();
            con.close();
            return true;
        }
        rs.closeO;
        s.close();
```

```
            con.close();
        }
        catch (ClassNotFoundException e) {
        System.out.println(e.toString());
        }
        catch (SQLException e) {
        System.out.println(e.toString());
        }
        catch (Exception e) {
            System.out.println(e.toString());
        }
        return false
    }
```

The LoginServlet Servlet

```
import javax.servlet.*;
import javax.servlet.http.*;
import java.io.*;
import java.util.*;

public class LoginServlet extends HttpServlet {
    public void doGet(HttpServletRequest request,
    HttpServletResponse response)
    throws ServletException, IOException {
    sendLoginForm(response, false);
}

private void sendl_oginForm(HttpServletResponse response,
    boolean withErrorMessage)
    throws ServletException, IOException {

    response. setContentType("text/html") ;
    PrintWriter out = response. getWriter( );
    out.println("<HTML>");
    out.println("<HEAD>");
    out.println("<TITLE>Login</TITLE>");
    out.println("</HEAD>");
    out.println(" <BODY> " ) ;
    out. println("<CENTER>") ;

    if (withErrorMessage)
        out. println( "Login failed. Please try again. <BR>") ;

    out.println("<BR>") ;
    out.println ("<BR><H2>Login Page</H2>") ;
    out.println("<BR>") ;
    out.println ("<BR>Please enter your user name and password.");
    out.print ln ("<BR>") ;
    out .println (" <BR><FORM METHOD=POST> " ) ;
    out.print ln ("<TABLE>") ;
    out. println ("<TR>");
    out.println("<TD>User Name:</TD>") ;
    out. println("<TD><INPUT TYPE=TEXT NAME=userName></TD>") ;
    out.println("</TR>");
```

```
        out.println("<TR>");
        out. print ln ("<TD>Password:</TD>") ;
        out.println("<TD><INPUT TYPE=PASSWORD NAME=password></TD>" ) ;
        out.println("</TR>") ;
        out.println("<TR>");
        out. println ("<TD ALIGN=RIGHT COLSPAN=2>" ) ;
        out. println ("<INPUT TYPE=SUBMIT VALUE=Login></TD>" ) ;
        out.println("</TR>");
        out.println("</TABLE>");
        out.println("</FORM>");
        out.println("</CENTER>");
        out.println("</BODY>");
        out.println("</HTML>");
}

public void doPost(HttpServletRequest request,
        HttpServletResponse response)
        throws ServletException, IOException {

        String userName = request. getParameter( "userName" );
        String password = request. getParameter( "password" );
        if (login(userName, password)) {
                RequestDispatcher rd =
                        request. getRequestDispatcher("AnotherServlet") ;
                rd. forward (request, response);
        }
        else {
                sendLoginForm(response, true);
        }
}

boolean login (String userName, String password) {
        try {
                Class. f orName ("sun. jdbc. odbc. JdbcOdbcDriver" ) ;
                Connection con =
                        DriverManager . getConnection ( " jdbc : odbc : JavaWeb" ) ;
                System. out. println(" got connection");

                Statement s = con.createStatement() ;
                String sql = "SELECT UserName FROM Users" +
                        "WHERE UserName='" + userName + " '" +
                        "AND Password='" + password + " ' " ;
                ResultSet rs = s.executeQuery(sql) ;
                if (rs.next()) {
                        rs. close () ;
                        s.close();
                        con.close();
                        return true;
                }
                rs.close();
                s. close () ;
                con.closeO;
        }
```

```
    catch (ClassNotFoundException e) {
        System. out .println(e.toString() ) ;
    }
    catch (SQLException e) {
        System.out.println ( e . toString ( ) ) ;
    }
    catch (Exception e) {
        System. out .println(e.toString() ) ;
    }
    return false;
}
```

The Single Quote Factor

The LoginServlet servlet in the previous section works fine, but not perfectly. The servlet passes the user name and password to the login method and the SQL statement is composed on the fly. For example, for user name equals "jeff" and password "Java", the SQL statement is as follows:

```
SELECT UserName
FROM Users
WHERE UserName='jeff'
AND Password='Java'
```

The values for the user name and the password are enclosed in the single quote pairs, as in 'jeff' and 'Java'. A username can be anything, including a name that includes an apostrophe, such as o'connor.This is also true for passwords. If the username is "o'connor" and password is "Java," the SQL statement will become the following:

```
SELECT UserName
FROM Users
WHERE UserName='o'connor'
AND Password='Java'
```

The database engine will encounter a single quote character after the o in "o'connor," and think it sees the end of the value for UserName, and expect a space after it. The engine finds a c instead, and this is confusing to the database server. An error will be thrown.

Does this mean you can't let users have a username that contains an apos-trophe? Fortunately, there is a work around. If a string value in an SQL state-ment contains an apostrophe, you just need to prefix the apostrophe with another apostrophe, resulting in the following SQL statement:

```
SELECT UserName
FROM Users
WHERE UserName='o''connor'
AND Password='Java'
```

Now everything works correctly. The LoginServlet servlet in the previous example should be fixed by passing all values entered by the user to a method called fixSQLFieldValue, which accepts a String. This method will replace every occurrence of a single quote character in the String with two single quote characters. The fixSQLFieldValue takes a String whose value may con-tain a single quote character and doubles every occurrence of it. The fixSQLFieldValue method uses a StringBuffer with an initial capacity of 1.1 times the length of the argument passed in String object. The 1.1 is based on the estimate that the resulting SQL statement will be at most 10 per-cent longer. Here is the code:

```
StringBuffer fixedValue = new StringBuffer((int) (length* 1.1));
```

The statement then tests each character of the String in a for loop. If a character is a single quote, it appends two single quote characters to the StringBuffer; otherwise, the character itself is appended, as you see here:

```
    for (int i=0; i<length; i++) {
        char c = value.charAt(i);
        if (c=='\'')
            fixedValue.append("''");
        else
            fixedValue.append(c);
    }
```

The fixSqlFieldValue method is given in the following listing.

The fixSqlFieldValue Method

```
public static String fixSqlFieldValue(String value) {
    if (value==null)
        return null;
    int length = value.length();
    StringBuffer fixedValue =
        new StringBuffer((int) (length * 1.1));
    for (int i=0; i<length; i++) {
        char c = value.charAt(i);
        if (c=='\'')
            fixedValue.append("''");
        else
            fixedValue.append(c);
    }
    return fixedValue.toString();
}
```

The login method in the above listings must pass the userName and password values before composing the SQL statement. With the fixSqlFieldValue method included in the LoginServlet servlet, the part that forms the SQL statement should be modified into the following line:

```
String sql = "SELECT UserName FROM Users" +
    "WHERE UserName='" + fixSqlFieldValue(userName) + "'" +
    "AND Password='" + fixSqlFieldValue(password) + "'";
```

Clearly, the fixSqlFieldValue method should be called every time you compose an SQL statement that includes a value from the user input.

```
import com.brainysoftware.java.StringUtil
StringUtil.fixSqlFieldValue(string)
```

To use this class, you need to create a directory structure under the /WEB-INF/classes directory under your application directory; that is, you need to create a directory called com under the classes directory, the brainysoftware directory under com, the Java directory under the brainysoftware directory, and copy the StringUtil.Java there.

Inserting Data into a Table with RegistrationServlet

Inserting a record into a table is a common database task. An example of this is a servlet named RegistrationServlet that takes user input to insert a record into a Users table. The table itself has columns as described in the following table.

The Users Table

Column Name	Column Type	Description
Id	Numeric	An autonumber that serves as the primary key and is auto-incremented by the database.
FirstName	String	The first name of the user.

LastName	String	The last name of the user.
UserName	String	The user name for the user to login. No duplicate allowed.
Password	String	The password for the user to login.

To insert a record containing a user s details, you need to retrieve the values entered by the user and compose an SQL Statement, such as the following:

```
INSERT INTO Users
(FirstName, LastName, UserName, Password)
VALUES
(firstName, lastName, userName, password)
```

The Id column is not included because it's an autonumber. A new unique number will be generated by the database server automatically for the Id col-umn when you insert a new record into the table. Also note that the UserName column value rrtust be unique. No two users can have the same user name.

The user can then fill in the necessary details: first name, last name, and a self-chosen user name and password. The user name must be unique. If the user enters a user name that's already taken, the servlet will display an error messag and present the same form again, with the previously typed in values retained.

If the servlet encounters an unexpected error, the error also will be displayed and the previously entered values retained.

On successful registration, the servlet will display the message "Successfully added one user" and present a clean form. The user then can enter another user s details if desired.

In the RegistrationServlet, the method to display the form is called sendRegistrationForm. The method accepts three arguments: an HttpServletRequest object, an HttpServletResponse object, and a boolean. The boolean indicates whether the previously entered values, if any, should be displayed. The sendRegistrationForm method is given in the listing. Note that when you need to display a value in an HTML input box, you need to call the encodeHtmlTag method (which is also part of the com.brainysoftware.java.StringUtil class) just in case the value contains one of the four HTML special characters. The RegistrationServlet therefore imports the com.brainysoftware.java.StringUtil class.

The sendRegistrationForm Method

```
private void sendRegistrationForm(HttpServletRequest request,
    HttpServletResponse response, boolean displayPreviousValues)
    throws ServletException, IOException {

    PrintWriter out = response.getWriter();
    out.println("<BR><H2>Registration Page</H2>");
    out.println("<BR>Please enter the user details.");
    out.println("<BR>");
    out.println("<BR><FORM METHOD=POST>");
    out.println("<TABLE>");
    out.println("<TR>");
    out.println("<TD>First Name</TD>");
    out.print("<TD><INPUT TYPE=TEXT Name=firstName");

    if (displayPreviousValues)
        out.print(" VALUE=\"" +
        StringIltil.encodeHtmlTag(firstName) + "\"");

    out.println("></TD>");
    out.println("</TR>");
```

```
out.println("<TR>");
out.println("<TD>Last Name</TD>");
out.print("<TD><INPUT TYPE=TEXT Name=lastName");

if (displayPreviousValues)
    out.print(" VALUE=\"" +
    StringUtil.encodeHtmlTag(lastName) + "\"");
    out.println("></TD>");
    out.println("</TR>");
    out.println("<TR>");
    out.println("<TD>User Name</TD>");
    out.print("<TD><INPUT TYPE=TEXT Name=userName");

if (displayPreviousValues)
    out.print(" VALUE=\"" + StringUtil.encodeHtmlTag(userName) + "\"");
    out.println("></TD>");
    out.println("</TR>");
    out.println("<TR>");
    out.println("<TD>Password</TD>");
    out.print("<TD><INPUT TYPE=PASSWORD Name=password");

if (displayPreviousValues)
    out.print(" VALUE=\"" +
    StringUtil.encodeHtmlTag(password) + "\"");

    out.println("></TD>");
    out.println("</TR>");
    out.println("<TR>");
    out.println("<TD><INPUT TYPE=RESET></TD>");
    out.println("<TD><INPUT TYPE=SUBMIT></TD>");
    out.println("</TR>");
    out.println("</TABLE>");
    out.println("</FORM>");
    out.println("<BR>");
    out.println("<BR>");
```

Unlike the form in the LoginServlet, however, the registration form does not include the <HTML> tag, the head section of the HTML, and the opening <BODY> tag. Nor does it contain the closing </ BODY> and </HTML> tags. Page headers can be sent to the browser by calling the sendPageHeader method, whose code is given in the following listing.

The sendPageHeader Method

```
private void sendPageHeader(HttpServletResponse response)
    throws ServletException, IOException {
    response.setContentType("text/html");
    PrintWriter out = response.getWriter();
    out.println("<HTML>");
    out.println("<HEAD>");
    out.println("<TITLE>Registration Page</TITLE>");
    out.println("</HEAD>");
    out.println("<BODY>");
    out.println("<CENTER>");
```

The page footer can be sent by calling the sendPageFooter method, listed in the following listing. .

The sendPageFooter Method

```
private void sendPageFooter(HttpServletResponse response)
    throws ServletException, IOException {
    PrintWriter out = response.getWriter();
    out.println("</CENTER>");
    out.println("</BODY>");
    out.println("</HTML>");
}
```

When first called, the servlet invokes its doGet method to send the registration form plus all necessary HTML tags to the browser. The doGet method is given in the following listing.

The doGet Method

```
public void doGet(HttpServletRequest request, HttpServletResponse
    response) throws ServletException, IOException {
        sendPageHeader(response);
        sendRegistrationFormfrequest, response, false);
        sendPageFooter(response);
    }
```

When the user submits the form, the doPost method is invoked. The doPost method first retrieves the values from the users, as follows:

```
firstName = request.getParameter("firstName");
lastName = request.getParameter("lastName");
userName = request.getParameter("userName");
password = request.getParameter("password");
```

After opening a Connection object and creating a Statement object, the method needs to determine whether the user name has already been taken. It does so by trying to retrieve a record whose UserName field value is userName, as follows:

```
Connection con =
    DriverManager.getConnection("j dbc:odbc:JavaWeb");
Statement s = con.createStatement();/
String sql = "SELECT UserName FROM Users" +
    "WHERE userName='" +
    StringUtil.fixSQLFieldValue(userName) + "'";
ResultSet rs = s.executeQuery(sql);
```

If the ResultSet object returned by the Statement interface's executeQuery method has a record, its next method will return true, indicating that the user name has already been taken; otherwise, false is returned.

If the user name has been taken, a message of "The user name... has been taken" is assigned to the message String and the error flag is set. If not, an SQL statement will be composed on the fly and the executeUpdate method of the Statement object will be called to insert a new record into the Users table.

```
if (rs.next()) {
    rs.close();
    message = "The user name <B>" +
        StringUtil.encodeHtmlTag(userName) +
        "</B> has been taken. Please select another name.";
    error = true;
}
else {
    rs.close();
    sql = "INSERT INTO Users" +
```

```
        " (FirstName, LastName, UserName, Password)" +
        " VALUES" +
        " ("" + StringUtil.fixSQLFieldValue(firstName) + "V +
        " " + StringUtil.fixSQLFieldValue(lastName) + "'," +
        " " + StringUtil.fixSQLFieldValue(userName) + "'," +
        " " + StringUtil.fixSQLFieldValue(password) + "')";
        int i = s.executeupdate(sql);
        if (i==!) {
            message = "Successfully added one user.";
        }
    }
```

The last part of the doPost method checks whether the message is null and whether the error flag has been set. If the message is not null, it is sent to the browser, as you see here;

```
    If (message I 'null) {
        PrintWriter out = response. getWriter();
        out.println("<B>* + message + "</B><BR>");
        out . println ( "<HR><BR> " ) ;
    }
```

If the error is set, the registration form is sent with the last argument equal to true, indicating the sendRegistrationForm to display the previously entered values. Otherwise, the sendRegistrationForm method is sent with the false value for the last argument, as you see here:

```
    if (error==true)
            sendflegistrationFormfrequest, response, true);
    else
            sendflegistrationFormf request, response, false);
    sendPageFooter(response) ;
    }
```

The doPost method is given in the following listing.

The doPost Method

```
    public void doPost(HttpServletflequest request,
        HttpServletResponse response)
        throws ServletException, IOException {
        sendPageHeaderfresponse);

        firstName = request.getParameter("firstName"J;
        lastName request.getParameter("lastName");
        userName == request.getParameter("userName");
        password * request.getParameterfpassword");

        boolean error = false;
        String message * null;
        try {

        Connection con = DriverManager.getConnection(" jdbc:odtic:JavaWeb") ;
        System. out. println(" got connection");

        Statement s = con.createStatement();

        String sql = "SELECT UserName FROM Users" +
            "WHERE userName='" + StringUtil.fixSQLFieldVaUie(userName) t "'";
        ResultSet rs = s.executeQuery(sql) ;
        if (rs.nextO) {
```

```
            rs. closet);
            message = "The user name <B>"
            * StringUtil.encodeHtmlTag(userName) +
            "</B> has been taken. Please select another name.";
            error = true;
        }
        else {
            rs.close();
            sql = "INSERT INTO Users" +
            " (FirstName, LastName, UserName, Password)" +
            "VALUES" +
            " ('" + StringUtil.fixSQLFieldValue(firstName) + "'," +
            " '" + StringUtil.fixSQLFieldValue(lastName) +
            "'," + " '" + StringUtil.fixSQLFieldValue(userName) +
            "'," + " '" + StringUtil.fixSQLFieldValue(password) + "')";
            int i = s.executeUpdate(sql) ;
            if (1==1) {
                message = "Successfully added one user.";
            }
        }
        s. closet);
        con. close();
    }
    catch tSOLException e) {
        message = "Error." + e.toStringO;
        error = true;
    }
    catch (Exception e) {
        message = "Error." •+• e.toString() ;
        error = true;
    }
    if (message!=null) {
        PrintWriter out = response. getWriter() ;
        out.println("<B>" + message + "</B><BR>");
        out . println ( "<HR><BR>" ) ;
    }
    if (error==true)
        sendRegistrationFormtrequest, response, true);
    else
        sendRegistrationForm(request, response, false);
    sendPageFooter(response) ;
    }
```

You may have noticed chat the code that loads the JDBC driver is missing from the doPost method. Instead, it has been moved to the init method of the servlet. The JDBC driver needs to be loaded only once, so it is appropriate to put it in the init method, which is only called once during the life cycle of the servlet. The init method is given in the following listing.

The init Method of the RegistrationServlet

```
public void init() {
    try {
        Class.forName("sun.jdbc.odbc.JdbcQdbcDriver";
        System.out.println("JDBC driver loaded");
    }
```

```
catch (ClassNotFoundException e) {
      System.out.println(e.toString());
   }
}
```

The complete code of the RegistrationServlet servlet is given in the following listing.

The RegistrationServlet

```
import javax. servlet.*;
import javax. servlet. http.*;
import java.io.*;
import java.util.*;
import java.sql.*;
import com.brainysoftware.java.StringUtil;

public class RegistrationServlet extends HttpServlet {

   private String firstName = "";
   private String lastName = "";
   private String userName = "";
   private String password = "";

   public void init() {
      try {
         Class. forName (" sun. jdbc. odbc. JdbcOdbcDriver");
         System, out. printing JDBC driver loaded");
      }
      catch (ClassNotFoundException e) {
         System.out.println(e.toString());
      }
   }

/**Process the HTTP Get request*/
public void doGetfHttpServletRequest request, HttpServletResponse response)
throws ServletException.IOException {
   sendPageHeader( response);
   sendRegistrationForm(request, response, false);
   sendPageFooter( response) ;

/**Process the HTTP Post request*/
public void doPost(HttpServletRequest request,
   HttpServletHesponse response)
   throws ServletException, IOException {
   sendPageHeader(response);

   firstName = request. getParameter("firstName");
   lastName = request, getParameter("lastName");
   userName = request. getParameter("userName") ;
   password = request. getParameter("password");

   boolean error = false;
   String message = null;
   try {
      Connection con = DriverManager.getConnection{"jdbc:odbc:Javaweb")
      System. out. println( "got connection") ;
```

```
    Statement s = con.createStatement() ;

    String sql = "SELECT UserName FROM Users" +
        "WHERE userName='" +• StringUtil.fixSQLFieldValue(userName) +
    "ResultSet rs = s.executeQuery(sql) ;
    if (rs.nextO) {
        rs. close();
        message = "The user name <B>"
            + StringUtil.encodeHtmlTag( userName) +
            </B> has been taken. Please select another name.";
        error = true;
    }
    else {
        rs. close() ;
        sql = "INSERT INTO Users" +
            " (FirstName, LastName, UserName, Password)" +
            "VALUES' +
            " ('" + StringUtil.fixSOLFieldValue(firstName) + "'," +
            " '" + StringUtil.fixSQLFieldValue(lastName) + "'," +
            " '" + StringUtil.fixSQLFieldValue(userName) + "'," +
            "'" + Stringutil.fixSQLFieldValue (password) + "')";
        int i = s.executeUpdate(sql);
        if (i«-1) {
            message = "Successfully added one user.";
        }
    }
    s. closef) ;
    con.close();
}
catch (SQLException e} {
    message = "Error." + e.toString() ;
    error = true;
}
catch (Exception e) {
    message = "Error." + e.toString() ;
    error = true;
}
if (message!=null) {
    PrintWriter out = response. getWriter( );
    out.println{"<B>" + message + "</B><BR>");
    out.println ( "<HR><BR> " ) ;
}
if (error—true)
    sendRegistrationForm(request, response, true);
else
    sendRegistrationForm(request, response, false);
    sendPageFooter(response) ;
}

/**
* Send the HTML page header, including the title
* and the <BODY> tag
* /
```

```
private void sendPageHeader(HttpServletResponse response)
    throws ServletException, IOException {
    response . setContentType( "text/html" ) ;
    PrintWriter out = response. getWriterf) ;
    out.println("<HTML>");
    out.println("<HEAD>") ;
    out. print ln ( "<TITLE>Regist ration Page</TITLE>") ;
    out.println("</HEAD>");
    out.println ("<BODY>") ;
    out.println("<CENTER>");
}

/**
* Send the HTML page footer, i.e. the </BODY>
* and the </HTML>
*/
private void sendPageFooter(HttpServletflesponse response)
    throws ServletException, IOException {
    PrintWriter out = response. getWriterf);
    out.println("</CENTER>");
    out.println("</BODY>");
    out.println("</HTML>");
}

/**Send the form where the user can type in
* the details for a new usef
*/
private void sendRegistrationForm(HttpServletRequest request,
    HttpServletResponse response, boolean displayPreviousValues)
    throws ServletException, IOException {

    PrintWriter out = response.getWriter();
    out.println("<BR><H2>Registration Page</H2>"); '
    out.println("<BR>Please enter the user details.");
    out.println("<BR>");
    out.println("<BR><FORM METHOD=POST>");
    out. println ("<TABLE>");
    out.println("<TR>");
    out.println("<TD>First Name</TD>");
    out.print("<TD><INPUT TYPE=TEXT Name=firstName");

    if (displayPreviousValues)
        out.print{"VALUE=\"" + StringUtil.encodeHtmlTag(firstName) + "\" ");

    out.println("></TD>");
    out.println("</TR>");
    out.println("<rR>");
    out.println("<TD>Last Name</TD>");
    out.print("<TD><INPUT TYPE=TEXT Name=lastName");

    if (displayPreviousValues)
        out.print(' VALUE=\"" + StringUtil.encodeHtmlTag(lastName) + "\"
```

```
        out.println("></TD>");
        out.println("</TR>");
        out.println("<TR>");
        out.println("<TD>User Name</TD>");
        out.print("<TD><INPUT TYPE=TEXT Name=userName");

        if (displayPreviousValues)
            out.print(" VALUE=\"" + StringUtil.encodeHtmlTag(userName) + "\"

            out.println("x/TD>");
            out.println("</TR>");
            out.println("<TR>");
            out.println("<TD>Password</TD>");
            out.print("<TD><INPUT TYPE=PASSWORD Name=password");

        if (displayPreviousValues)
            out.print(" VALUE=\"" + StringUtil.encodeHtmlTag(password) + "\"
            out.println("></TD>");
            out.println("</TR>");
            out.println("<TR>");
            out.println("<TD><INPUT TYPE=RESET></TD>");
            out.println("<TD><INPUT TYPE=SUBMIT></TD>");
            out.println("</TR>");
            out.println("</TABLE>");
            out.println("</FORM>");
            out.println("<BR>");
            out.println("<BR>"};
        }
}
```

Displaying All Records

Displaying data is inevitable when you program with a database. You can achieve this task easily with the knowledge of JDBC you have learned so far. Which data you need to display depends on your application. In this section and the next, you will display data and format it in an HTML table. This sec-tion displays all records in the Users table, and the next section implements a Search Page that displays data selectively.

Displaying all records could not be simpler. You need only to open a Connection, create a Statement object, and execute a query to produce a ResultSet object. You then iterate each record in a while loop. The code is very straightforward and is given in the following listing.

Displaying All Records in the Users Table

```
    import javax.servlet.*;
    import javax.servlet.http.*;
    import java.io.*;
    import java.sql.*;
    import com.brainysoftware.Java.StringUtil;

    public class DataViewerServlet extends HttpServlet {
        /**Load JDBC driver*/
        public void init() {
            try {
                Class.forName("sun.jdbc.odbc.JdbcOdbcDriver"};
```

```
			System.out.printlnf'JDBC driver loaded");
		}
		catch (ClassNotFoundException e) {
			System.out.println(e.toString());
}
/**Process the HTTP Get request*/
public void doGet(HttpServletRequest request, HttpServletResponse response)
	throws ServletException, IOException {
	response.setContentType("text/html");

	PrintWriter out = response.getwriterf);
	out.println("<HTML>");
	out.println{"<HEAD>");
	out.println("<TITLE>Display All Users</TITLE>");
	out.println("</HEAD>");
	out.println("<BODY>");
	out.println("<CENTER>");
	out.println("<BR><H2>Displaying All Users</H2>");
	out.println("<BR>");
	out.println("<BR>");
	out.println("<TABLE>');
	out.println("<TR>");
	out.println("<TH>First Name</TH>");
	out.println("<TH>Last Name</TH>"};
	out.println{"<TH>User Name</TH>");
	out.println("<TH>Password</TH>");
	out.println("</TR>"};

	String sql = "SELECT FirstName, LastName, UserName, Password" +
		"FROM Users";
	try {
		Connection con = DriverManager,getConnection("jdbc:odbc:JavaWeb");
		System.out.println("got connection");

		Statement s = con.createStatement();
		ResultSet rs = s.executeOuery(sql);

		while (rs.next()) {
			out.println("<TR>");
			out.println("<TD>" + StringUtil.encodeHtmlTag(rs,getString(1)) + "</TD>");
		out.println("<TD>" + Stringlttil.encodeHtmlTag{rs,getstring(2)) + "</TD>");
		out.println("<TD>" + StringUtil.encodeHtmlTag(rs,getString(3)) + "</TD>");
		out.println("<TD>" + StringUtil.encodeHtmlTag(rs.getstring(4)) + "</TD>");
		out.println{"</TR>");
	}
	rs.close();
	s.close();
	con.close();
}
	catch (SQLException e) {
	}
	catch (Exception e) {
	}
```

```
        out.println("</TABLE>");
        out.println("</CENTER>");
        out.println("</BODY>");
        out.println("</HTML>");
    }
    public void doPostfHttpServletRequest request, HttpServletResponse response)
        throws ServletException, IOException {
        doGet(request, response);
        }
    }
```

Search Page

In many cases, displaying all records is not preferable. More often than not, you need only to display data selectively. For example, someone might try to find the details of a person called Jones without knowing the person's first name. Displaying all the users does not help much, especially if you have a large number of records in the table. The page is more useful if you can let the user specify a keyword and display all records that match the keyword.

In this example, you will write a SearchServlet that lets the user enter a keyword that could be the first name, the last name, or even part of the first name or last name.

The main part of this servlet is the sendSearchResult method that connects to the database, executes the following SQL statement, and sends the result to the browser.

```
SELECT FirstName, LastName, UserName, Password
FROM users
WHERE FirstName LIKE '%keyword%'
OR LastName LIKE ' %keyword%'
```

The code is given in the following listing.

The SearchServlet Servlet

```
import  javax.servlet.*;
import  javax.servlet.http.*;
import  java.io.*;
import  java.util.*;
import  java.sql.*;
import  com.brainysoftware.Java.StringUtil;

public class SearchServlet extends HttpServlet {
private String keyword = "";

public void init() {
    try {
        Class. forNamef" sun. jdbc.odbc.JdbcOdbcDriver"
        System. out. println("JDBC driver loaded");
    }
    catch (ClassNotFoundException e) {
        System. out. print ln(e.toString()) ;
    }
}

/**Process the HTTP Get request*/
public void doGet(HttpServletRequest request,
HttpServletResponse response) throws ServletException,
IOException {
```

```
        sendPageHeader(response);
        sendSearchForm(response) ;
        sendPageFooter(response);

    /**Process the HTTP Post request*/
    public void doPost(HttpServletRequest request,
    HttpServletResponse response) throws ServletException,
    IOException {

        keyword = request. getParameter( "keyword" );
        sendPageHeader(response);
        sendSearchForm(response);
        sendSearchResult(response) ;
        sendPageFooter(response);
    }

    void sendSearchResult(HttpServletResponse response)
        throws IOException {
        PrintWriter out = response. getWriter() ;
        try {
            Connection con = DriverManager.getConnection{" jdbc:odbc:JavaWeb"
            System. out. println("got connection");

            Statement s = con.createStatement() ;

            out.println("<TABLE>");
            out.println("<TR>");
            out.printin("<TH>First Name</TH>");
            out.println("<TH>Last Name</TH>") ;
            out.println("<TH>User Name</TH>");
            out.println ("<TH>Pas.sword</TH>" ) ;
            out.println("</TR>");
            String sql = "SELECT FirstName, LastName, UserName, Password" +
            " FROM Users" + 11 WHERE FirstName LIKE '%
                + StringUtil.fixSqlFieldValue(keyword) + "%" +
                ' OR LastName LIKE 'V -t- StringUtil.fixSqlFieldValue(keyword)
                + " % " + ;
            ResultSet rs = s.executeQuery(sql);
            while (rs.next()) {
            out.println("<TR>"); out.println("<TO>" + StringUtil. encodeHtmlTag (rs.getString(l)) + "</
    TD> ");
            out. print In ("<TD>" + StringUtil. encodeHtmlTag(rs.getString(2}) + "</TD> ");
            out.println("<TD>° + StringUtil. encodeHtmlTag(rs.getString(3)) + "</TD> ");
            out.println("<TD>" + StringUtil. encodeHtmlTag(rs.getString(4)) + "</TD> ");
            out.println("</TR>");
        }
        s. close ();
        con.closef);
    }
    catch (SQLException e) {
    }
    catch (Exception e) {
    }
```

```
Out.println('</TABLE>");

/**
* Send the HTML page header, including the title
* and the <BODY> tag
*/
private void sendPageHeader(HttpServletResponse response)
    throws ServletException, IOException {
    response. setContentType(" text /html") ;
    PrintWriter out = response. getWriter( );
    out.println("<HTML>");
    out.println("<HEAD>");
    out.println("<TITLE>Displaying Selected Record(s)</TITLE>");
    out.println("</HEAD>");
    out.println("<BODY>");
    out.println("<CENTER>");
}

/**
* Send the HTML page footer, i.e. the </BODY>
* and the </HTML>
*/
private void sendPageFooter(HttpServletResponse response)
    throws ServletException, IOException {
    PrintWriter out - response. getWriter();
    out.println{"</CENTER>");
    out.println("</BODY>");
    out.println("</HTML>");
}

/**Send the form where the user can type in
* the details for a new user
*/
private void sendSearchForm(HttpServletResponse response)
    throws IOException {

    PrintWriter out = response. getwriter();
    out.println( "<BR><H2>Search Form</H2>" ) ;
    out.println ("<BR>Please enter the first name, last name or part of any.") ;
    out.println("<BR>") ;
    out.println ( " <BR><FORM METHOD=POST>' ) ;
    out.print("Name: <INPUT TYPE=TEXT Name=keyword") ;
    out.println(" VALUE=\"" + StringUtil.encodeHtmlTag(keyword)
    out.println(">") ;
    out.println("<INPUT TYPE=SUBMIT>");
    out.println("</FORM>");
    out.println("<BR>");
    out.println ("<BR>"};
    }
}
```

An Online SQL Tool

When working with a database, you may want to change data in a database for testing. When connecting to a database from a servlet, however, it is often the case that the database resides somewhere beyond your physical reach. Oftentimes, the only way to connect to the database is through the database client program. Of course, the client program needs to be installed on the computer you are using to connect to the database. This is not good if you need to be mobile, or if you need to be able to connect to the database from different locations. Because the database is available to a servlet, you can also make it available online.

The program given in this application is an online SQL tool that lets you manipulate your data or data structure by using a web browser. You can type in any SQL statement and send it to the server to be executed by the database server. If the database server returns a ResultSet, the tool displays it in the form of a table. If the database server sends an error message, that also is displayed in the form.

The user can type an SQL statement in the box.

When first called, the servlet will invoke the doGet method that calls the sendSqlForm. The latter sends the SQL form to the browser. The doGet method is given in the following listing.

The doGet Method of the SQLToolServlet

```
public void doGet(HttpServletRequest request,
    HttpServletResponse response)
throws ServletException, IOException {
    sendSqlForm(request, response);
}
```

The sendSqlForm method does what its name implies: it sends the SQL form. The method is given in the following listing.

The sendSqlForm Method

```
private void sendSqlForm(HttpServletRequest request,
    HttpServletResponse response)
    throws ServletException, IOException {

    response.setContentType("text/html");
    PrintWriter out = response.getwriter();
    out.println("<HTML>");
    out.println("<HEAD>");
    out.println("<TITLE>SQL Tool Servlet</TITLE>");
    out.println ( "</HEAD> " ) ;
    out.println (" <BODY> " ) ;
    out.println("<BR><H2>SQL Tool</H2>");
    out.println("<BR>Please type your SQL statement in the following box.");
    out.println("<BR>");
    out.println ( "<BR><FORM METHOD=POST> " ) ;
    out.println("<TEXTAREA NAME=sql COLS=80 ROWS=8>"};
    String sql = request. getParameter("sql");

    // Reprint tne previously entered SOL in the TextArea
    if (sql!=null)
        out.println(sql);

    out.println("</TEXTAREA>");
    out.println("<BR>");
    out.println("<INPUT TYPE=SUBMIT VALUE=Execute>") ;
    out. println ("</FORM>");
    out.println("<BR>");
    out.println("<HR>"};
```

```
        out.println("<BR>");

        if (sql!=null) {
            executeSql{ sql. trim () , response);
        }

        out.println("</BODY>");
        out.println("</HTML>");
    }
```

The sendSqlForm method retrieves the sql value from the HttpServletRequest object. When this method is called from the doGet method, the sql parameter is null. When called from the doPost method, however, the sql parameter is not null. When the latter is the case, the executeSql method is called.

The executeSql method accepts a String containing the SQL statement, which it tries to execute, and a HttpServletResponse object that it can write to for sending text to the browser. The executeSql method is given in the following listing.

The executeSql Method

```
    public void executeSql(String sql, HttpServletResponse response)
        throws ServletException, IOException {

        PrintWriter out = response.getWriter();
        try {
            //Class.forName("sun.j dbc.odbc.JdbcOdbcDriver");
        Connection con = DriverManager.getConnection("jdbc:odbc:JavaWeb");
        System. out. println(" got connection") ;

        Statement s = con.createStatement() ;

        if (sql.toUpperCase().startsWith("SELECT"}) {
            out.println("<TABLE BORDER=1>");
            ResultSet rs = s.executeQuery(sql);
            ResultSetMetaData rsmd = rs.getMetaData();
            // Write table headings
            int columnCount = rsmd.getColumnCount();
            out.println("<TR>");
            for (int 1*1 ; i<=columnCount; i++) {
                out.println("<TD><B>" + rsmd.getColumnName(1) + "</B></TD>\n");
        }
        out.println("</TR>");
        while (rs.next(J) {
            out.println("<TR>") ;
            for (int i=1; i<=columnCount; 1++( {
                out.println( "<TD>" + StringUtil.encodeHtmlTag(rs.getString(1}) + "</TD>" );
                }
                out.println("</TR>");
            }
            rs.close();
            out.println("</TABLE>");
            }
            else {
                int i = s.executeUpdate(sql);
                out.println("Record(s) affected: " + i);
```

```
        }
        s.close();
        con. close( );
        out.println("</TABLE>");
    }
    catch (SQLException e) {
    out.println("<B>Error</B>");
    out.println("<BR>");
    out. printing. toString()) ;
    }
    catch (Exception e) {
    out.println("<B>Error</B>") ;
    out.println( "<BR>") ;
    out.println{e.toString());
    }
}
```

The complete code of the SQLToolServlet is given in the following listing.

The SQLToolServlet Servlet

```
import  javax.servlet.*;
import  javax.servlet. http.*;
import  java.io.*;
import  java.util.*;
import  java.sql.*;
import  com. brainysoftware. Java. StringUtil;

public class SQLToolServlet extends HttpServlet {

    /**Load the JDBC driver*/
    public void init() {
        try {
            Class.forName ( "sun . jdbc . odbc . JdbcOdbcDriver" ) ;
            System. out. print ln ("JDBC driver loaded"};
        }
        catch (ClassNotFoundException e) {
        System. out. println(e.toString() ) ;
        }
    }

    /**Process the HTTP Get request*/
    public void doGet(HttpServletRequest request,
        HttpServletResponse response)
    throws ServletException, IOException {
        sendSqlForm( request, response);
    }

    /**Process the HTTP Post request*/
    public void doPost(HttpServletRequest request,
    HttpServietResponse response) throws ServletException, IOException {
        sendSq!Form( request,  response);
    }

    /**Send the form where the user can type in
    * an SOL statement to be processed
```

```
*/
private void sendSq!Form(HttpServletRequest request,
      HttpServletResponse  response)
      throws ServletException, IOException {

response. setContentTyps( "text /html");
PrintWriter out = response. getwriter() ;
out.println("<HTML>");
out.println("<HEAD>") ;
out.println("<TITLE>SQL Tool Servlet</TITLE>") ;
out.println("</H£AD>");
out.println ( " <BODY>" ) ;
out. println ("<BR><H2>SQL Tool</H2>");
out.println{"<BR>Please type your SQL statement in the following box.");
out.println("<BR>");
out.println ( " <BR><FORM METHOD=POST> " ) ;
out.println ("<TEXTAREA NAME=sql COLS=80 ROWS=8>");
String sql = request, getParameter{ "sql" );

// Reprint the previously entered SQL in the TextArea
if (sqll=null)
      out.println (sql);

out.println ( "</TEXTAREA>" ) ;
out.println("<BR>");
out.println ("<INPUT TYPE=SUBMIT VALUE=Execute>") ;
out.println("</FORM>");
out.println("<BR>");
out.println("<HR>");
out.println("<BR>");

if  (sql!=null)  {
      executeSql(sql.trim() ,  response) ;
}

out.println("</BODY>");
out.println("</HTML>");
}

/**execute the SQL */
public void executeSql(String sql, HttpServletResponse response)
      throws ServletException, IOException {

      PrintWriter out = response. getWriter( );
      try {
            //Class. forName( "sun. jdbc.odbc.JdbcOdbcDriver");
            Connection con = DriverManager.getConnection("jdbc:odbc:JavaWeb");
            System. out. println ("got connection");

            Statement s = con.createStatement() ;

            if (sql.toUpperCase().startsWith("SELECT")) {
                  out.println("<TABLE BOROER=1>");
```

```
                    ResultSet rs = s.executeQuery(sql);
                    ResultSetMetaData rsmd = rs.getMetaData();
                    // Write table headings
                    int columnCount = rsmd. get ColumnCount() ;
                    out.println{"<TR>");
                    for (int i=1; i<=columnCount; i++) {
                        out.println("<TD><B>" + rsmd.getColumnName(i) + "<B>"<>/TD>\n);
                    }
                out.println("</TR>");
                while (rs.next() {
                    out.println("<TR>");
                    for (int i=1; i<=columnCount; i++) {
                        out.println("<TD>" StringUtil.encodeHtmlTag(rs.getString(i)) + "</TD>");
                    }
                    out.println("</TR>");
                }
                rs.closet);
                out.printlnt"</TABLE>");
            }
            else {
                int i = s.executeUpdate(sql);
                out.println{"Record(s) affected; " + i);
            }
            s. closet);
            con.closet);
            out.println("</TABLE>");
        }
        catch (SQLException e) {
            out.println("<B>Error</B>");
            out.println("<BR>"};
            out.println(e.toString());
        }
        catch (Exception e) {
            out.println("<B>Error</B>");
            out.println("<BR>");
            out.println{e.toString());
            }
        }
```

Should I Keep the Connection Open?

All the examples given up to this point in this chapter have closed the Connection object after the servlet services the request. However, you might ask the following question: If the servlet is called more than once and it opens a Connection object with the same arguments (url, user, and password) each time, wouldn't it better to leave the Connection object open for future use; that is, for .when the servlet is called again?

In fact, opening a Connection object is one of the most expensive opera-tions in database programming. Being able to save CPU cycles by keeping the Connection object is very tempting. In many reference books, however,including those of non-Java web programming, you are always told to close the Connection object as soon as it's no longer needed. The recommendation is so intense that some people have the impression that keeping the connec-tion open is forbidden. Why?

Actually, an open connection is not taboo. Whether to leave the Connection object open really depends on the application you are building. First, however; you should consider that it is true that the response time will be faster if you can access the database without having to open a Connection object. For example, you can make the object variable of the Connection object class level so that it can be accessed from anywhere in the servlet and put the code that opens the Connection object in the init method. This way, the Connection object will be opened when the servlet first initializes. On subsequent requests, the Connection is already open, so the user will experience faster response.

A servlet can be accessed by multiple users at the same time, however. The bad news is that the underlying architecture of the Connection object will allow only one query to be processed at a time. As a result, only one user can use the connection at a time. The others will have to queue. Considering that modern databases allow multiple connections, this is really a waste.

The conclusion is that if your servlet is accessed only by a single user at a time, leaving the Connection object open makes the response time faster. If you can't guarantee that there will not be simultaneous accesses to the servlet, however, response rime will start to decrease because the second and subse-quent users have to wait until the first user gets serviced.

That's why, except in some rare cases, you should always close your Connection object. Rather than having the Connection object open, a technique called *connection pooling* is implemented by the driver to manage the Connection objects to a database so that multiple connections can be pooled. A connection can be used and, when finished, returned to the pool to be used by another request.

Transactions

A *transaction* is a unit of work that can comprise one or more SQL operations. In the examples you have seen so far, every SQL operation is executed and the change is committed (made permanent) on the database right after the SQL statement is executed.

In some cases, however, a set of SQL operations must succeed or fail as a whole. If one of the SQL statements fails, the other SQL statements in the group must also be rolled back. Consider the following scenario.

In an online store a customer purchases several books. When the customer checks out and pays using a credit card, the following things must happen:

1. A record must be added to the Orders table specifying the order, including the delivery address and the credit card details. This operation results in an OrderId used to identify each item in the OrderDetails table.
2. A record must be inserted into the OrderDetails table for each purchase item. Each item is linked to the Orders table using the OrderId value returned by the previous SQL operation.

Now, if everything goes smoothly, both SQL statements will be executed successfully in the database.Things can go wrong, however. Say, for example, the first operation is committed successfully but the second SQL statement fails. In this case, the order details are lost and the customer won't get anything. The customer's credit card will still be debited by the amount of the purchase, however, because a record is added to the Orders table.

In this case, you want both SQL statements to succeed or fail as a whole. You can do this using a transaction. Upon a failed transaction, you can notify the customer so that he or she can try to check out again.

By default, a Connection object's auto-commit state is true, meaning that the database is updated when an SQL statement is executed. If you want to group a set of SQL statements in one transaction, you must first tell the Connection object not to update the change until it is explicitly notified to do so. You do this by calling the setAutoCommit method and passing false as its argument, as follows:

```
connection.setAutoCommitffalse};
```

Then, you can execute all the SQL statements in the group normally, using the executeQnery and the updateQuery methods. After the last call to the executeQuery or the updateQuery method, you call the commit method of the Connection object to make the database changes permanent, such as connection.commit();

If you don't call the commit method within a specified time, all SQL state-ments will be rolled back after calling the setAutoCommit method. Alternatively, you can roll back the transaction explicitly by calling the Connection object's rollback method:

```
connection.rollback();
```

Connection Pooling

Creating a Connection, object is one of the most expensive operations in data-base programming. This is especially true when the Connection object is used for only one or two SQL operations, as demonstrated in the servlets discussed in this chapter.

To make the use of Connection objects more efficient, those objects are not pooled. When the application starts, a certain number of Connection objects are created and stored in a pool. When a database client, such as a servlet, needs to use a Connection object, it does not create the object but instead requests one from the pool. When the client is finished with it, the Connection object is returned to the pool.

You can create a pool of Connection objects programmatically; however, there is an easier and more robust way: by using the connection pooling fea-ture of the JDBC Optional Package. The good thing about this feature is that the connection pooling is transparent. There is no single line of code that the programmer needs to change to use it.

To use the connection pooling feature in the javax.sql package, you need to connect to the data source using the javax.sql,DataSource interface. The code is as follows:

```
Context context = new InitialContext();
DataSource ds = (DataSource)context.lookup("jdbc/myDB");
Connection connection = ds.getConnection(user, password)
```

Revision

1. Most web applications use a database. Database accessing and programming therefore play a significant role in web development.
2. The Java.sql package provides the API for accessing and processing data in a data source.
3. The Driver interface is implemented by every JDBC driver class.
4. The Connection interface represents a connection to the database.
5. You use the statement interface method to execute an SQL statement and obtain the results that are produced.
6. The ResultSet interface represents a table-like database result set.
7. The PreparedStatement interface extends the Statement interface and represents a precompiled SQL statement.
8. Before you can manipulate data in a database, you need to connect to that database server.
9. Database servers have their own proprietary "languages" for communication.
10. Because JDBC drivers normally come in a .jar file, the first thing you need to do after you get the driver is tell Tomcat where to find it by copying the jar file into the WEB-INF\lib directory under your application directory.

11. After you register a JDBC driver with the DriverManager, you can use it to get a connection to the database. In JDBC, a database connection is represented by the Java.sql.Connection interface.
12. After you have a Connection object, your SQL skill takes over. Basically, you can pass any SQL statement that your database server understands.
13. A ResultSet is the representation of a database table that is returned from a Statement object.
14. When the user types the URL in the Location or Address box and first calls the servlet, its doGet method is invoked.
15. The values for the user name and the password are enclosed in the single quote pairs, as in 'jeff' and 'Java'.
16. Inserting a record into a table is a common database task.
17. Displaying data is inevitable when you program with a database.
18. In many cases, displaying all records is not preferable. More often than not, you need only to display data selectively.
19. When working with a database, you may want to change data in a database for testing.
20. A *transaction* is a unit of work that can comprise one or more SQL operations.

Multiple Choice Questions

1. __________ acts as a mediator between the Java Application and the database..
   (a) ODBC (b) JDBC
   (c) RPC
2. __________ manages a list of database drivers..
   (a) DriverManager (b) Driver
   (c) ODBC
3. Most JDBC drivers should accept properties:
   (a) user (b) password
   (c) hostname (d) all above
4. ________ interface with all methods for contacting a database.
   (a) Database (b) Driver
   (c) Connection
5. Use the ______________ method of interface java.sql.DatabaseMetaData to probe the database for stored procedures.
   (a) getMethod() (b) getProcedures()
   (c) any one of them

Answers

1. b 2. a 3. d 4. c 5. b

Subjective Questions

1. What is JDPC?
2. Desribe the various steps involved in getting the databae.
3. What is meant by A Database-Based Login Servlet?

4. What is The Single Quote Factor?
5. How would you insert data into a table with RegistrationServlet?
6. How would you search a page?
7. What are Transactions?
8. What is Connection Pooling?
9. Write short note on the following:

   The Driver Interface

   The Connection Inerface

   The Statement Interface

   The ResultSet Interface

   The PreparedStatement Interface

   The ResultSetMetaData Interface

11 Common Gateway Information (CGI), Practical Extraction and Review Language (PERL)

Introduction

The Common Gateway Interface (CGI) is a standard for interfacing external applications with information servers, such as HTTP or Web servers. A plain HTML document that the Web daemon retrieves is static, which means it exists in a constant state: a text file that doesn't change. A CGI program, on the other hand, is executed in real-time, so that it can output dynamic information. For example, let's say that you wanted to "hook up" your Unix database to the World Wide Web, to allow people from all over the world to query it. Basically, you need to create a CGI program that the Web daemon will execute to transmit information to the database engine, and receive the results back again and display them to the client. This is an example of a *gateway,* and this is where CGI, currently version 1.1, got its origins.

The database example is a simple idea, but most of the time rather difficult to implement. There really is no limit as to what you can hook up to the Web. The only thing you need to remember is that whatever your CGI program does, it should not take too long to process. Otherwise, the user will just be staring at their browser waiting for something to happen.

Overview

A CGI program can be written in any language that allows it to be executed on the system, such as:

- C/C++
- Fortran
- PERL
- TCL
- Any Unix shell
- Visual Basic
- AppleScript

It just depends what you have available on your system. If you use a programming language like C or Fortran, you know that you must compile the program before it will run. If you look in the /cgi-src directory that came with the server distribution, you will find the source code for some of the CGI programs in the /cgi-bin directory. If, however, you use one of the scripting languages instead, such as PERL, TCL, or a Unix shell, the script itself only needs to reside in the /cgi-bin directory, since there is no associated source code. Many people prefer to write CGI scripts instead of programs, since they are easier to debug, modify, and maintain than a typical compiled program.

Get the Form data from

As you now know, there are two methods which can be used to access your forms. These methods are GET and POST. Depending on which method you used, you will receive the encoded results of the form in a different way.

- The GET method

  If your form has METHOD="GET" in its FORM tag, your CGI program will receive the encoded form input in the environment variable QUERY_STRING.
- The POST method

  If your form has METHOD="POST" in its FORM tag, your CGI program will receive the encoded form input on stdin. The server will NOT send you an EOF on the end of the data, instead you should use the environment variable CONTENT_LENGTH to determine how much data you should read from stdin.

A simple CGI script

The first task is to write a very simple peri program which just prints out an HTML document. This should include all the usual <head> ... </head> and <title> ... </title> commands. However, there are two things to note.

First of all, the first two lines of the print out are special. The first line must be

Content-type: text/html

so that the browser knows what kind of document it is and therefore how to display it. In this case it is a text document of the subclass HTML. You can also send back plain ASCII documents by using text/ plain instead. Sound samples and images also have to have the content type specified appropriately. The second line must be blank (ie it must contain just a line feed). This line is reserved for future development.

The second point is really a tip: Strings in peri programs can span over several lines. This means the.main part of your peri program can just be one enormous **print** statement

Forms

Usually the information supplied by the QUERY_STRING variable should come from the user press-ing buttons and entering text in the HTML document. It is this information we would like to package up and send to the CGI script. Each group of buttons and text boxes is called a form, and forms are enclosed between the HTML tags <form> ... </form>. You also have to tell it the URL to send the information to, and how the information is sent. The result is something like this:

```
<form action="http:/www/comp.leeds.ac.uk/sam-cgi/answerme"
    method="GET">
Some text in here
It can anything except another form.
</form>
```

The action tag is the URL of the CGI script. The method GET tells it to use the QUERY_STRING method of sending information. As indicated, almost anything can go between the form tags, including text and various types of input devices. In particular we can have...

Submit buttons

A submit button is the input device that actually calls the URL. It has a value which is the mes-sage that appears on the button. Here is the code for a form with just a submit button in it. When you click on the submit button the URL specified in the form's action is called.

```
<form action="http//www.comp.leeds.ac.uk/cgi-bib/Perl/environment-example"
    method="GET">
<input type="submit" value="Click me">
</form>
```

The result is a form which looks like this.

Top of Form

Click me

Bottom of Form

If you click the submit button then the URL will be called. However the QUERY_STRING variable will be null because no information was specified. The answer is to use...

Checkboxes

A checkbox is a simple on/off button. A checkbox has a name (its key) and a value that this key has when the box is clicked. As an example, here is the HTML code for a form with a checkbox and a submit button in it.

The result of this code is the following form

Top of Form

Do it

Bottom of Form

Now if the submit button is clicked when the box is checked then the information lights=on is packaged into QUERY_STRING. However if the box is not checked then no information is packaged into QUERY_STRING and it remains empty. Notice also that the checkbox does not appear with a message. This is something you have to add yourself as ordinary HTML text.

Here is example HTML code for a form with two checkboxes and a message for each.

```
<form action="http://www.comp.leeds.ac.uk/cgi-bin/Perl/environment-example"
    method="GET">
<input type="checkbox" name="lights" value="on"> Lights
<input type="checkbox" name="camera" value="on"> Camera
<input type="submit" value="Do it">
</form>
```

The result of this code is the following form

Top of Form

Lights **Cameras** **Do it**

Bottom of Form

Click the submit button with various combinations of checked boxes and watch how the QUERY_STRING environment variable changes. If both boxes are checked then the names are separated by an & sign, as we saw earlier.

Radio Buttons

Radio buttons are just like checkboxes except they are grouped together and only one button in the group may be selected at a time. All the buttons in a group must have the same name and each one should have a different value. You can also specify which buttons (if any) are checked initially. When the submit button is clicked the name and the value of the selected button are packaged up for QUERY_STRING.

Here is some example code for five such buttons. They are all of type radio, and are in the group named cert. The 15 button is checked initially.

```
<form action="http://www.comp.leeds.ac.uk/cgi-bin/Perl/environment-example"
    method="GET">
<input type="radio" name="cert" value="u"> U
<input type="radio" name="cert" value="pg"> PG
<input type="radio" name="cert" value="12"> 12
```

```
<input type="radio" name="cert" value="15" checked > 15
<input type="radio" name="cert" value="18"> 18
<input type="submit" value="Certify">
</form>
```

The result of this HTML code is the following form.

OU OPG O12 O 15 O 18 Certify

Again, try this out for yourself and watch QUERY_STRING change. Notice that the value of cert for the U and PG buttons are in lowercase because this is what we specified with the value tag.

Text boxes

Finally we deal with text input devices. These are simply boxes into which the user can enter some text which is then packaged up under a particular name/Here is some example code for two text boxes and a submit button. The <br> tag causes a line break.

```
<form action="http://www.comp.leeds.ac.uk/cgi-bin/Perl/environment-example"
    method="GET">
Director: <input type="text" name="dir"> <br>
Producer: <input type="text" name="prod">
<input type="submit" value="Fire">
</form?
```

The result is the following form

Director : [ ]

Producer : [ ] Fire

Recall that spaces are encoded as + signs and some other characters are encoded in their hexadecimal form. Try entering signs like &, + and % in particular.

PERL (PRACTICAL EXTRACTION AND REVIEW LANGUAGE)

A Basic Program

Here is the basic peri program that we'll use to get started.

```
# ! /usr/local/bin/perl
#
# Program to do the obvious
#
print "Hello world";
# Print a message
```

Each of the parts will be discussed in turn.

The first line

Every perl program starts off with this as its very first line:

```
#!/usr/local/bin/perl
```

although this may vary from system to system. This line tells the machine what to do with the file when it is executed (i.e., it tells it to run the file through Perl).

Comments and Statements

Comments can be inserted into a program with the # symbol, and anything from the # to the end of the line is ignored (with the exception of the first line). The only way to stretch comments over several lines is to use a # on each line.

Everything else is a Perl statement which must end with a semicolon, like the last line above.

Simple Printing

The print function outputs some information. In the above case it prints out the the literal string Hello world, and of course the statement ends with a semicolon.

Scalar Variables

The most basic kind of variable in Perl is the scalar variable. Scalar variables hold both strings and numbers, and are remarkable in that strings and numbers are completely interchangable. For example, the statement

$priority = 9;

sets the scalar variable Spriority to 9, but you can also assign a string to exactly the same variable:

$priority = 'high';

Perl also accepts numbers as strings, like this:

$priority = '9 ';
$default = '0009' ;

and can still cope with arithmetic and other operations quite happily.

In general variable names consists of numbers, letters and underscores, but they should not start with a number and the variable $_ is special, as we'll see later. Also, Perl is case sensitive, so $a and $A are different.

Operations and Assignment

Perl uses all the usual C arithmetic operators:

```
$a = 1 + 2;        # Add 1 and 2 and store in $a
$a = 3 – 4;        # Subtract 4 from 3 and store in $a
$a = 5 * 6;        # Multiply 5 and 6
$a = 7 / 8;        # Nine to the power of 10
$a = 9 ** 10;      # Add 1 and 2 and store in $a
$a = 5 % 2;        # Remainder of 5 divided by 2
++$a;              # Increment $a and then return it
$a++;              # Return $a and then increment it
–$a;               # Decrement $a and then return it
$a–;               # Return $a and then decrement it
```

and for strings Perl has the following among others:

```
$a = $b . $c;      # Concatenate $b and $c
$a = $b x $c;      # $b repeated $c times
```

To assign values Perl includes

```
$a = $b;           # Assign $b and $a
$a += $b;          # Add $b to $a
$a –= $b;          # Subtract $b from $a
$a .= $b;          # Append $b onto $a
```

i ll

Note that when Perl assigns a value with $a = $b it makes a copy of $b and then assigns that to $a. Therefore the next time you change $b it will not alter $a.

Other operators can be found on the **perlop** manual page. Type **man perlop** at the prompt.

Interpolation

The following code prints *apples and pears* using concatenation:

$a = 'apples';

```
$b = 'pears';
print $a.' and '.Sb;
```

It would be nicer to include only one string in the final print statement, but the line

```
print '$a and $b';
```

prints literally *$a and $b* which isn't very helpful. Instead we can use the double quotes in place of the single quotes:

```
print "$a and $b";
```

The double quotes force *interpolation* of any codes, including interpreting variables. This is a much nicer than our original statement. Other codes that are interpolated include special characters such as newline and tab. The code \n is a newline and \t is a tab.

Array Variables

A slightly more interesting kind of variable is the *array variable* which is a list of scalars (ie numbers and strings). Array variables have the same format as scalar variables except that they are prefixed by an @ symbol. The statement

```
@food = ("apples", "pears", "eels");
@music = ("whistle", "flute");
```

assigns a three element list to the array variable @food and a two element list to the array variable @music.

The array is accessed by using indices starting from 0, and square brapkets are used to specify the index. The expression

```
$food[2]
```

returns *eels.* Notice that the @ has changed to a $ because *eels* is a scalar.

Array Assignments

As in all of Perl, the same expression in a different context can produce a different result. The first assignment below explodes the @music variable so that it is equivalent to the second assignment.

```
@moremusic = ("organ", @music, "harp");
©moremusic = ("organ", "whistle", "flute", "harp"),-
```

This should suggest a way of adding elements to an array. A neater way of adding elements is to use the statement

```
push(@food, "eggs");
```

which pushes eggs onto the end of the array @food. To push two or more items onto the array use one of the following forms:

```
push(@food, "eggs", "lard");
push(@food, ("eggs", "lard")),
push(@food, (Jmorefood) ;
```

The **push** function returns the length of the new list.

To remove the last item from a list and return it use the pop function. From our original list the pop function returns *eels* and @food now has two elements:

```
$grub = pop(@food); # Now $grub = "eels"
```

It is also possible to assign an array to a scalar variable. As usual context is important. The line

```
$f = ©food;
```

assigns the length of @food, but

```
$f = "©food";
```

turns the list into a string with a space between each element. This space can be replaced by any other string by changing the value of the special $" variable. This variable is just one of Peri's many special variables, most of which have odd names.

Arrays can also be used to make multiple assignments to scalar variables:

```
($a, $b) = ($c, $d);        # Return $a and then decrement it
($a, $b) = @food;           # $a and $b are the first two items of @food
($a, @somefood) = @food;    # $a is the first item of @food
                            # @somefood is a list of the others
($somefood, $a) = @food;    # @somefood is @food and @a is undefined
```

The last assignment occurs because arrays are greedy, and @somefood will swallow up as much of @food as it can. Therefore that form is best avoided.

Finally, you may want to find the index of the last element of a list. To do this for the @food array use the expression

```
$#food
```

Displaying arrays

Since context is important, it shouldn't be too surprising that the following all produce different results:

```
print @food; # By itself
print "@food";          # Embedded in double quotes
print @food. " ";       # In a scalar context
```

File Handling

Here is the basic perl program which does the same as the UNIX cat command on a certain file.

```
#! /usr/local/bin/perl
#
# Program to open the password file, read it in.
# print it, and close it again,
$file = '/etc/passwd';          #   Name the file
open(INFO, $file);              #   Open the file
@lines = <INFO>;                #   Read it into an array
close(INFO);                    #   Close the file
print @lines;                   #   Print the array
```

The **open** function opens a file for input (i.e. for reading). The first parameter is the filehandle which allows Perl to refer to the file in future. The second parameter is an expression denoting the filename. If the filename was given in quotes then it is taken literally without shell expansion. So the expression '-/notes/todolist' will not be interpreted successfully. If you want to force shell expansion then use angled brackets: that is, use <-/notes/todolist> instead.

The **close** function tells Perl to finish with that file.

There are a few useful points to add to this discussion on filehandling. First, the **open** statement can also specify *a* file for output and for appending as well as for input. To do this, prefix the filename with a > for output and a >> for appending:

```
open(INFO, $file);              #   Open for input
open(INFO, ">$file");           #   Open for output
open(INFO, ">>$file");          #   Open for appending
open(INFO, ">>$file");          #   Also open for input
```

Second, if you want to print something to a file you've already opened for output then you can use the print statement with an extra parameter. To print a string to the file with the INFO filehandle use

```
print INFO "This line goes to the file.Xn";
```

Third, you can use the following to open the standard input (usually the keyboard) and standard output (usually the screen) respectively:

```
open(INFO, '-');            # Open standard input
open(INFO, '>-');           # Open standard output
```

In the above program the information is read from a file. The file is the INFO file and to read from it Perl uses angled brackets. So the statement

```
@lines = <INFO>;
```

reads the file denoted by the filehandle into the array @lines. Note that the <INFO> expression reads in the file entirely in one go. This because the reading takes place in the context of an array variable. If ©lines is replaced by the scalar $lines then only the next one line would be read in. In either case its newline character at the end.

Conditionals

Of course Perl also allows if/then/else statements. These are of the following form:

```
if ($a)
{
    print "The string is not empty\n";
}
else
{
    print "The string is empty\n";
}
```

For this, remember that an empty string is considered to be false. It will also give an "empty" result if $a is the string 0.

It is also possible to include more alternatives in a conditional statement:

```
if (!$a)                        #  The ! is the not operator
{
        print "The string is empty\n";
}
elseif (length($a) == 1)        #  If above fails, try this
{
        print "The String has one character\n";
}
elseif (length($a) == 2)        #  If that fails, try this
{
        print "The string has two characters\n";
}
else                            #  Now, everything has failed
{
        print "The string has lots of characters\n";
}
```

In this, it is important to notice that the elsif statement really does have an "e" missing.

String Matching

One of the most useful features of Perl (if not the most useful feature) is its powerful string manipulation facilities. At the heart of this is the regular expression (RE) which is shared by many other UNIX utilities.

Regular expressions

A regular expression is contained in slashes, and matching occurs with the =~ operator. The following expression is true if the string the appears in variable $sentence.

```
$sentence =~ /the/
```

The RE is case sensitive, so if

```
$sentence = "The quick brown fox";
```

then the above match will be false. The operator!- is used for spotting a non-match. In the above example

```
$sentence !~ /the/
```

is true because the string *the* does not appear in $sentence.

The $_ special variable

We could use a conditional as

```
if ($sentence =~ /undeer/)
{
    print "We're talking about rugby\n";
}
```

which would print out a message if we had either of the following

```
$sentence = "Up and under";
$sentence = "Best winkles in Sunderland";
```

But it's often much easier if we assign the sentence to the special variable $_ which is of course a scalar/If we do this then we can avoid using the match and non-match operators and the above can be written simply as

```
if (/under/)
{
    print "We're talking about rugby\n";
}
```

The $_ variable is the default for many Perl operations and tends to be used very heavily.

More on REs

In an RE there are plenty of special characters, and it is these that both give them their power and make them appear very complicated. It's best to build up your use of REs slowly; their creation can be something of an art form.

Here are some special RE characters and their meaning

```
.       # Any single character except a newline
^       # The beginning of the line or string
$       # The end of the line or string
*       # Zero or more of the last character
+       # One or more of the last character
?       # Zero or one of the last character
```

and here are some example matches. Remember that should be enclosed in /.../ slashes to be used.

```
t.e     # t followed by anything followed by e
        # This will match  the
        #                  tre
        #                  tle
        # but not te
        #                  tale
~f      # f at the beginning of a line
```

```
~ftp          # e at the end of a line
e$            # tle at the end of a line
tle$          # un followed by zero or more d characters
              # This will match  un
              #                  und
              #                  undd
              #                  unddd (etc)
              # Any string without a newline. This is because
              # the . matches anything except a newline and
              # the * means zero or more of these.
^$            # A line with nothing in it.
```

There are even more options. Square brackets are used to match any one of the characters inside them. Inside square brackets a - indicates "between" and a ^ at the beginning means "not".

```
[qjk]         # Either q or j or k
[^qjk]        # Neither q or j or k
[a–z]         # Anything from a to z inclusive
[^a–z]        # No lower case letters
[a–zA–Z]      # Any letter
[a–z]+        # Any non-zero sequence of lower case letters
```

At this point you can probably skip to the end and do at least most of the exercise. The rest is mostly just for reference.

A vertical bar | represents an "or" and parentheses (...) can be used to group things together:

```
jelly|cream   # Either jelly or cream
(eg|le)gs     # Either eggs or legs
(da)+         # Either da or dada or dadada or . . .
```

Here are some more special characters:

```
\n            # A newline
\t            # A tab
\w            # Any alphanumeric (word) character
              # The same as [a–zA–Z0–9_]
\W            # Any non-word character.
              # The same as [^a–zA–Z0–9_]
\d            # Any digit. The same as (0 – 9)
\D            # Any non-digit. The same as [^0 – 9]
\s            # Any whitespace character; space,
              # tab, newline, etc.
\S            # Any non-whitespace character
\b            # A word boundary, outside [ ] only
\B            # No word boundary
```

Clearly characters like $, |, [,), \, / and so on are peculiar cases in regular expressions. If you want to match for one of those then you have to preceed it by a backslash. So:

```
\|            # Vertical bar
\[            # An open square bracket
\)            # A closing parenthesis
\*            # An asterisk
\^            # A carat symbol
\/            # A slash
\\            # A backslash
```

Some Example REs

As was mentioned earlier, it's probably best to build up your use of regular expressions slowly. Here are a few examples. Remember that to use them for matching they should be put in /.../ slashes

```
[01l          # Either "0" or "1"
\/0           # A division by zero; "/0"
\/ 0          # A division by zero with a space; "/ 0"
\/ s0         # A division by zero with a whitespace
              # "/ 0" where the space may be a tab etc.
\/ *0         # A division by zero with possibly some
              # spaces; "/0" or "/ 0" or "/  0" etc.
\/ \s*0       # A divison by zero with possibly some whitespace
\/ \s*0\,0*   # As the previous one, but with decimal
              # point and maybe some 0s after it. Accepts
              # "/0." and "/0.0" and "/0.00" etc and
              # "/ 0." and "/  0.0" and "/       0.00" etc.
```

Substitution and Translation

As well as identifying regular expressions Perl can make substitutions based on those matches. The way to do this is to use the s function which is designed to mimic the way substitution is done in the vi text editor. Once again the match operator is used, and once again if it is omitted then the substitution is assumed to take place with the $_ variable.

To replace an occurrence of *london* by *London* in the string Ssentence we use the expression

$sentence =~ s/ london/ London/

and to do the same thing with the $_ variable just

s/l ondon /London /

Notice that the two regular expressions *(london* and *London) are* surrounded by a total of three slashes. The result of this expression is the number of substitutions made, so it is either 0 (false) or 1 (true) in this case.

Options

This example only replaces the first occurrence of the string, and it may be that there will be more than one such string we want to replace. To make a global substitution the last slash is followed by a g as follows:

***s/l* ondon / London / g**

which of course works on the $_ variable. Again the expression returns the number of substitutions made, which is 0 (false) or something greater than 0 (true).

If we want to also replace occurrences of *lOndon, lonDON, LoNDoN* and so on then we could use

s/[Ll] [Oo] [Nn] [Dd] [Oo] [Nn] /London/g

but an easier way is to use the i option (for "ignore case"). The expression

s/ london/ London/ gi

will make a global substitution ignoring case. The i option is also used in the basic/.../ regular expres-sion match.

Remembering Patterns

It's often useful to remember patterns that have been matched so that they can be used again. It just so happens that anything matched in parentheses gets remembered in the variables $1 ,...,$9. These strings can also be used in the same regular expression (or substitution) by using the special RE codes \1,...,\9. For example

```
$_ = "Lord Whopper of Fibbing";
s/([A-Z] }/:\l:/g;
print "$_\n";
```

will replace each upper case letter by that letter surrounded by colons. It will print *:L:ord Whopper of :F:ibbing.* The variables $1 ,...,$9 are read-only variables; you cannot alter them yourself.

As another example, the test

```
if ( / (\b. + \b) \1/)
{
    print "Found $1 repeatedVn";
}
```

will identify any words repeated. Each \b represents a word boundary and the .+ matches any non-empty string, so \b.+\b matches anything between two word boundaries. This is then remembered by the parentheses and storecl as \1 for regular expressions and as $1 for the rest of the program.

The following swaps the first and last characters of a line in the $_ variable:

```
s/"{.)(.*)(.)$/\3\2\1/
```

The ^ and $ match the beginning and end of the line. The \1 code stores the first character; the \2 code stores everything else up the last character which is stored in the \3 code. Then that whole line is replaced with \1 and \3 swapped round.

After a match, you can use the special read-only variables $' and $& and $' to find what was matched before, during and after the seach. So after

```
$_= "Lord Whopper of Fibbing";
/PP/;
```

all of the following are true. (Remember that eq is the string-equality test.)

```
$' eq "Lord Wo";
$& eq "pp";
$' eq "er of Fibbing";
```

Finally on the subject of remembering patterns it's worth knowing that inside of the slashes of a match or a substitution variables are interpolated. So

```
$search = "the";
s/$search/xxx/g;
```

will replace every occurrence of *the* with xxx. If you want to replace every occurence of *there* then you cannot do **s/$searchre/xxx/** because this will be interpolated as the variable Ssearchre. Instead you should put the variable name in curly braces so that the code becomes

```
$search = "the";
s/${search}re/xxx/;
```

Translation

The **tr** function allows character-by-character translation. The following expression replaces each a with e, each *b* with *d,* and each c with *f* in the variable $sentence. The expression returns the number of substitutions made.

```
$sentence =~ tr/abc/edf/
```

Most of the special RE codes do not apply in the tr function. For example, the statement here counts the number of asterisks in the $sentence variable and stores that in the $count variable.

```
$count = ($sentence =~ tr/*/*/);
```

However, the dash is still used to mean "between". This statement converts $_ to upper case.

```
tr/a-z/A-Z/;
```

Type in the example program using a text editor, and save it. Emacs is a good editor to use for this because it has its own Perl mode which formats lines nicely when you hit tab (use 'M-x peri-mode'). But as ever, use whichever you're most comfortable with.

After you've entered and saved the program make sure the file is executable by using the com-mand

chmod u+x *progname*

at the UNIX prompt, where *progname* is the filename of the program. Now to run the program just type any of the following at the prompt.

```
peri progname
. /progname
progname
```

If something goes wrong then you may get error messages, or you may get nothing. You can always run the program with warnings using the command

peri -w *progname*

at the prompt. This will display warnings and other (hopefully) helpful messages before it tries to execute the program. To run the program with a debugger use the command

peri -d *progname*

When the file is executed Perl first compiles it and then executes that compiled version. So after a short pause for compilation the program should run quite quickly. This also explains why you can get compilation errors when you execute a Perl file which consists only of text.

Make sure your program works before proceeding. The program's output may be slightly unex-pected - at least it isn't very pretty. We'll look next at variables and then tie this in with prettier printing.

Control Structures

More interesting possiblities arise when we introduce control structures and looping. Perl supports lots of different kinds of control structures which tend to be like those in C, but are very similar to Pascal, too. Here we discuss a few of them.

Foreach

To go through each line of an array or other list-like structure (such as lines in a file) Perl uses the foreach structure. This has the form

```
foreach $morsel (@food)       # Visit each item in turn
                              # and call it $morsel
{
    print "$morsel\n";        # Print the item
    print "Yum yim\n";        # That was nice
}
```

The actions to be performed each time are enclosed in a block of curly braces. The first time through the block $morsel is assigned the value of the first item in the array @food. Next time it is assigned the value of the second item, and so until the end. If @food is empty to start with then the block of statements is never executed.

Testing

The next few structures rely on a test being true or false. In Perl any non-zero number and non-empty string is counted as true. The number zero, zero by itself in a string, and the empty string are counted as false. Here are some tests on numbers and strings.

```
$a == $b                      # Is $a numerically equal to $b?
                              # Beware: Don't use the - operator;
$a ! = $b                     # Is $a numerically unequal to $b?
$a eq $b                      # Is $a string-equal to $b
```

```
$a ne $b                    # Is $a string-unequal to $b?
```

You can also use logical and, or and not:

```
($a && $b)          # Is $a and $b true?
($a | $b)           # Is either $a or $b true?
! ($a)              # is $a false?
```

For

Perl has *a* for structure that mimics that of C. It has the form

```
for (initialise; test; inc)
{
        first_action;
        second_action;
        etc
}
```

First of all the statement *initialise* is executed. Then while *test* is true the block of actions is executed. After each time the block is executed *inc* takes place. Here is an example for loop to print out the numbers 0 to 9.

```
for ($i = 0; $i < 10; ++$i)        # Start with $i = 1
                                   # Do it while $i < 10
                                   # Increment $i before repeating
{
        print "$i\n";
}
```

While and Until

Here is a program that reads some input from the keyboard and won't continue until it is the correct password

```
#!/usr/local/bin/perl
do
{
        "Password? ";              # Ask for input
        $a = <STDIN>;              # Get input
        chop $a;                   # Chop off newline
}
while ($a ne "fred")               # Redo while wrong input.
```

Revision

1. The Common Gateway Interface (CGI) is a standard for interfacing external applications with information servers, such as HTTP or Web servers.
2. A CGI program can be written in any language that allows it to be executed on the system, such as:

   C/C++

   Fortran

   PERL

   TCL

   Any Unix shell

Visual Basic

AppleScript

3. Usually the information supplied by the QUERY_STRING variable should come from the user pressing buttons and entering text in the HTML document.
4. A submit button is the input device that actually calls the URL.
5. A checkbox is a simple on/off button. A checkbox has a name (its key) and a value that this key has when the box is checked.
6. Radio buttons are just like checkboxes except they are grouped together and only one button in he group may be selected at a time.
7. Finally we deal with text input devices. These are simply boxes into which the user can enter some text which is then packaged up under a particular name.
8. The most basic kind of variable in Perl is the scalar variable.
9. A slightly more interesting kind of variable is the array variable which is a list of scalars (i.e., numbers and strings). Array variables have the same format as scalar variables except that they are prefixed by an @ symbol.
10. One of the most useful features of Perl (if not the most useful feature) is its powerful string manipulation facilities. At the heart of this is the regular expression (RE) which is shared by many other UNIX utilities.

Multiple Choice Questions

1. __________ is a standard for interfacing external applications with information servers, such as HTTP or Web servers.

   (a) CGI (b) PERL
   (c) SP (d) JAVA

2. A CGI program can be written in _______ language.

   (a) C/C++ (b) Fortran
   (c) PERL (d) ANY

3. If your form has _________ in its FROM tag, your CGI program will receive the encoded form input in the environment variable QUERY_STRING.

   (a) METHOD="GET" (b) METHOD="POST"
   (c) EITHER A OR B (d) NONE

4. If your form has ________ in its FROM tag, your CGI program will receive the encoded form input on stdlin.

   (a) METHOD="GET" (b) METHOD="POST"
   (c) EITHER A OR B (d) NONE

5. If you click the _______ button then the URL will be called.

   (a) SUBMIT (b) CHECK
   (c) RADIO (d) NONE

6. QUERY_STRING variable will be ________ because no information was specified.

   (a) ZERO (b) 0
   (c) NUL (d) NONE

7. Comments can be inserted into a program with the ______ symbol.

   (a) % (b) #
   (c) @ (d) $

8. Array variables have the same format as scalar variables except that they are prefixed by an _____ symbol.

(a) % (b) #

(c) @ (d) $

9. To close the file we use __________.

(a) close (INFO); (b) exit()

(c) quit() (d) none

10. A regular expression is contained in slashes, and matching occurs with the _____ operator.

(a) =~ (b) !=

(c) ~= (d) none

Answers

1. a	2. d	3. a	4. b	5. a
6. c	7. b	8. c	9. a	10. a

Subjective Questions

1. What is the procedure for writing a simple CGI script?
2. Write a code for using submit buttons.
3. What is the difference between checkbox and readio button?
4. Write a basic program in Perl and explain.
5. What is the syntax for declaring scalar variable?
6. What are array variables in Perl?
7. What do you mean by interpolation?
8. What is the syntax for declaring scalar variable?
9. How would you handle files in Perl?
10. Explain string matching in Perl.
11. What is control structures in Perl? Explain.
12. Write a program to find sum of n numbers in Perl.
13. Write a program to find the average of marks of five subjects in Perl.
14. Write a program in Perl to calculate simple interest.

12 RMI, VBScript & COM/DOM

INTRODUCTION

One of the central and unique features of RMI is its ability to download the definition of an object's class if the class is not defined in the receiver's Java virtual machine. Like any other Java application, a distributed application built by using Java RMI is made up of interfaces and classes. The interfaces declare methods.

An Overview of RMI Applications

RMI applications often comprise two separate programs, a server and a client. A typical server program creates some remote objects, makes references to these objects accessible, and waits for clients to invoke methods on these objects. A typical client program obtains a remote reference to one or more remote objects on a server and then invokes methods on them. RMI provides the mechanism by which the server and the client communicate and pass information back and forth. Such an application is sometimes referred to as a *distributed object application.*

Distributed object applications need to do the following:

- **Locate remote objects.** Applications can use various mechanisms to obtain references to remote objects. For example, an application can register its remote objects with RMI's simple naming facility, the RMI registry. Alternatively, an application can pass and return remote object references as part of other remote invocations.
- **Communicate with remote objects.** Details of communication between remote objects are handled by RMI. To the programmer, remote communication looks similar to regular Java method invocations.
- **Load class definitions for objects that are passed around.** Because RMI enables objects to be passed back and forth, it provides mechanisms for loading an object's class definitions as well as for transmitting an object's data.

Advantages of Dynamic Code Loading

One of the central and unique features of RMI is its ability to download the definition of an object's class if the class is not defined in the receiver's Java virtual machine. All of the types and behavior of an object, previously available only in a single Java virtual machine, can be transmitted to another, possibly remote, Java virtual machine. RMI passes objects by their actual classes, so the behavior of the objects is not changed when they are sent to another Java virtual machine. This capability enables new types and behaviors to be introduced into a remote Java virtual machine, thus dynamically extending the behavior of an application. The compute engine example in this trail uses this capability to introduce new behavior to a distributed program.

Remote Interfaces, Objects, and Methods

Like any other Java application, a distributed application built by using Java RMI is made up of interfaces and classes. The interfaces declare methods. The classes implement the methods declared in the interfaces and, perhaps, declare additional methods as well. In a distributed application, some implementations might reside in some Java virtual machines but not others. Objects with methods that can be invoked across Java virtual machines are called *remote objects.*

An object becomes remote by implementing a *remote interface,* which has the following characteristics:

- A remote interface extends the interface j ava. rmi. Remote.
- Each method of the interface declares j ava. rmi. RemoteException in its throws clause, in addition to any application-specific exceptions.

RMI treats a remote object differently from a non-remote object when the object is passed from one Java virtual machine to another Java virtual machine. Rather than making a copy of the implementation object in the receiving Java virtual machine, RMI passes a remote *stub* for a remote object. The stub acts as the local representative, or proxy, for the remote object and basically is, to the client, the remote reference. The client invokes a method on the local stub, which is responsible for carrying out the method invocation on the remote object.

A stub for a remote object implements the same set of remote interfaces that the remote object implements. This property enables a stub to be cast to any of the interfaces that the remote object implements. However, only those methods defined in a remote interface are available to be called from the receiving Java virtual machine.

Creating Distributed Applications by Using RMI

Using RMI to develop a distributed application involves these general steps:

1. Designing and implementing the components of your distributed application.
2. Compiling sources.
3. Making classes network accessible.
4. Starting the application.

Designing and Implementing the Application Components

First, determine your application architecture, including which components are local objects and which components are remotely accessible. This step includes:

- **Defining the remote interfaces.** A remote interface specifies the methods that can be invoked remotely by a client. Clients program to remote interfaces, not to the implementation classes of those interfaces. The design of such interfaces includes the determination of the types of objects that will be used as the parameters and return values for these methods. If any of these interfaces or classes do not yet exist, you need to define them as well.
- **Implementing the remote objects.** Remote objects must implement one or more remote interfaces. The remote object class may include implementations of other interfaces and methods that are available only locally. If any local classes are to be used for parameters or return values of any of these methods, they must be implemented as well.
- **Implementing the clients.** Clients that use remote objects can be implemented at any time after the remote interfaces are defined, including after the remote objects have been deployed.

Compiling Sources

As with any Java program, you use the j avac compiler to compile the source files. The source files contain the declarations of the remote interfaces, their implementations, any other server classes, and the client classes.

Making Classes Network Accessible

In this step, you make certain class definitions network accessible, such as the definitions for the remote interfaces and their associated types, and the definitions for classes that need to be downloaded to the clients or servers. Classes definitions are typically made network accessible through a web server.

Starting the Application

Starting the application includes running the RMI remote object registry, the server, and the client. The rest of this section walks through the steps used to create a compute engine.

Building a Generic Compute Engine

This trail focuses on a simple, yet powerful, distributed application called a *compute engine.* The compute engine is a remote object on the server that takes tasks from clients, runs the tasks, and returns any results. The tasks are run on the machine where the server is running. This type of distributed application can enable a number of client machines to make use of a particularly powerful machine or a machine that has specialized hardware.

The novel aspect of the compute engine is that the tasks it runs do not need to be defined when the compute engine is written or started. New kinds of tasks can be created at any time and then given to the compute engine to be run. The only requirement of a task is that its class implement a particular interface. The code needed to accomplish the task can be downloaded by the RMI system to the compute engine. Then, the compute engine runs the task, using the resources on the machine on which the compute engine is running.

The ability to perform arbitrary tasks is enabled by the dynamic nature of the Java platform, which is extended to the network by RMI. RMI dynamically loads the task code into the compute engine's Java virtual machine and runs the task without prior knowledge of the class that implements the task. Such an application, which has the ability to download code dynamically, is often called a *behavior-based application.* Such applications usually require full agent-enabled infrastructures. With RMI, such applications are part of the basic mechanisms for distributed computing on the Java platform.

Writing an RMI Server

The compute engine server accepts tasks from clients, runs the tasks, and returns any results. The server code consists of an interface and a class. The interface defines the methods that can be invoked from the client. Essentially, the interface defines the client's view of the remote object. The class provides the implementation.

Designing a Remote Interface

This section explains the Compute interface, which provides the connection between the client and the server. You will also learn about the RMI API, which supports this communication.

Implementing a Remote Interface

This section explores the class that implements the compute interface, thereby implementing a remote object. This class also provides the rest of the code that makes up the server program, including a main method that creates an instance of the remote object, registers it with the RMI registry, and sets up a security manager.

Designing a Remote Interface

At the core of the compute engine is a protocol that enables tasks to be submitted to the compute engine, the compute engine to run those tasks, and the results of those tasks to be returned to the client. This protocol is expressed in the interfaces that are supported by the compute engine.

Each interface contains a single method. The compute engine's remote interface, Compute, enables tasks to be submitted to the engine. The client interface, Task, defines how the compute engine executes a submitted task.

The compute.Compute interface defines the remotely accessible part, the compute engine itself. Here is the source code for the Compute interface:

```
package compute;
import java.rmi.Remote;
import java.rmi.RemoteException;
public interface Compute extends Remote {
    <T> T executeTask (Task<T> t) throws RemoteException;
}
```

By extending the interface java.rmi. Remote, the compute interface identifies itself as an interface whose methods can be invoked from another Java virtual machine. Any object that implements this interface can be a remote object.

As a member of a remote interface, the executeTask method is a remote method. Therefore, this method must be defined as being capable of throwing a Java.rmi.RemoteException. This exception is thrown by the RMI system from a remote method invocation to indicate that either a communication failure or a protocol error has occurred. A RemoteException is a checked exception, so any code invoking a remote method needs to handle this exception by either catching it or declaring it in its throws clause.

The second interface needed for the compute engine is the Task interface, which is the type of the parameter to the executeTask method in the Compute interface. The compute.Task interface defines the interface between the compute engine and the work that it needs to do, providing the way to start the work. Here is the source code for the Task interface:

```
package compute;
public interface Task<T> {
    T execute();
}
```

The Task interface defines a single method, execute, which has no parameters and throws no exceptions. Because the interface does not extend Remote, the method in this interface doesn't need to list j ava . rmi . RemoteException in its throw clause.

The Task interface has a type parameter, T, which represents the result type of the task's computation. This interface's execute method returns the result of the computation and thus its return type is T.

The Compute interface's executeTask method, in turn, returns the result of the execution of the Task instance passed to it. Thus, the executeTask method has its own type parameter, T, that associates its own return type with the result type of the passed Task instance.

RMI uses the Java object serialization mechanism to transport objects by value between Java virtual machines. For an object to be considered serializable, its class must implement the java. io. Serializable marker interface. Therefore, classes that implement the Task interface must also implement Serializable, as must the classes of objects used for task results.

Different kinds of tasks can be run by a Compute object as long as they are implementations of the Task type. The classes that implement this interface can contain any data needed for the computation of the task and any other methods needed for the computation.

Here is how RMI makes this simple compute engine possible. Because RMI can assume that the Task objects are written in the Java programming language, implementations of the Task object that were previously unknown to the compute engine are downloaded by RMI into the compute engine's Java virtual machine as needed. This capability enables clients of the compute engine to define new kinds of tasks to be run on the server machine without needing the code to be explicitly installed on that machine.

The compute engine, implemented by the ComputeEngine class, implements the Compute interface, enabling different tasks to be submitted to it by calls to its executeTask method. These tasks are run using the task's implementation of the execute method and the results, are returned to the remote client.

Implementing a Remote Interface

This section discusses the task of implementing a class for the compute engine. In general, a class that implements a remote interface should at least do the following:

1. Declare the remote interfaces being implemented
2. Define the constructor for each remote object
3. Provide an implementation for each remote method in the remote interfaces

An RMI server program needs to create the initial remote objects and *export* them to the RMI runtime, which makes them available to receive incoming remote invocations. This setup procedure can be either encapsulated in a method of the remote object implementation class itself or included in another class entirely. The setup procedure should do the following:

1. Create and install a security manager
2. Create and export one or more remote objects
3. Register at least one remote object with the RMI registry (or with another naming service, such as a service accessible through the Java Naming and Directory Interface) for bootstrapping purposes

The complete implementation of the compute engine follows. The engine. ComputeEngine_class implements the remote interface Compute and also includes the main method for setting up the compute engine. Here is the source code for the ComputeEngine class:

```
package engine;
import java.rmi.RemoteException;
import java.rmi.registry.LocateRegistry;
import java.rmi.registry.Registry;
import java.rmi.server.UnicastRemoteObject;
import compute.Compute;
import compute.Task;

public class ComputEngine implements Compute {

    public ComputeEngine() {
        super();
    }

    public <T> T executeTask(Task<T> t) {
        return t.execute();
    }

    public static void main(String[ ] args) {
        if (System.getSecurityManager() == null) {
            System.setSecurityManager (new.SecurityManager());
    }
    try {
        String name = "Compute";
        Compute engine = new ComputeEngine();
```

```
        Compute stub =
            (Compute)UnicastRemoteObject.exportObject (engine, 0);
        Registry registry = LocateRegistry.getRegistry();
        registry.rebind(name, stub);
        System.out.println("ComputeEngine bound");
    } catch (Exception e) {
        System.err.println("ComputeEngineexception:");
        e.printStackTrace();
        }
    }
}
```

The following sections discuss each component of the compute engine implementation.

Declaring the Remote Interfaces Being Implemented

The implementation class for the compute engine is declared as follows:

```
public class ComputeEngine implements Compute
```

This declaration states that the class implements the Compute remote interface and therefore can be used for a remote object.

The ComputeEngine class defines a remote object implementation class that implements a single remote interface and no other interfaces. The ComputeEngine class also contains two executable program elements that can only be invoked locally. The first of these elements is a constructor for ComputeEngine instances. The second of these elements is a main method that is used to create a ComputeEngine instance and make it available to clients.

Defining the Constructor for the Remote Object

The ComputeEngine class has a single constructor that takes no arguments. The code for the constructor is as follows:

```
public ComputeEngine() {
    super();
}
```

This constructor just invokes the superclass constructor, which is the no-argument constructor of the Object class. Although the superclass constructor gets invoked even if omitted from the ComputeEngine constructor, it is included for clarity.

12.2.5 Providing Implementations for Each Remote Method

The class for a remote object provides implementations for each remote method specified in the remote interfaces. The compute interface contains a single remote method, executeTask, which is implemented as follows:

```
public <T> T execueTask(Task<T> t) {
    return t.execute();
}
```

This method implements the protocol between the ComputeEngine remote object and its clients. Each client provides the ComputeEngine with a Task object that has a particular implementation of the Task interface's execute method. The ComputeEngine executes each client's task and returns the result of the task's execute method directly to the client.

Passing Objects in RMI

Arguments to or return values from remote methods can be of almost any type, including local objects, remote objects, and primitive data types. More precisely, any entity of any type can be passed to or from a remote method as long as the entity is an instance of a type that is a primitive data type, a remote object, or a *serializable* object, which means that it implements the interface java.io.serializable.

Some object types do not meet any of these criteria and thus cannot be passed to or returned from a remote method. Most of these objects, such as threads or file descriptors, encapsulate information that makes sense only within a single address space. Many of the core classes, including the classes in the packages j ava.lang and java.util, implement the Serializable interface.

The rules governing how arguments and return values are passed are as follows:

- Remote objects are essentially passed by reference. A *remote object reference* is a stub, which is a client-side proxy that implements the complete set of remote interfaces that the remote object implements.
- Local objects are passed by copy, using object serialization. By default, all fields are copied except fields that are marked static or transient. Default serialization behavior can be overridden on a class-by-class basis.

Passing a remote object by reference means that any changes made to the state of the object by remote method invocations are reflected in the original remote object. When a remote object is passed, only those interfaces that are remote interfaces are available to the receiver. Any methods defined in the implementation class or defined in non-remote interfaces implemented by the class are not available to that receiver.

For example, if you were to pass a reference to an instance of the ComputeEngine class, the receiver would have access only to the compute engine's executeTask method. That receiver would not see the ComputeEngine constructor, its main method, or its implementation of any methods Of java.lang .Object.

In the parameters and return values of remote method invocations, objects that are not remote objects are passed by value. Thus, a copy of the object is created in the receiving Java virtual machine. Any changes to the object's state by the receiver are reflected only in the receiver's copy, not in the sender's original instance. Any changes to the object's state by the sender are reflected only in the sender's original instance, not in the receiver's copy.

The most complex method of the ComputeEngine implementation is the main method. The main method is used to start the ComputeEngine and therefore needs to do the necessary initialization and housekeeping to prepare the server to accept calls from clients. This method is not a remote method, which means that it cannot be invoked from a different Java virtual machine. Because the main method is declared static, the method is not associated with an object at all but rather with the class ComputeEngine.

Creating and Installing a Security Manager

The main method's first task is to create and install a security manager, which protects access to system resources from untrusted downloaded code running within the Java virtual machine. A security manager determines whether downloaded code has access to the local file system or can perform any other privileged operations.

If an RMI program does not install a security manager, RMI will not download classes (other than from the local class path) for objects received as arguments or return values of remote method invocations. This restriction ensures that the operations performed by downloaded code are subject to a security policy.

Here's the code that creates and installs a security manager:

```
if (System.getSecurityManager() == null) {
    System.setSecurityManager(new SecurityManager());
}
```

Making the Remote Object Available to Clients

Next, the main method creates an instance of ComputeEngine and exports it to the RMI runtime with the following statements:

```
Compute engine = new ComputeEngine();
Compute stub =
    (Compute) UnicastRemoteObject.exportObject(engine, 0);
```

The static UnicastRemoteObject.exportobj ect method exports the supplied remote object so that it can receive invocations of its remote methods from remote clients. The second argument, an int, specifies which TCP port to use to listen for .incoming remote invocation requests for the object. It is common to use the value zero, which specifies the use of an anonymous port. The actual port will then be chosen at runtime by RMI or the underlying operating system. However, a non-zero value can also be used to specify a specific port to use for listening. Once the exportobject invocation has returned successfully, the ComputeEngine remote object is ready to process incoming remote invocations.

The exportob j ect method returns a stub for the exported remote object. Note that the type of the variable stub must be Compute, not ComputeEngine, because the stub for a remote object only implements the remote interfaces that the exported remote object implements.

The exportob j ect method declares that it can throw a RemoteException, which is a checked exception type. The main method handles this exception with its try/catch block. If the exception were not handled in this way, RemoteException would have to be declared in the throws clause of the main method. An attempt to export a remote object can throw a RemoteException if the necessary communication resources are not available, such as if the requested port is bound for some other purpose. Before a client can invoke a method on a remote object, it must first obtain a reference to the remote object. Obtaining a reference can be done in the same way that any other object reference is obtained in a program, such as by getting the reference as part of the return value of a method or as part of a data structure that contains such a reference.

The system provides a particular type of remote object, the RMI registry, for finding references to other remote objects. The RMI registry is a simple remote object naming service that enables clients to obtain a reference to a remote object by name. The registry is typically only used to locate the first remote object that an RMI client needs to use. That first remote object might then provide support for finding other objects.

The java.rmi.registry.Registry remote interface is the API for binding (or registering) and looking up remote objects in the registry. The j ava. rmi. registry. LocateRegistry class provides static methods for synthesizing a remote reference to a registry at a particular network address (host and port). These methods create the remote reference object containing the specified network address without performing any remote communication. LocateRegistry also provides static methods for creating a new registry in the current Java virtual machine, although this example does not use those methods. Once a remote object is registered with an RMI registry on the local host, clients on any host can look up the remote object by name, obtain its reference, and then invoke remote methods on the object. The registry can be shared by all servers running on a host, or an individual server process can create and use its own registry.

The ComputeEngine class creates a name for the object with the following statement:

```
String name = "Compute";
```

The code then adds the name to the RMI registry running on the server. This step is done later with the following statements:

```
Registry registry = LocateRegistry.getRegistry();
registry.rebind(name, stub);
```

This rebind invocation makes a remote call to the RMI registry on the local host. Like any remote call, this call can result in a RemoteException being thrown, which is handled by the catch block at the end of the main method.

Note the following about the Registry. rebind invocation:

1. The no-argument overload of LocateRegistry. getRegistry synthesizes a reference to a registry on the local host and on the default registry port, 1099. You must use an overload that has an int parameter if the registry is created on a port other than 1099.
2. When a remote invocation on the registry is made, a stub for the remote object is passed instead of a copy of the remote object itself. Remote implementation objects, such as instances of ComputeEngine, never leave the Java virtual machine in which they were created. Thus, when a client performs a lookup in a server's remote object registry, a copy of the stub is returned. Remote objects in such cases are thus effectively passed by (remote) reference rather than by value.
3. For security reasons, an application can only bind, unbind, or rebind remote object references with a registry running on the same host. This restriction prevents a remote client from removing or overwriting any of the entries in a server's registry. A lookup, however, can be requested from any host, local or remote.

Once the server has registered with the local RMI registry, it prints a message indicating that it is ready to start handling calls. Then, the main method completes. It is not necessary to have a thread wait to keep the server alive. As long as there is a reference to the ComputeEngine object in another Java virtual machine, local or remote, the ComputeEngine object will not be shut down or garbage collected. Because the program binds a reference to the ComputeEngine in the registry, it is reachable from a remote client, the registry itself. The RMI system keeps the ComputeEngine's process running. The ComputeEngine is available to accept calls and won't be reclaimed until its binding is removed from the registry *and* no remote clients hold a remote reference to the ComputeEngine object.

The final piece of code in the ComputeEngine. main method handles any exception that might arise. The only checked exception type that could be thrown in the code is RemoteException, either by the UnicastRemote object.exportobject invocation or by the registry rebind invocation. In either case, the program cannot do much more than exit after printing an error message. In some distributed applications, recovering from the failure to make a remote invocation is possible. For example, the application could attempt to retry the operation or choose another server to continue the operation.

Creating a Client Program

The compute engine is a relatively simple program: it runs tasks that are handed to it. The clients for the compute engine are more complex. A client needs to call the compute engine, but it also has to define the task to be performed by the compute engine.

Two separate classes make up the client in our example. The first class, computePi, looks up and invokes a Compute object. The second class, Pi, implements the Task interface and defines the work to be done by the compute engine. Tof the Pi class is to compute the value of *n* to some number of decimal places.

The non-remote Task interface is defined as follows:

```
package compute;
public interface Task<T> {
    T execute() ;
}
```

The code that invokes a Compute object's methods must obtain a reference to that object, create a Task object, and then request that the task be executed. The definition of the task class Pi is shown later. A Pi object is constructed with a single argument, the desired precision of the result. The result of the task execution is a java.math.bigDecimal representing *n* calculated to the specified precision.

Here is the source code for client.ComputePi, the main client class:

```
package client;

import java.rmi.registry.LocateRegistry;
import java.rmi.registry.Registry;
import java.math.BigDecimal;
import compute.Compute;

public class ComputePi {
    public static void main(String args()) {
        if (System.getSecurityManager() == null) {
            System.setSecurityManager (new SecurityManager());
        }
        try {
            String name = "Compute";
            Registry registry = LocateRegistry.getRegistry(args[0]);
            Compute comp = (Compute) registry.lookup(name);
            Pi task = new Pi (Integer.parseInt(args[1]));
            BigDecimal pi = comp.executeTask(task);
            System.out.println(pi);
        } catch (Exception e) {
            System.err.println("ComputePi exception:");
            e.printStackTrace();
        }
    }
}
```

Like the computeEngine server, the client begins by installing a security manager. This step is necessary because the process of receiving the server remote object's stub could require downloading class definitions from the server. For RMI to download classes, a security manager must be in force.

After installing a security manager, the client constructs a name to use to look up a compute remote object, using the same name used by ComputeEngine to bind its remote object. Also, the client uses the LocateRegistry. getRegistry API to synthesize a remote reference to the registry on the server's host. The value of the first command-line argument, args[0], is the name of the remote host on which the Compute object runs. The client then invokes the lookup method on the registry to look up the remote object by name in the server host's registry. The particular overload of LocateRegistry. getRegistry used, which has a single String parameter, returns a reference to a reaistrv at the named host and the default registry port, 1099. You must use an overload that has an int parameter if the registry is created on a port other than 1099.

Next, the client creates a new Pi object, passing to the Pi constructor the value of the second command-line argument, args [1], parsed as an integer. This argument indicates the number of decimal places to use in the calculation. Finally, the client invokes the executeTask method of the Compute remote object. The object passed into the executeTask invocation returns an object of type BigDecimal, which the program stores in the variable result. Finally, the program prints the result. The following figure depicts the flow of messages among the ComputePi client, the rmiregistry, and the ComputeEngine.

The Pi class implements the Task interface and computes the value of *n* to a specified number of decimal places. For this example, the actual algorithm is unimportant. What is important is that the algorithm is computationally expensive, meaning that you would want to have it executed on a capable server.

Here is the source code for client. Pi. the class that implements the Task interface:

```
package client;

import compute.Task;
import java.io.Serializable;
import java.math.BigDecimal;

public class Pi implements Task<BigDecimal>, Serializable {

    private static final long serialVersionUID = 227L;

    /** constants used in pi computation */
    private static final BigDecimal FOUR =
        BigDecimal.valueOf (4);

    /** rounding mode to use during pic computation */
    private static final int roundingMode =
        BigDecimal.ROUND_HALF_EVEN;

    /** digits of precision after the decimal point*/
    private final int digits;

    /**
    * Construct a task to calculate pi to be specified
    * precision
    * /
    public Pi(int digits) {
        this digits = digits;
    }

    /**
    * Calculate pi.
    */
    public BigDecimal execute() {
        return computePi(digits);
    }

    /**
        * Compute the value of pi to the specified number of
        * digits after the decimal point. The value is
        * computed using Machin's formula;
        *
        *     pi/4 = 4*arctan(1/5) – arctan(1/239)
        *
        * and a power series expansion of arctan(x) to
        * sufficient precsion.
        * /
    public static BigDecimal computePi(int digits) {
        int scale = digits + 5;
        BigDecimal arctan1_5 = arctan(5, scale);
        BigDecimal arctan1_239 = arctan(235, scale);
```

```
            BigDecimal pi = arctan1_5.multiply(FOUR).subtract{
                            arctan1_239).multiply(FOUR);
            return pi.setScale(digits,
                            BigDecimal.ROUND_HALF_UP);
    }
    /**
        * Compute the value, in radians, of the arctangent of
        * the inverse of the supplied integer to the specified
        * number of digits after the decimal point. The value
        * is computed using the power series expansion for the
        * arc tangent;
        *
        * arctan(x) = x – (x^3)/3 + (x^5)/5 – (x^7)/7 +
        * (x^9)/9
        * /
    public static BigDecimal arctan(int inverseX, int scale)
    {
        BigDecimal result. numer, term;
        BigDecimal invX = BigDecimal.valueOf(inverseX);
        BigDecimal invX2 =
        BigDecimal.valueOf(inversesX);

        numer = BigDecimal1.ONE.divide(invX,
                                scale, roundingMode);

        result = numer;
        int i = 1;
        do {
            numer = numer.divide(invX2, scale, roundingMode);
            int denom = 2 * i + 1;
            term =
                numer.divide(BigDecimal.valueOf(denom), scale roundingMode);
            if ((i % 2) ! = 0) {
                result = result.subtract(term);
            } else {
            result = result.add(term);
            }
            i++1
        } while (term.compareTo(BigDecimal.ZERO) ! = 0);
        return result
        }
    }
```

Note that all serializable classes, whether they implement the Serializable interface directly or indirectly, must declare a private static final field named serialVersionUiD to guarantee serialization compatibility between versions. If no previous version of the class has been released, then the value of this field can be any long value, similar to the 22 7L used by Pi, as long as the value is used consistently in future versions. If a previous version of the class has been released without an explicit serialVersionUiD declaration, but serialization compatibility with that version is important, then the default implicitly computed value for the previous version must be used for the value of the new version's explicit declaration. The serialver tool can be run against the previous version to determine the default computed value for it.

The most interesting feature of this example is that the Compute implementation object never needs the Pi class's definition until a Pi object is passed in as an argurrjent to the executeTask method.ile At that point, the code for the class is loaded by RMI into the Compute object's Java virtual machine, the execute method is invoked, and the task's code is executed. The result, which in the case of the Pi task is a BigDecimal object, is handed back to the calling client, where it is used to print the result of the computation.

The fact that the supplied Task object computes the value of Pi is irrelevant to the ComputeEngine object. You could also implement a task that, for example, generates a random prime number by using a probabilistic algorithm. That task would also be computationally intensive and therefore a good candidate for passing to the ComputeEngine, but it would require very different code. This code could also be downloaded when the Task object is passed to a compute object. In just the way that the algorithm for computing *n* is brought in when needed, the code that generates the random prime number would be brought in when needed. The compute object knows only that each object it receives implements the execute method. The Compute object does not know, and does not need to know, what the implementation does.

Compiling the Example Programs

In a real-world scenario in which a service such as the compute engine is deployed, a developer would likely create a Java Archive (JAR) file that contains the Compute and Task interfaces for server classes to implement and client programs to use. Next, a developer, perhaps the same developer of the interface JAR file, would write an implementation of the compute interface and deploy that service on a machine available to clients. Developers of client programs can use the Compute and the Task interfaces, contained in the JAR file, and independently develop a task and client program that uses a Compute service.

Starting the Server

Before starting the compute engine, you need to start the RMI registry. The RMI registry is a simple server-side bootstrap naming facility that enables remote clients to obtain a reference to an initial remote object. It can be started with the rmiregistry command. Before you execute rmiregistry, you must make sure that the shell or window in which you will run rmiregistry either has no CLASSPATH environment variable set or has a CLASSPATH environment variable that does not include the path to any classes that you want downloaded to clients of your remote objects.

To start the registry on the server, execute the rmiregistry command. This command produces no output and is typically run in the background. For this example, the registry is started on the host zaphod.

By default, the registry runs on port 1099. To start the registry on a different port, specify the port number on the command line. Do not forget to unset your CLASSPATH environment variable.

Microsoft Windows

```
start rmiregistry 2001
```

Solaris OS or Linux

```
rmiregistry 2001 &
```

Once the registry is started, you can start the server. You need to make sure that both the compute .jar file and the remote object implementation class are in your class path. When you start the compute engine, you need to specify, using the Java. rmi . server. codebase property, where the server's classes are network accessible. In this example, the server-side classes to be made available for downloading are the Compute and Task interfaces, which are available in the compute.jar file in the public_html \classes directory of user ann. The compute engine server is started on the host zaphod, the same host on which the registry was started.

Microsoft Windows:

```
java - cp c:\home\ann\src; c:\home\ann\public_html\classes\compute.jar
    –Djava.rmi.server.codebase=file:/c:/home/ann/public_html/classes/compute.jar
    –Djava.rmi.server.hostname=zaphod.east.sun.com
    –Djava.security.policy=server.policy
        engine.ComputeEngine
Solaris OS or Linux:
java -cp /home/ann/arc: /home/ann/public_html/classes/compute.jar
    –Djava.rmi.server.codebase=http://zaphod/-ann/classes/compute.jar
    –Djava.rmi.server.hostname=zaphod.east.sun.com
    –Djava.security.policy=server.policy
        engine.ComputeEngine
```

The above java command defines the following system properties:

1. The java.rmi.server. codebase property specifies the location, a codebase URL, from which the definitions for classes originating from this server can be downloaded. If the codebase specifies a directory hierarchy (as opposed to a JAR file), you must include a trailing slash at the end of the codebase URL.
2. The j ava.rmi.server.hostname property specifies the host name or address to put in the stubs for remote objects exported in this Java virtual machine. This value is the host name or address used by clients when they attempt to communicate remote method invocations. By default, the RMI implementation uses the server's IP address as indicated by the j ava. net. InetAddress. getLocalHost API. However, sometimes, this address is not appropriate for all clients and a fully qualified host name would be more effective. To ensure that RMI uses a host name (or IP address) for the server that is routable from all potential clients, set the j ava. rmi. server. hostname property.
3. The java. security.policy property is used to specify the policy file that contains the permissions you intend to grant.

Starting the Client

Once the registry and the compute engine are running, you can start the client, specifying the following:

1. The location where the client serves its classes (the Pi class) by using the java.rmi.server.codebase property.
2. The java.security.policy property, which is used to specify the security policy file that contains the permissions you intend to grant to various pieces of code
3. As command-line arguments, the host name of the server (so that the client knows where to locate the Compute remote object) and the number of decimal places to use in the *n* calculation.

Start the client on another host (a host named ford, for example) as follows:

```
Microsoft Windows:
java - cp c:\home\jones\src; c:\home\jones\public_html\classes\compute.jar
    –Djava.rmi.server.codebase=file:/c:/home/jones/public_html/classes/
    –Djava.security.policy=client.policy
        client.ComputePi zaphod.east.sun.com 45
Solaris OS or Linux:
java -cp /home/jones/arc: /home/jones/public_html/classes/compute.jar
    –Djava.rmi.server.codebase=http://ford/-jones/classes/
    –Djava.security.policy=client.policy
        engine.ComputePi zaphod.east.sun.com 45
```

Note that the class path is set on the command line so that the interpreter can find the client classes and the JAR file containing the interfaces. Also note that the value of the j ava. rmi. server. codebase property, which specifies a directory hierarchy, ends with a trailing slash.

After you start the client, the following output is displayed:

```
3.141592653589793238462643383279502884197169399
```

When the ComputeEngine server binds its remote object reference in the registry, the registry downloads the Compute and Task interfaces on which the stub class depends. These classes are downloaded from either the ComputeEngine server's web server or file system, depending on the type of codebase URL used when starting the server.

Because the ComputePi client has both the compute and the Task interfaces available in its class path, it loads their definitions from its class path, not from the server's codebase.

Finally, the Pi class is loaded into the ComputeEngine server's Java virtual machine when the pi object is passed in the executeTask remote call to the ComputeEngine object. The Pi class is loaded by the server from either the client's web server or file system, depending on the type of codebase URL used when starting the client.

Introduction to VB Script

We have seen how to say hello to the world in many different ways using HTML. Now let us say the same thing using Visual Basic Script, or sometimes called VB Script. Here, when you load the Web page, the script will run and pop up a dialog box telling us Hello. You will notice a big difference between this example and the one used in HTML and XML. HTML is primarily utilized for text output, whereas Visual Basic Script is utilized for programming behind the scenes.

Here is the actual example:

```
<HTML>
<title>Beginning to learn Visual Basic Script</title>
<h1><Font size=3 Colour=maroon>
<b><i>Hello World Visual Basic Script></I></b></font></h1>
<hr>
<Body Language="vbScript" onLoad=StartSub>
<SCRIPT LANGUAGE="VBSript">
'StartSub is our sub routine that is first loaded
Sub StartSub
    ' The MsgBox function with the first
    ' argument as "Hello World!" to define the message
    MsgBox "Hello World!"
End Sub
</SCRIPT>
<br>This is to test VB Script</br>
</BODY>
</HTML>
```

This will give you on screen the message, as shown on the next page.

Adding VB Script code to HTML

It was easy to see that we can use both HTML and VBScript for Web pages. But, can we use them together. Let's see that. In the above example we make some changes to include both HTML and VBScript codes. The resultant text now becomes:

```
<HTML>
<title>Beginning to Learn Visual Basic Script</title>
<h1><font size=6 colour=blue>
<b><i>Hello World Visual Basic Script</I></b></font></h1>
```

```
<hr>
<INPUT TYPE=BUTTON VALUE="Click Here!" NAME="Hello">
<SCRIPT LANGUAGE="VBScript">
'This subroutine is run when the button is selected
Sub Hello_OnClick
  ' Msgbox function with three parameters. The first is the
  ' message, the second indicates what buttons to show and
  ' the last is the title of the msgbox
  MsgBox "Hello World!", 0, "HTML and VBScript combined"
End Sub
</SCRIPT>
</BODY>
</HTML>
```

This would result in the figure shown next.

The HTML section begins with the usual <SCRIPT></SCRIPT> tags. The key is defining a way for our code to run whenever the button is clicked. The button has an OnClick event so that if we define the event in our script, the specified code will run.

In this example, the Hello_OnClick subroutine in the script will be run whenever the button is clicked. And in this case, we have a message box for our "Hello World!" message. An event is run whenever an event takes place in the browser. Thus, in our example, the click of a button fires off the OnClick event.

Adding a Script to Your Documents

You have seen small examples tags. The first tag will indicate the script language and the end tag will end the scripting section. Following is an example:

```
<SCRIPT LANGUAGE="VBScript">
<!—
...
...
...
—>
</SCRIPT>
```

You need to specify the language, since Microsoft Internet Explorer Version 3.0 can support other

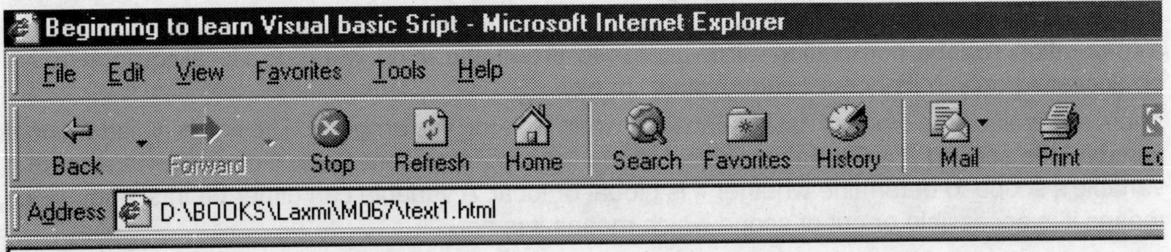

scripting languages such as JavaScript. If you are deploying your HTML pages on the Internet for anyone to see, there is a good chance that someone who is using a browser that does not support Visual Basic Script may venture across your Web page. The HTML comment tags, <!—and —> placed as shown, will prevent these browsers from displaying the script code.

Where you place script code in a page can be at your discretion. You can even break the script code up into different sections of the document. Most likely though, it will be easier to keep all your code together in one section of your HTML document.

You can attach Visual Basic Script code to your document in order ways, though keeping your code together in one section is generally recommended. One way of add short sections of code (inline) is in the tag defining an HTML element. For example, the following HTML input button has an OnClick even that shows a message box:

```
<INPUT NAME="A button" TYPE=BUTTON
    VALUE="A Button" OnClick='MsgBox "I was clicked!"'>
```

You can place the Visual Basic Script code right in the declaration of the input element. The event name, in this case OnClick, is followed by the Visual Basic Script code in single quotes. In our example, we will call the MsgBox function. Multiple Visual Basic Script statements can be included as long as they are separated with colons (:).

You can also write a <SCRIPT> tag so that it applies only to a particular event for a specific HTML input element. In the following example, we have the same code for the AButton HTML button defined.

```
<INPUT NAME="AButton" TYPE=BUTTON VALUE="A Button">
<SCRIPT LANGUAGE="VBScript" EVENT="OnClick" FOR="AButton">
<!—
    MsgBox "I was clicked!"
—>
</SCRIPT>
```

In this case, the sample script will just be run whenever the AButton is clicked by the user. Notice that in the script definition we have defined which event we want the script to run FOR, in this case the OnClick event for the button.

Data Types of Visual Basic

One of the first things to understand about using variables in Visual Basic Script is how long the life of a variable is. Just because you may be wrong with variable Monica in one section of your code does not mean the value of Monica is available in another area of your code. You can think of this like a corporation. A global corporation will be accessed all over the world, whereas a local corporation will be accessed only in its home country, state, or city.

There are really three types of data to consider when working in a program. The first is called a **Local** variable, the second is a **Global** variable and the third is **Persistent** Data. Generally, we refer to a variable's scope to determine whether it is global or local. A variable can either be in scope, which means it is accessible or out of scope which means it is not.

Global Variables

Let's take a look at a simple Global variable example.

```
<!—Start the Script —>
<SCRIPT LANGUAGE=:VBScript">
Dim OurGlobal
StartSub
Sub StartSub
....OurGlobal = 5
....CheckGlobal
....MsgBox OurGlobal
End Sub
Sub CheckGlobal
....MsgBox OurGlobal
....OurGlobal = 10
End Sub
</SCRIPT>
```

Local Variables

Now let's take a look at utilizing Local variables in script code.

```
<!—Start the Script —>
<SCRIPT LANGUAGE=:VBScript">
StartSub
Sub StartSub
....Dim StartLocal
....StartLocal = 10
....DemoSub
....MsgBox StartLocaL
...MsgBox DemoLocaL
....DemoSub
End Sub
Sub DemoSub
....Dim DemoLocaL
....MsgBox "Demo Local = " & DemoLocaL
....DemoLocaL = 30
....MsgBox "StartLocaL = " & StartLocal
End Sub
</SCRIPT>
```

In this example, our startup routine is StartSub, the set of code that will be run first when the browser starts the script. It initializes the StartLocaL variable to 10. We then run DemoSub. In DemoSub, we have another Local variable, DemoLocaL. To ensure that DemoLocal is empty, we display the value. We then set its value to 30. We also display the value of StartLocal. Since DemoSub has no reference to StartLocal, the value is empty. We then return to StartSub and this displays both the value of StartLocal and DemoLocal. Since Startsub has not yet finished, StartLocal is still 10 and of course

DemoLocal is empty because StartSub has no access to its value. Finally, we run DemoSub again. Now, you might think the value of DemoLocal is still 30. But, our first running of DemoSub has finished; that means that the instance of DemoLocal we had then is no longer around. Thus, we are starting with a newly initialized value of DemoLocal and it will be empty initially. Again, try this code in the browser and play around with it.

Strings in Script

All variables contain variant data. Variants in their basic form contain either numeric data or strings (alpha, numeric and special characters). Strings are very important in working with data. For example, if you want someone to enter his or her name and address information, the data will be stored in a string. But, if you want the user to enter his or her weight, then that will be stored in numeric subtype.

The first thing to point out about working with strings in Visual Basic Script is that they are always surrounded by double quotes (for example, "Hello"). The following is an example of how to set a variable to a string:

```
Dim strTest
strTest = "This is a String in Quotes!"
```

In this example, the variable strTest is set to contain the string: This is a String in Quotes! You will always need to ensure that you utilize the double quotes or Visual Basic Script will politely inform you that you have an error.

Visual Basic Script provides a wide range of functions for working with strings. These include, but are not limited to Left, Right, Mid and Trim. Let's take a look at a few quick and simple example of working with strings to help get some basic string manipulation ideas down. Let's first look at basic string concatenation (or addition) using the & operator. To get started, let's review a visual example.

Now, let's look at a Visual Basic Script example that builds on our above graphical example. The following code will build the string "HelloWorld".

```
Dim strTest1
Dim strTest2
Dim strTest3
strTest1 = "Hello"
strTest2 = "World"
strTest3 = strTest1 & strTest2
```

You can think of the & symbol as acting like addition and in reality it concatenates (combines) two strings. Note that the new value is "HelloWorld" with no space between Hello and World. Taking from our visual example, the following example will build the string with a space.

```
Dim strTest1
Dim strTest2
Dim strTest3
strTest1 = "Hello"
strTest2 = "World"
strTest3 = strTest1 & " " & strTest2
```

In this case, we don't have a variable that defines the space; we simply place a literal string of "" into our string concatenation equation. The result is that strTest3 has "Hello World" with a space dividing the two words.

Simply adding strings together is not all we can do. We have access to powerful string manipulation functions including Left, Right and Instr for parsing strings. Let's set up an example of splitting off the extension of a filename. In our example, the filename will be "very wonderful.htm". In order to strip off the two parts surrounding the "-", we will have to find the location of the "-" in this string. Knowing where this character is located will help us determine where the filename ends and the extension begins. Fortunately, we have many built-in utilities to help us tear this string apart.

```
Dim intPos
Dim strFileName
Dim strLeft
Dim StrRight
StrFileName = " very wonderful.htm"
intpos = Instr(1, strFileName, ".")
```

Instr first wants to know where in the string we would like to start searching. In this case we will start in position 1. It then wants to know what string we are going to look in which in this case, is the filename stored in strFileName. Last we have to tell it what string to search for, which is ".". In this case intPos will have a value of 15, which is the location of the ".". Now let's get all of the characters to the left of the period using the Left function:

```
strLeft = Left(strFileName, intPos – 1)
```

This will return the string "very wonderful". We know the "-" is at position 15, so if we get the left 14 characters of the string, we will have "very wonderful". Now, let's get the extension of the filename:

```
strRight = Right(strFileName, Len(StrFileName) – intPos)
```

In order to do this, we have to utilize one more string function, Len. This will return the length of the filename, which is 18. So, if we get the right of characters of the string for a length of 3, we will have "htm". Since the "-" is at 15 and the length is 18, we can subtract the "-" position from the length to get the extension (18-15).

Visual Basic Script is rich in string manipulation tools; this has always been one of the strengths of the Visual Basic language.

Let's first take a look at the use of the ASC and Char functions in the ConvertAsc_ChrDemo subroutine of the code. Every character in a string has a corresponding value of 65. The ASC function will return the ASCII value of a character and the CHR function will convert as ASCII value into a character (string) representation. The first set of code in the subroutine converts integer 65 into a string, A.

```
....' Get the character representation
....' For ASCII 65
....strWork = chr(65)
```

The next set of code then converts the capital A back into a value of 65.

```
.....' Retrieve the ASCII value for the
.....' the character
...... intWork = asc(strWork)
```

The second subroutine, PaperStringDemo, demonstrates parsing or taking apart, a string using a wide variety of functions. Let's review how a couple of these work. First, the mid function us utilized as follows:

```
......' Set out working string
.......strWork = -ABCDEFGHIJKL"
.......' Get the Mid
......StriTemp = Mid(strWork,1,3)
```

The Mid function is a powerful string parser. It allows you to select which part of a string you would like to work with. In the above code example, we will get the section of the "ABCDEFGHIJKL" starting with visualize some of the Mid, Left, Right, and Inter functions.

Numbers in Script

Let's move on to working with numbers. As mentioned in the last session, all variables are of a variant data type, which means they can contain any subtype whether it be a string, number or date. Often, you will want to work with just variables that are numeric. So, if you are building a Web page to help to work with just variables that are numeric. So, if you are building a Web page to help millionaires track their money, you would want a long data type. If you are a science wizard calculating the nth root of the pie in the sky of the earth, you would want to use the double data type.

Of course when working with numeric data, there is not just one numeric subtype. In fact there are several, including Boolean, Byte, Integer, Long, Single and Double.

Using these numeric data types comes down to how precisely you need to define your number. If it is a simple yes or no, then using a Double would be a little bit of an overkill. Likewise, if you are doing scientific work, an Integer or even a Long will most likely not cut it. But, fortunately, Visual Basic Script will handle adjusting the data subtype for your.

Visual Basic Script also provides a wide range of functions for working with numbers. These functions loosely fall into three categories:

Standard Operators
Logical Operators and
Math Functions.

Standard Operators consists of those familiar math operations such as *, -, *, / and =. But this also includes some operators that may not be as familiar and obvious. For example, the integer division operator, \ is similar to the division operator, /, but performs a little different function. Let's look at the following example:

```
Dim intVar1
Dim intVar2
intVar1 = 17
intVar2 = intVar1 / 3
intVar3 = intVar1 \ 3
```

With normal division, intVar2 will have a value of 5.66667(17/3). With integer division, intVar3 will have a value of 5. If you only need an integer after a division operation and want the remainder to be discarded, the \ integer division operator is a valuable tool. Another function to utilize when working with division remainders is the Mod function. Mod performs modules arithmetic on two numbers with return the remainder after the division of the two numbers. Let's look at another example:

```
Dim intVar1
Dim intVar2
intVar1 = 17
intVar2 = intVar1 mod 3
```

In this case, intVar2 will have a value of 2 (3 goes into 175 times with 2 left over) which is the remainder of the division. Logical operators perform logical with operations on two variables. Let's look at a simple example of performing a logical AND operation.

```
Dim intVar1
intVar1 = 1 & 0
intVar1 = 1 & 1
intVar1 = 0 & 0
```

In the first case, intVar1 is going to return 0. In the second, the return value is 1. In the last case, the return value is 0. The logical AND of two values will only be true if both values are 1 or TRUE. The OR operator combines the values differently as outlined as follows:

```
Dim intVar1
intVar1 = 1 & 0
intVar1 = 1 & 1
intVar1 = 0 & 0
```

In the first and second case, 1 will be returned; in the last, 0 will be returned.

There are four sections to the script: Standard Operators, then Logical Operators followed by Match Functions and the last section showing several miscellaneous functions. Let's take a look at two of these miscellaneous functions: Int and Fix. The code for following with these is as follows:

```
...... Dim varWork
......' Set the initial value
......varWork = -57.98
......' Show the integer of the value
......document.write "Int(varWork) = " & Int(varWork) & "<br>"
......' show the 'fix' of the value which
......' Will round down
......document.write "Fix(varWork) = " & Fix(varWork) & "<br>"
```

The Int and Fix functions return the Integer portion of a number. But, if a number is negative, Int returns the first negative integer less than or equal to the number. Fix returns the first negative integer greater than or equal to the number. Review the other code examples to see the effect of the different operators and functions. Most of these will be as familiar as your high school math. Some will only be useful if you are a heavy duty math person.

Dates and Time in Script

This session is good for single people to review since we will use Visual Basic Script to show you how to get a date and make time to be with that date. Visual Basic Script provides a variable subtype of Date that can be utilized for working with both dates and time. If you need to work with dates or time in your Visual Basic Script pages, you will be glad this date type is built into the language. There are many powerful built-in functions for manipulating both. Let's first look at how we can retrieve the current date and time using the Date, Time, and Now function:

```
Dim dtVar1
Dim dtVar2
Dim dtVar3
dtVar1 = Date
dtVar2 = Time
dtVar3 = Now
```

The dtVar1 variable will contain the current system date in the format of "1/1/96". The dtVar2 variable will contain the current system time in the format of "10:37:30 PM". Finally, the dtVar3 variable will contain both the current system data and time in the format of "5/10/96 to 10:37:30 PM".

The following table briefly overview the Date functions provided in Visual Basic Script

Function	*Description*
Date	Returns the current system date
Time	Returns a Variant of subtype Date indicating the current system time.
DateSerial	Returns a Variant of subtype Date for a specified year, month and day.
DateValue	Returns a Variant of subtype Date.
Day	Returns a whole number between 1 and 31, inclusive, representing the day of the month.
Month	Returns a whole number between 1 and 12, inclusive, representing the month of the year.
Weekday	Returns a whole number representing the day of the week.
Year	Returns a whole number representing the year.
Hour	Returns a whole number between 0 and 23, inclusive, representing the hour of the day.
Minute	Returns a whole number between 0 and 59, inclusive, representing the minute of the hour.

Second	Returns a whole number between 0 and 59 inclusive representing the second of the minute.
Now	Returns the current date and time according to the setting of your computer's system date and time.
TimeSerial	Returns a Variant of subtype Date containing the time for a specific hour, minute and second.
TimeValue	Returns a Variant of subtype Date containing the time.

Fortunately, Visual Basic Script provides solid, built-in functionality for working with dates and times. These functions are easy and straightforward to use.

Arrays in Script

So far we have been working with variables known as scalars. Scalar variables only allow you to assign a single value to the variable. Many times, however, you will want to easily track many values at one time. Let's say you have a Web page that, depending on a class number, shows all of the students in the class. And, this Web page will allow you to enter the grade for each student. There are several challenges to making this happen. First, each class might have a different number of students. Second, it would be very unwieldy to try to declare and work with enough variables to define each student and grade. Just imagine having variables intStudent1 through intStudent50 to try to do this.

Fortunately, there is a special type of variable in Visual Basic Script called an Array that allows you to easily work with large amounts of data in a convenient fashion. Dimensioning an Array variable is very easy, as follows:

```
Dim arrVar(5)
```

The array arrVar has a single dimension with six elements. Note that there are six elements and not five because the first element of the array starts at zero, not one. Let's look at an example defining an array of students from our class example.

```
<SCRIPT LANGUAGE="VBSript">
Dim arrStudents(50)
arrStudents(0) = "Akash Chopra"
arrStudents(1) = "Virendra Sehwag"
arrStudents(2) = "Rahul Dravid"
arrStudents(3) = "V V S Laxman"
arrStudents(4) = "Saurav Ganguly"
arrStudents(5) = "Parthiv Patel"
</SCRIPT>
```

In this example, we have an array of students, arrStudents(50) that has 51 elements. We have defined the first six elements to have the student's names by using a numerical index into the array. Note that the remaining elements remain empty. To access the value of an element in the array, simply use the index as follows:

```
strTestVar = arrStudents(1)
```

In this case strTestVar will be equal to Johny Anderson. Arrays can also have two dimensions. Let's implement the ability to also store the students' grades along with their names as follows:

```
dim  arrStudents(50,1)
......arrStudents(0,0) = "Akash Chopra"
......arrStudents(0,1) = 5

.....arrStudents(1,0) = "Virendra Sehwag"
.....arrStudents(1,1) = 15
```

```
....arrStudents(2,0) = "Rahul Dravid"
....arrStudents(2,1) = 25
```

In this example, our arrStudents variable has two dimensions. The best way to think of two-dimensional arrays is as a table with columns and rows. You can have multidimensional arrays for working with all kinds of data relationship.

You may have noticed that in our examples with arrStudent, we have dimensioned 51 elements but are only using a few of these. Let's look at another way to dimension and work with array size.

```
<SCRIPT LANGUAGE="VBScript">
Dim arrStudents()
Dim intNumsStudents
IntNumStudents = GetNumbersStudents
ReDim arrStudents(intNumStudents)
...
...
...
intNumStudents = GetNumberStudents
ReDim Preserve arrStudents(intNumStudents)
</SCRIPT>
```

You can also declare an array whose size changes at any time your script is running. This is called a dynamic array. The array is initially declared locally within a procedure or globally using the Dim statements as with any other array. The difference is that no size or number of dimensions is placed inside the parentheses. To use a dynamic array, you must subsequently use ReDim to determine the number of dimensions and the size of each. In the above example, ReDim to determine the number of dimensions and the size of each. In the above example, ReDim sets the initial size of the dynamic array to the value returned from a function called GetNumberStudents. Now let's say that later we need to check again how many students there are in a class, in case some have been added or deleted. A subsequent ReDim statement resizes the array to the new number of students, but uses the Preserve keyboard to preserve the contents of the array. Without the Preserve keyword, all data in the array would have been erased.

The topic of our code example for this session is working with several of the array functions provided in Visual Basic Script. Utilizing these functions will help you in your endeavors with arrays. Arrays are powerful tools for dealing with large amounts of data. Remember that an array's elements can be of any data subtype and allow you to mix and match what type of data you will be working with, as in our student name and grade example.

VB Script Operators

Here is the list of operators which are supported by VB Script.

Operator	+
Syntax	`result = expression1+expression2`
Description	It is used to sum two numbers. In the above syntax: result can be any numeric variable, and expression1 and expression2 can be any expressions. You can use the + operator to concatenate two strings together, but the preferred operator for this is &. Also, the + operator can be used to add two date variable together.
Operator	And
Syntax	`result = expression1 And expression2`
Description	It is used to perform a logical conjunction on two expressions. If and only if, both expressions evaluate to True, the result is True. If either expression evaluates to False, the result is False.

Operator &

Syntax `result = expression1 & expression2`

Description It is used to force string concatenation of two expressions. Whenever an expression is not a string, it is converted to a String subtype.

Operator /

Syntax `result = number1/number2`

Description It is used to divide two numbers and return a floating point result.

Operator Eqv

Syntax `result = expression1 Eqv expression2`

Description It is used to perform a logical equivalence on two expressions. The result is determined according to the following table:

Expression1	***Expression2***	***Return Value***
True (1)	True (2)	True (1)
True (1)	False (0)	False (0)
False (0)	True (1)	False (0)
False (0)	False (0)	True (1)

Operator ^

Syntax `result = number ^ exponent`

Description It is used to raise a number to the power of an exponent.

Operator Imp

Syntax `result = expression1 Imp expression2`

Description It is used to perform a logical implication on two expressions. The following table illustrates how result is determined.

Expression1	***Expression2***	***Return Value***
True	True	True
True	False	False
True	Null	Null
False	True	True
False	False	True
False	Null	True
Null	True	True
Null	False	Null
Null	Null	Null

The Imp operator performs a bitwise comparison of identically positioned bits in two numeric expressions and sets the corresponding bit in result according to the following truth table:

Expression1	***Expression2***	***Return Value***
0	0	1
0	1	1
1	0	0
1	1	1

Operator	\
Syntax	`result = number1 \ number2`
Description	It is used to divide two numbers and return an integer result. Before division is performed, numeric expressions are rounded to Byte, Integer, or Long subtype expressions.
Operator	Is
Syntax	`result = object1 Is object2`
Description	It is used to compare two object reference variables. If object1 and object2 both refer to the same object, the result is True; if they do not, the result is False.
Operator	Mod
Syntax	`result = number1 Mod number2`
Description	It is used to divide two numbers and return only the remainder.
Operator	*
Syntax	`result = number1 * number2`
Description	It is used to raise a number to the power of an exponent.
Operator	–
Syntax	`result = number1 - number2`
Description	It is used to find the difference between two numbers or to indicate the negative value of a numeric expression.
Operator	Not
Syntax	`result = Not expression`
Description	It is used to perform logical negation on an expression. The following table illustrates how result is determined:

Expression	***Return Value***
True (1)	False (0)
False (0)	True (1)
Null	Null

Operator	Or
Syntax	`result = expression1 or expression2`
Description	It is used to perform a logical disjunction on two expressions. If either or both expressions evaluate to True, the result is True. The following table illustrates how result is determined:

Expression1	***Expression2***	***Return Value***
True	True	True
True	False	True
True	Null	True
False	True	True
False	False	False
False	Null	Null
Null	True	True
Null	False	Null
Null	Null	Null

The Or operator also performs a bitwise comparison of identically positioned bits in two numeric expressions and sets the corresponding bit in result according to the following truth table:

Expression1	***Expression2***	***Return Value***
0	0	0
0	1	1
1	0	1
1	1	1

Operator Xor

Syntax `result = expression1 Xor expression2`

Description It is used to perform a logical exclusion on two expressions. If one, and only one, of the expressions evaluates to True, the result is True. When neither expression is Null, the result is determined according to the following table:

Expression1	***Expression2***	***Return Value***
True	True	False
True	False	True
False	True	True
False	False	False

The Xor operator also performs a bitwise comparison of identically positioned bits in two numeric expressions and sets the corresponding bit in result according to the following truth table:

Expression1	***Expression2***	***Return Value***
0	0	0
0	1	1
1	0	1
1	1	0

Getting the Message Across

With new tools provided in Visual Basic Script, a whole new set if opportunities is available for communicating directly with your users. Now that we have the ability to manipulate HTML buttons, the windows status bar and so on, the information displayed in these items can be dynamically controlled.

For example, from Visual Basic Script can change the caption of an HTML button with a simple statement, as follows:

button.Value = "new value"

Simply by setting a new value for the button control, we have new options, for making our controls state dependent. An example would be playing a game. While the game is stopped, the button value could be Play. While the game is in play, the button value could be stop. Thus, this one HTML button can now be used for different purposes.

Another example of available options is the ability to set the status bar text of the Internet Explorer browser by referencing the status property of the Window object. We can set this status based on user actions in our Web document.

When the button is clicked on, the code reference the caption value of the label control and its caption is updated to display the new text, "Updated!". The following code demonstrates the changes in the button values using visual Basic Script.

```
Sub Button1_OnClick()
......Click1Count = Click1Count + 1
......Button1.Value = "Clicked " & Click1Count & " times"
End Sub
```

In this example, there is a global counter that is incremented each time the button is clicked. We then display the new button value by concatenating the "Clicked" text with the counter value and then the "times" text. Each button will reflect the number of times each was clicked.

Finally, the browser status bar is set by clicking on the other three buttons using the following code:

```
Sub Status1_OnClick()
.....SetStatus "You clicked the first status button"
End Sub
Sub SetStatus(Status)
.....window.status = status
End Sub
```

A general subroutine, SetStatus is provided to set the window status value based on the Status argument. Each click event for the buttons calls this subroutine and passes in the text to be displayed. Remember that the browser also uses the status bar based on various events the user may initiate on the Web page.

Confirming a User's Action

Earlier, the MsgBox function has been used to display various values in our code examples and we have purposely glossed over some of the available features of the function. In this session we will take a closer look at how the MsgBox function can be utilized to query the user to have the user provide feedback confirmation and then take appropriate action from our code.

There are many argument options available for confirming the display of the dialog box to the user. Specifically, the second MsgBox argument allows the developer to configure what icons and buttons are displayed. The following table overviews these argument settings:

Value	*Description*
0	OK Button only
1	OK and Cancel button
2	Abort, Retry and Ignore buttons
3	Yes, No and Cancel buttons
4	Yes and No buttons
5	Retry and Cancel buttons
16	Critical Message icon
32	Warning Query icon
48	Warning Message icon
64	Information Message icon
0	First button is the default
256	Second button is the default
512	Third button is the default
768	Fourth button is the default
0	The message box is model and the user must respond to the message box before continuing work in the current application.
4096	The message box is system-model and no other system interaction can take place until the user responds to the messagebox.

As you can see, there are many different configuration options. Keep in mind that you can combine these settings by adding up the values when setting the second argument. For example, the following code will show a message box with Yes and No buttons and the second button will be the default:

```
ReturnValue = msgbox("Prompt", 4 + 256)
```

Also note that one does not have to have each argument of MsgBox defined to call the function. And a value does not have to be returned as shown earlier. If a value is returned from the function, then that value indicates which buttons was selected. The following table describes what each return value means:

Value	*Button*
1	OK
2	Cancel
3	Abort
4	Retry
5	Ignore
6	Yes
7	No

Using an Imagine Map for Navigation

Well, now that we have played around with changing captions, values and popping up message dialog boxes all over the place, we are now ready to move on to some more graphical examples. Image maps are useful for providing graphical hot links to other Web documents and other parts of your Web pages. But, we can use our Visual Basic Script toolbox to provide more capabilities to work with an image map.

The following HTML code sets up the image map:

```
<!—Build a Link with no actual jump document —>
<A id="ImageLink" href="">
<!—Show the bitmap image —>
<IMG SRC="6-3.bmp" WIDTH=176 HEIGHT=163 BORDER=1 NAME=Letters>
</A>
```

First, there is a link setup name "ImageLink". In this case, there is no reference document to jump to when the image is clicked on. Next, the image is shown using the <IMG> tag. An HTML input text box element is placed on the Web page. This is where feedback will be provided based on the mouse pointer location over the image. Then, the script starts as follows;

```
<!—Start the script, —>
<SCRIPT Language="VBScript">
' The Mouse event of the ImageLink
Sub ImageLink_MouseMove(Shift, Button, X, Y)
```

The first subroutine in the script is the MouseMove event of the link we set up in the document. This event will be fired off every time the mouse moves over the image. The four parameters are the status of the shift key, mouse button. X location and Y location. In this session, we will be working with the X and Y location of the mouse over the image.

In the following code, a check is done to see which quadrant of the graphic the mouse is in. The width of the image is 175 pixels and the height is 163 pixels. Next in the code is the InRect subroutine which is being called to check if the X and Y location of the mouse is in the specified bounding box for each character. If the mouse is in the bounding box, then an appropriate message is displayed in the HTML input text element.

```
' We check to see if the mouse is over the 'A'
```

```
if (inRect(X, Y, 0, 0, 175/2, 163/2)=true) then
......' Indicate we are over the A
.......HelpText.value = "A is the Letter"
Else
.....' Check to see if we are over the 'C'
......If (InRect(X, Y, 0, 163/2, 175/2, 163)=true) then
.... Else
....' Check to see if we are over the 'B'
......If (InRect(X, Y, 175/2, 0, 175, 163/2)=true) then
......' Indicate we are over the B
.......HelpText.value = "B is the Letter"
........Else
.....' Check to see if we are over the 'D'
......If (InRect(X, Y, 175/2, 163/2, 175, 163)=true) then
......' Indicate we are over the D
.......HelpText.value = "D is the Letter"
......End If
.....End If
...End If
End If
End Sub
```

The InRect function takes in the X and Y mouse location and the coordinates to check the location. A set of AND logic is done to check each coordinate. If all are true, then InRect will return true; otherwise, it will return false.

```
Function InRect(X, Y, X1, Y1, X2, Y2)
.........' We check to see if the X coordinate
......' is between the two X boundaries. We also
      ' check to see if the Y coordinate is between
      ' the two boundaries. InRect will either return
      ' true or false.
      InRect = X >= X1 AND X<= X2 AND Y >= Y1 and Y <= Y2
End Function
</script>
```

The final result is that with each mouse move over the image, an appropriate message about that section of the image is displayed. Having this type of capability opens up many options for providing interactive graphics in your documents.

Creating and Calling Subroutines

In Visual Basic Script, the most common procedure type is the Subroutine or Sub for short. A Sub is a number of functionally related lines of code bundled together and given a name. The name is important because this is the identifier that will be used to call the Sub into action when it is needed. Precisely where the Sub starts and where it ends is determined by the Sub....End Sub delimiters.

```
Sub name (parameter List)
    statements
        Exit Sub
End Sub
```

The subroutine is defined by the two delimiting statements Sub SubName....EndSub. The code between the subroutine delimiters will be run whenever this subroutine is called from elsewhere in the code.

This is how a Visual Basic Script Sub called Square a might look.

```
Sub Square
```

```
        Dim X
        X = X * X
    End Sub
```

If you look carefully at the code above, you can see that this procedure is flawed. The variable x has been created, dimensioned, inside the Sub and is therefore a local variable. In this example, the variable x can only exist while the code inside Sub Square is actually being processed. The scope of the variable extends only as far as the delimiters of the sub, so the variable is created, then destroyed, every time the Sub Square is run.

As you know, when Visual Basic Script variables are first dimensioned, they are assigned a default value. For X, since it is being used to hold number values, the value will be zero. Since x is zeroed every time the Sub is run, the result of X = X * X will always be zero, which is not much use of anybody. What is needed is the ability to pass a value for X from outside the world of Sub Square and then give it to X after it has been dimensioned. The programmer needs to decide when they want Sub Square to run and what data Sub Square might need.

In the example below, the subroutine called Square shows how the value of x can passed into the subroutine from elsewhere.

```
    Sub Square(X)
            X = X * X
    End Sub
```

In order to call a subroutine, you just write the sub's name and possibly some values or variables. When the name is read by the Visual Basic Script interpreter. It will not be confused; it keeps a record of all the procedures in a program. It uses the sub's name as a label to indicate where it should jump to find the sub's code. To call Sub Square and pass it a value, you would type the sub's name and beside it the value or variable holding the value, you wish to pass. In our case it is Sub Square and the variable X, which has been given a value of 10.

```
    Sub Square
        Dim X
        X = 10
        Square X
    End Sub
```

To be able to receive a value, Sub Square must be told to expect a variable from the calling procedure. In this case the value of X passed from Sub Math is popped into the Sub Square variable called SqX. Note the use of the keyword ByVaL placed directly before each parameter the Sub Square is expecting to receive. This is required only when the value of the parameter is to be altered, as in this case. If the value is for reference only then there is no need to use any prefix keyword.

ByRef is always assumed if ByRef is not used; it is the default mode of passing variables. The example below illustrates the ByVal method of passing variables to a procedure

```
    Sub Square (ByVaL X)
        X = X * X
        MsgBox X
    End Sub
```

Only use the ByVaL keyword when the variable being sent is to have its value altered within the called procedure. In the case of Sub Square, the value of X is altered by the processing and so the ByVal statement is used. Also, this example shows the first use of a new command, MsgBox. This simple command causes a message box to appear displaying whatever message you like. In this case the value of X will be displayed. For simple message boxes, the syntax for the MsgBox command is just MsgBox message. The message can be a string of variable holding a string or a number.

With subroutines such as these you might create a whole library of mathematical functions to be called at your whim. Code organized and structured into subroutines is easy to read and maintain, but, most importantly, the same subroutine can be called from anywhere in your code as many times as needed. Since the code is reused time and again, it makes for enormous savings in time and cost. Subroutines also allow the programmer to concentrate on perfecting one piece of code at a time, which reduces the overall time spent debugging the code.

Functions

If a subroutine is just code with structure applied, then a function is just a subroutine with a little more structure applied. The functional difference between the two lies in the function's ability to pass a data value back to the calling code. This value can be anything you like. It may be the result of successful calculation or perhaps an error code indicating why a function went wrong. Whatever they are used for, functions provide enhanced communication between the procedures that compromise a program.

Defining a Function: Defining a function is no different from defining a sub, except you use the word Function instead of Sub in the procedure's declaration.

```
Syntax
Function name (parameter List)
    statements
        name = return_value
        Exit Function
    statements
        name = return_value
End Function
```

Function fSquare below returns the value it calculates back to the call by making the name of the function equal to the value or variable that needs returning. Here is function fSquare returning the value of the variable X.

```
Function fSquare(X)
    X = (X * X)
    FSquare = X
End Function
```

It is important to note that setting the return value of a function does not terminate the function. This statement is processed like any other and left behind. Although you can set the function's return value as often as you want, the value won't actually be returned until the function terminates at the End Function delimiter or the Exit Function command. It is good coding practice to set the return value of the function only when there is no possibility of further relevant processing – in other words, processing that could alter the value being returned. In simple functions such as fSquarel, the equation that sets the return variable is commonly the last line to be processed.

Calling a Function: Like a subroutine, a function can receive values passed by the call, but unlike a sub, a function can also return a value to the call. In order to be capable of receiving a value the call must be in the form of an equation, which always takes the form:

```
Empty Variable = function Name
```

Just as with subroutines, if the function requires data to work with, you can send some winging its way by placing it (value or variable) to the right of the function name. There is a minor technicality at this point and that is remembering to enclose the value to be sent to a function in brackets; subs do not need brackets, but functions do. In Visual Basic Script, the call to function fSquare, passing the value of X, looks like this X = fSquare(X). Sub Math below shows the result returned from fSquare being displayed via a message box.

```
Sub Math
    Dim X
```

```
        X = 10
        X = fSquare(X)
        MsgBox X
    End Sub
```

In the example above you can see that the procedure Math is using the function fSquare to perform the arithmetic for it. This helps keep Sub Math nice and simple. If any other procedures need to have a number squared, then they can call the same fSquare function which helps to prevent unnecessary duplication of code.

Forcing an Exit From a Sub: A Sub or Function comes to an end when the End Sub or End Function delimiter is encountered. Here the processor returns to the point where the call was made and continues with the next line in the calling procedure. Occasionally it becomes imperative to stop processing mid-procedure; for example, of incorrect data is sent to a procedure and this is detected, then there is little point in carrying on with the rest of the processing.

In other cases it may simply be more efficient to leave the procedure before the End delimiter is reached. For example, a text search procedure might scan onscreen text for the first instance of a particular word. If successful, it would probably want to stop and return something meaningful (like the position of the word) as soon as it found the first instance of the word. Continuing to process all the remaining text after the procedure has been successful is just a waste of time. Being able to exit a procedure at just the right moment can be as important as knowing just when to call a procedure. How it's done is simplicity itself. The command is Exit Sub or Exit Function, depending on the procedure type you wish to exit from.

If....Then....Else

The If…Then…..Else construct allows the programmer to test data values and respond by running either one piece of code or another. A program with many If statements can react to a wide range of data and is not rigidly fixed into one particular processing pathway. Precisely which subset of the program's code actually runs depends on the depends on the data. When only one line of code is dependent on the If condition being true, then it is fine to use the shorthand syntax shown below.

For more complex statements it is better to use the full syntax:

```
If Condition1 Then
    statements
Else Condition2 Then
    statements
Else
    statements
End If
```

Decision Time

In procedures, decisions need to be made all the time. It is how the logic flows that determines the code's functionality. Such decisions can be defined by asking questions and then supplying instructions depending on the answers. The example below uses the If construct to translate the traveller's dilemma into Visual Basic Script.

```
Sub TakeMeHome
    If RightForm = Home Then
        GoRightFork
    Else
        GoLeftFork
    End If
End Sub
```

The code below shows that translating the full question into Visual Basic Script does not present any problems.

```
Sub TakeMeHome
    If RightFork = Home Then
        GoRightFork
    Elseif LeftFork = Home Then
        GoLeftFork
Else
    GoBackwards
    End If
End Sub
```

Here are some sample If structures from Visual Basic Script. The first is the simplest.

```
If X<10 Then X=X+1
```

X will only be incremented if its value is less than 10. When only one statement is dependent on the condition, in this case X<10, then it is acceptable to use this shorthand version of the If syntax. With the single-line syntax, it is possible to have multiple statements executed, but they must all be on the same line and separated by colons, as in the following example.

```
If X>10 Then X=X+1: Y=Y+1: Z+Z+1
```

It is usual practice not to format an If statement with multiple dependent line sin the manner shown above, even though it is perfectly legal. The recognized way of formatting exactly the same statement is as follows.

```
If X<10 Then
    X=X+1
    Y=Y+1
    Z=Z+1
End If
```

In both these examples, the dependent lines that increment X, Y and Z will only be run if the value of X is greater than 10 when x is tested by the If statement. What would run if X was less than 10? For these examples, the answer is nothing at all, but it raises an interesting point. If you are testing a condition to see if it is true, you are in fact making two tests: the first to see if it is true and also an implicit second test, to see if it is false, the logical opposite of true.

Else: We use the statement Else to delimit lines that are dependent on the condition being tested not resolving to true. In the following sample you can see the Else statement being used to take care of all logical possibilities other than the condition being tested in the If statement above it.

```
If X<10 Then
    X=X+1
Else
    X=X-1
End If
```

X will be incremented if its value is less than 10; however, if X is not less than 10, then the value of X is decremented. The Else statement acts as the logical opposite to whatever tests are being applied. What it really says is, "If none of the above conditions are true....". The Else statement is not a specific test, since it represents the logical opposite of the tests being applied. There is a way to specific an Else statement to ensure that, in the event of the first condition failing, there are secondary conditions to apply. The ElseIf statement allows a second and third (as many as you want) condition to be built into the If…Then…Else construct.

ElseIf: It is used to define any further conditions required by the If block. Each ElseIf condition is dependent on all the conditions above it being false. If the ElseIf condition evaluates to true, then its dependent code is run. In the following example you can see an ElseIf statement being used to specify a second condition. It is important to remember that although ElseIfs look like they are only testing one condition, in logical terms they represent a retest of all the conditions that went before, in addition to the condition the ElseIf specifies. In order to even test the ElseIf condition, we know the first If condition must have failed.

```
If X<10 Then
    X=X+1
ElseIf X>20 Then
    X=X-1
Else
    MsgBox "X > 10 and < 20"
End If
```

X will be incremented if its value is less than 10. X will be decremented if its value exceeds 20. However, if neither of the conditions are true, then the code dependent on the Else statement is run and a message box displays an appropriate string. This shows how the logic of each failed condition is assumed for the next tested condition. The equivalent logic without using ElseIf or Else could be achieved like this:

```
If X<10 Then X=X+1
If not (X<10) And X>20 Then X=X-1
If not (X<10 And X>20) Then MsgBox "X is greater than 10 but Less than 20"
```

This form of logic may appear to be the same at first glance, but it would potentially give very different results. Using the If…then….Else construct, only one of the conditional branches can be run; using the three line logic shown above means this effect is not guaranteed. The Else and ElseIf are both optional. You can have as many ElseIf clauses as you want, but none can appear after an Else statement.

Complex logical structures can be built easily using the If…End If control blocks. This lends the code a great deal of flexibility. All situations and all circumstances can be catered to. The ElseIf component allows each ElseIf branch to be a specific. Yes – No question in its own right; you can have more than one ElseIf if required, so all valid possibilities can be checked for. In reasonably complex if structures, this Else code is often used "In Case of Emergency"; this is where error processing code is commonly found. If you have checked for all valid contingencies and none of them have resolved to true, then what is left over must be invalid. In the following example all valid conditions are checked and appropriate action taken if the value being tested turns out to be invalid.

```
If X=42 Then
    MsgBox "This is the Magic Number"
ElseIf X>0 And X<1000
    MsgBox "X is Valid but not magic"
Else
    MsgBox "ERROR: X is an Invalid number"
End If
```

You can see that x has three logical states: magic, valid or invalid. The first If condition tests to see if X is the magic number 42. If it is not the magic number, then the second test is performed to make sure that X is at least valid; that is, it lies between 0 and 1000.

Failing both of these then, it is an iron-clad certainty that X lies outside the valid range and is certainly not the magic number. In this instance, the Else instance, appropriate error trapping takes place.

Nesting Ifs

Nesting means placing one thing inside another, like a Dancing doll. Any code be made dependent on an If statement, even another If. The example below shows two Ifs, one nested inside the other.

```
If X< 10 Then
```

```
        If Y < 10 Then MsgBox "X and Y are both Less than 10"
    End if
```

The limit for nested If statements is so high it will never concern you. Here is an example for a triple-nested If statement.

```
    If X< 10 Then
        If Y < 10 Then
            If z<10 Then
                MsgBox "X, Y, Z  are both Less than 10"
            End if
        End If
    End if
```

The only line of code that actually does anything is the line that displays the message box, by nesting the If statements together, another form or compound test is being created. Nesting Ifs produces results very similar to those created by using ElseIf.

It is common to see a new nested If being created when a new test is being applied and ElseIfs being used when the test is a variation on the If condition. This is not a syntax rule, but it does make sense. In this final complex example, you see all the Ifs components and formatting in action.

```
    If X< 10 Then
        MsgBox "X=10"
    Else
        If Y < 10
            MsgBox "X not=10 and Y<10"
        ElseIf Y=10
            If Z=5 Then
                MsgBox "X not=10 and Y=10 and Z=5"
            End If
        End If
    End if
```

Don't worry if the example above looks a little daunting; just start at the top with first If condition and work your way down and to the right. If you imagine some simple values for X,Y and X then you can follow the logic-making decisions about which dependent code should run whenever a condition is being tested. The message boxes show the code's current logical position.

The For...Next Loop

Techniques involved in repeating certain lines of code and over again is called looping; the same lines of code are run repeatedly, in a loop, until a condition is met and the loop stops looping. The For....Next construct is used to show which lines are to be repeated and very importantly, supplies a condition that can stop the loop when it is supposed to.

Loop: A group of lines within a body of code that can be selectively repeated a specified number of times.

```
    For counter = start To end [Step step]
        statements
        [Exit For]
        statements
    Next
```

The For delimiter is made up of a number of a components that help to define the specific nature of the loop. The For Counter component sets up a predeclared variable to be used as a counter for the loop. The start and end components store the number at which to start looping and the number at which to end looping. And the number at which to end looping. A loop of this type works by setting the counter variable to the start number and then running the code in the box of the loop before jumping back to the top line again. After each run of the loop code, the counter variable is automatically incremented.

This process carries on until the counter variable has the same value as the End variable. The additional and optional component Step holds the increment value used by the loop. If Step is omitted from the statement, then an increment value of + 1 is assumed. If there is no Step component, then the number of times a loop will repeat itself is equal to Start variable minus the End variable.

The Next delimiter shows whether the block of code to be repeated ends. All the lines of code contained between the For and Next delimiters are part of the body of the loop and will therefore be repeated. When the processor reaches the Next delimiter, it jumps back up to the For statement and checks to see if it is okay to do another loop. Here is an example of a simple loop in Visual Basic Script.

```
For Counter = 1 to 10
    MsgBox Counter
Next
```

This code will display 10 message boxes, each displaying a number one larger than the previous message box. The number will start at 1 and the loop will end after the tenth message box is displayed.

The Step statement allows the counter to increment in values other than one. If the value of the Step variable is two, then the counter will increase it by two for each iteration of the loop. If the Step value is negative, then the counter will count down instead of up. In many circumstances, counting down can be just as useful as counting up.

```
For CountDown = 10 to 1 Step -1
    MsgBox CountDown
Next
```

In Formula One racing it is a common sight to see men with headphones learning out over the railing at the edge of the track, holding aloft a large board with various numbers on it. Each time the driver comes around, he recognizes the stretch of track and for an instance, glances up at the board. Now he knows his position, how many laps there are until the finish line and other useful things. For the rest of the lap the driver need pay no attention to such details, concentrating all his energies into going as fast as possible. Only when he comes around again need the driver give up any of his resources to refresh his lap information. Such a system works well for racing drivers because it is a very efficient use of their resources.

This is what a For...Next car driving loop might look like in Visual Basic Script.

```
Sub GrandPrix
    Dim Lap
    Dim EndRace
    EndRace = 10
    For Lap = 1 to EndRace
        MsgBox "This is Lap Number " & Lap
    Next
    MsgBox "The Race is Over"
End Sub
```

The Do Loop Construct

A new loop construct, the Do loop, will show how to create loops that will repeat until or while a condition, any condition chosen by the programmer, is found to be true. This adds enormous flexibility to the looping concept.

Do Loop: Code found inside the Do loop delimiters will be repeated while a condition is true or until a condition becomes true.

The syntax will be:

```
Do While | Until condition
    statements
        Exit Do
    statements
Loop
```

Top check: A Do loop that checks its condition before entering the loop. The conditional statement is found to the right of the Do keyword. If the condition resolves to true on this first check, then the loop code will not be run.

Do While or Do Until?

Do loops are used when you need to repeat a block of code, not for a set number of times but for as many times as it takes to complete a task. Remember, a condition is simply an equation that can be resolved to either true or false. Conditions are tested and used to decide whether or not to stop the loop. The form of Do loop that will continue until a condition is true is called a Do Until loop: Do Until condition = True. The logical opposite of that statement is Do While. Visual Basic Script allows both logical types to be used so that the conditional tests are kept simple. Here are some Do loop examples.

```
X=0
Do While Not(X=10)
    X=X+1
Loop

X=0
Do UntiL X = 10
    X=X+1
Loop
```

Both these examples will loop for 10 iterations. You can see that the simplest code will be achieved by using the Do UntiL. The Do While is more complex because it uses the extra Not. Although the logic is the same, the clarity of the code is not.

Do Until isn't always the simpler method. Here are two more examples.

```
X=0
Do While X<10
    X=X+1
Loop

X=0
Do Until Not(X<10)
    X=X+1
Loop
```

As the first set of Do loops, this pair is logically identical and both will loop 10 times. From a computer's point of view, there is nothing to choose between them. From a Visual Basic Scripter's point of view, the While is much simpler to understand.

```
Do
    statements
        Exit Do
    statements
Loop while | UntiL Condition
```

Definition

Bottom Check: A Do loop that checks its condition only after running through the loop code once.

The conditional statement is found to the right of the Loop keyboard. Even through the conditional statement might be true before the loop begins, the loop code will run once before this is checked.

Top/Bottom Checking

Another flexible feature of the Do loop is its ability to be either a top-checking loop or a bottom-checking loop. The top-checking check its condition beside the Do keyword whereas the bottom-checking loop its condition beside the Loop keyword at the bottom .if you remember how a computer reads one line at a time from the top of the code to the bottom, then you can see that a bottom-checking will have all of its code run at least once before the condition has a chance of being checked. Compare this to the top-checking loop, which always tests its condition before any loop dependent code is run. For a top-checking loop, the exit condition must be false to start with before the loop is run at all. In contrast, a bottom-checking loop will always run the loop code at least once.

Each of the two types is used depending on the logic you want to express. Here are two examples of apparently equivalent loop logic.

```
X=10
Do While X<10
    X=X+1
Loop

X=10
Do
    X=X+1
Loop While X<10
```

The actions of the loops will not be the same though. In the first example X will not be incremented by one because, although the exit condition has been met before the loop starts running, this is not tested until the loop has run once. It is more likely that correct loop structure for this case would be a top-checking loop, but as these next two example show, that is not always the case.

```
X=10
Do While X>10 and X<20
    X=X+1
Loop

X=-10
Do
    X=X+1
Loop While X>10 and X<20
```

In the first example the loop simply will never run. In the second example the loop is allowed to run once before testing the value of its condition. X has been incremented and the loop continues for another nine iterations.

Loops can go for Ever

The main asset of Do loops, their conditional testing, is also the source of their greatest danger. When you are running a loop that will stop only when its exit condition is met, what happens if its exit condition cannot be met? Consider the following example.

```
Do until True = False
    X=X+1
Loop
```

This is perfectly valid Visual Basic Script. You can type it in now if you want to, but it would not be advisable. If such a loop were to appear, the processor would treat it like any other. X would be incremented until True = False, which is just a logical way of saying the loop would never stop.

Writing a loop like this is not as hard as it might seem. Here is an example of an innocuous-looking loop that would stop your code in its tracks.

```
X=10
Do
    X=X+1
Loop UntiL X=10
```

In this case the loop will stop only when X is equal to 10. X is equal to 10 at the outset, but the loop is a bottom checker. This means the loop code will be run before the value of X is tested. Unfortunately, the loop code increases the value of X to 11, and so the loop condition, when it is finally tested, is found to be false. The loop will keep on looping, adding one to X and hopelessly rechecking the condition, but now things are out of control. X will never be equal to 10 and the loop will never exit.

The Select Case Construct

As your experience grows and the programming tasks you tackle increase in logical complexity, it is common to find large and unwieldy blocks of code emerging. One of the main culprits for this effect is the multiple nesting of If statements. The Select Case construct is much neater way of packaging a series of conditional tests that direct the program flow. In many cases, it is the construct of choice simply because the resulting code is so much clearer and not because it adds extra functionality over a complex If statement.

Select Case: The Select Case statement will execute one of several groups of statements. The statements to be run will depend on the value of the expression tested by Select Case. If the value of the expression tested matches any more one Case branch, only the statements following the first match are executed.

```
Select Case testexpression
    Case expressionlist-1
        statements
    Case expressionlist-n
        statements
    Case Else
        statements
End Select
```

Testing Multiple Values

The power of a Select Case statement lies in its ability to test any number of values for a given expression and then run the appropriate, dependent code block. Each case branch can have more than one possible value, which will trigger the case-dependent code. Each value is separate by a comma. To illustrate this, consider the simple Select Case statement below.

```
X=5
Select Case X
    Case 1
        MsgBox "X is equal to 1"
    Case 2,3,4
        MsgBox "X is either 2 or 3 or 4"
    Case 5
        MsgBox "X is equal to 5"
End Select
```

If this example was run, then the message "X is equal to 5" would be displayed in a message box. The Select Case statement first evaluates the value of X. It then tests each Case statement in turn. If the case has multiple values separated by commas, each value is tested in turn. This process continues until one of the values on one of the Case branches matches the test expression value. The code that is dependent on this Case statement is then run. After a case is found to be true, the processor ignores all the other case branches and restarts processing on the first line after the End Select statement. In the example, only the Case statement Case 5 resolves to true since x, the test expression, was made equal to five before the Select Case was run.

Select Case statements can evaluate any type of data, not just numbers. The example below shows a Select Case statement testing the variable MyName to determine the appropriate code to run.

```
MyName = "Sachin"
Select Case MyName
    Case "Sachin"
        MsgBox "Hello Sachin"
    Case "Rahul", "Saurav"
        MsgBox "Hello Rahul or Saurav?"
End Select
```

You can see that this example will display the message "Hello Sachin" but what would have happened if the variable MyName had been set to a name that was not mentioned in any of the Case branches? In this situation, the Select Case would have simply ignored all the dependent code and nothing would have happened. Sometimes it is useful to have some default code to deal with just such an eventuality. Sometimes it is useful to have some default code to deal with just such an eventuality. This default ability is provided by the Case Else statement. Like the Else components of an If....Then...Else construct, the Case Else stands for: if none of the above are true. Here is the same example as before, but this time it has default code to deal with any names it does not recognize.

```
MyName = "Madhuri"
Select Case MyName
    Case "Sachin"
        MsgBox "Hello Sachin"
    Case "Rahul", "Saurav"
        MsgBox "Hello Rahul or Saurav?"
    Case Else
        MsgBox "Howdy Stranger"
End Select
```

Now running the example would result in a welcoming "Howdy Stranger" message should MyName contain any name, indeed any string, other than those specified in the preceding Case statements. The code dependent on the Case Else statement will only be run if none of the previous conditions have been met.

COM (Component Object Model)

History of COM

One of the first methods of interprocess communication in Windows was DDE that allowed sending and receiving messages in so-called "conversations" between applications. This technology was the base of OLE (Object Linking and Embedding) that in turn was one of the leading technologies introduced with Windows 3.0. Text conversations and Windows messages proved not to be as flexible as to allow sharing application features in a robust and extensible way. By the time version 3.1 of Windows was released, COM was created as a new foundation, and OLE changed to OLE2.

Visual Basic introduced the VBX (packaged extensions in the form of dynamically linked libraries) that allowed the creation of objects to be graphically placed in a form and manipulated by properties and methods. These were later adapted for use by other languages such as Visual C++. OLE custom controls (OCXs) would be the replacement for the widely popular VBX.

Antony Williams, one of the more notable thinkers involved in the creation of the COM architecture, distributed a couple of internal papers in Microsoft that embraced the concept of software components; Object Architecture: Dealing With the Unknown -or- Type Safety in a Dynamically Extensible Class Library in 1988 and On Inheritance: What It Means and How To Use It in 1990. These provided the foundation of many, if not all, of the ideas behind the basics of COM.

From many of these ideas spawned Microsoft's first object-based framework, OLE, which is the abbreviation of Object Linking and Embedding. OLE was built on top of dynamic data exchange (DDE) arid designed specifically for compound documents. It was introduced with Word for Windows arid Excel in 1991, and was later included with Windows, starting with version 3.1 in 1992. An example of a compound document is a spreadsheet embedded in a Word for Windows document; as changes are made to the spreadsheet within Excel, they appear automatically inside the Word document.

In 1991, Microsoft introduced Visual Basic Extensions (VBX) with Visual Basic 1.0.

In 1993, Microsoft released OLE 2 with its underlying object model. While OLE 1 was focused on compound documents, COM and OLE 2 were designed to address software components in general. In 1994 OLE controls (OCX) were introduced as the successor to VBX controls. At the same time, Microsoft stated that OLE 2 would just be known as "OLE", and that OLE was no longer an acronym, but a name for all of the company's component technologies.

In early 1996, Microsoft found a new use for OLE Custom Controls, expanding their Web browser's capability to present content, renamed some parts of OLE relating to the Internet **ActiveX,** and gradually renamed all OLE technologies to ActiveX, except the compound document technology that was used in Microsoft Office. Later that year, DCOM was introduced as an answer to CORBA.

COM

Component Object Model (COM) is an interface standard for software componentry introduced by Microsoft in 1993. It is used to enable interprocess communication and dynamic object creation in any programming language that supports the technology. The term COM is often used in the software development industry as an umbrella term that encompasses the OLE, OLE Automation, ActiveX, COM+ and DCOM technologies.

The essence of COM is a language-neutral way of implementing objects that can be used in environments different from the one they were created in, even across machine boundaries. For well-authored components, COM allows reuse of objects with no knowledge of their internal implementation, as it forces component implementers to provide well-defined interfaces that are separate from the implementation. The different allocation semantics of languages are accommodated by making objects responsible for their own creation and destruction through reference-counting. Casting between different interfaces of an object is achieved through the QueryInterfaceQ function. The preferred method of inheritance within COM is the creation of sub-objects to which method calls are delegated.

Although the interface standard has been implemented on several platforms, COM is primarily used with Microsoft Windows. COM is expected to be replaced at least to some extent by the Microsoft .NET framework, and support for Web Services through the Windows Communication Foundation (WCF). However, COM objects can still be used with all .NET languages without problems. Networked DCOM uses binary proprietary formats, while WCF encourages the use of XML-based SOAP messaging. COM is very similar to other component software interface standards, such as CORBA and Java Beans, although each has its own strengths and weaknesses. It is likely that the characteristics of COM make it most suitable for the development and deployment of desktop applications, for which it was originally designed.

COM is a software architecture that allows applications and systems to be built from the components supplied by different software vendors. It is a set of binary and network standards that allows any software to communicate with each other regardless of the hardware, the operating system, and the programming language used for development. COM is not a programming language but a specification that defines how components can communicate with each other.

In a component based system, components interact with each other by calling methods and passing data. COM ensures that there is a standard method of interaction between the components. All the COM objects need to follow these standards when prqviding the functionality. Standards are as important to component architectures as they are to any system with interchangeable parts. For example, the signal that a television or radio receives is specified in a standard. Without standards, nothing will function togather.

Benefits Of COM

Any programming language that can create structures of pointers and call functions through pointers can create and use component objects. Component objects can be implemented in a number of programming languages and can be used for client software written in different programming languages. This is because COM provides a binary and not source code standard. This is the fundamental benefit of COM over object oriented language.

Thus COM can be simply defined as 'an island of objects in a sea of applications'.

DCOM

DCOM(Distributed Component Object Model) is a model as COM but it is specially designed for distributed applications. While there is a need of using the applications as distributed then DCOM becomes in picture.

Distributed Component Object Model (DCOM) is a proprietary Microsoft technology for communication among software components distributed across networked computers. DCOM, which originally was called "Network OLE", extends Microsoft's COM, and provides the communication substrate under Microsoft's COM+ application server infrastructure. It has been deprecated in favor of Microsoft .NET.

The addition of the "D" to COM was due to extensive use of DCE/RPC - more specifically Microsoft's enhanced version, known as MSRPC.

In terms of the extensions it added to COM, DCOM had to solve the problems of

- Marshalling - serializing and deserializing the arguments and return values of method calls "over the wire".
- Distributed garbage collection - ensuring that references held by clients of interfaces are released when, for example, the client process crashed, or the network connection was lost.

One of the key factors in solving these problems is the use of DCE/RPC as the underlying RPC mechanism behind DCOM. DCE/RPC has strictly defined rules regarding marshalling and who is responsible for freeing memory.

DCOM was a major competitor to CORBA. Proponents of both of these technologies saw them as one day becoming the model for code and service-reuse over the Internet. However, the difficulties involved in getting either of these technologies to work over Internet firewalls, and on unknown and insecure machines, meant that normal HTTP requests in combination with web browsers won out over both of them. Microsoft, at one point, attempted and failed to head this off by adding an extra http transport to DCE/RPC called ncacn_http(Network Computing Architecture, Connection-based, over HTTP). This was later resurrected to support an Exchange 2003 connection over HTTP.

Revision

1. Visual Basic Script is also called VB Script.
2. The first tag indicates the script language and the end tag ends the scripting section.
3. There are really three types of data to consider when working in a program. The first is called a **Local** variable, the second is a **Global** variable and the third is **Persistent** Data.
4. All variables contain variant data.
5. Visual Basic Script provides a wide range of functions for working with strings.
6. Visual Basic Script is rich in string manipulation tools; this has always been one of the strengths of the Visual Basic language.
7. Visual Basic Script also provides a wide range of functions for working with numbers.
8. Visual Basic Script provides solid, built-in functionality for working with dates and times.
9. Operator + is used to sum two numbers.
10. And is used to to perform a logical conjunction on two expressions.
11. & is used to force string concatenation of two expressions.
12. / is used to divide two numbers and return a floating point result.
13. ^ is used to raise a number to the power of an exponent.
14. \ is used to divide two numbers and return an integer result.
15. Is is used to compare two object reference variables.
16. Mod is used to divide two numbers and return only the remainder.
17. * is used to raise a number to the power of an exponent.
18. – is used to find the difference between two numbers or to indicate the negative value of a numeric expression.
19. Not is used to perform logical negation on an expression.
20. Or is used to perform a logical disjunction on two expressions.
21. Xor is used to perform a logical exclusion on two expressions.
22. Visual Basic Script can change the caption of an HTML button with a simple statement
23. SetStatus is provided to set the window status value based on the Status argument.
24. MsgBox function can be utilized to query the user to have the user provide feedback confirmation.
25. Image maps are useful for providing graphical hot links to other Web documents and other parts of your Web pages.
26. In Visual Basic Script, the most common procedure type is the Subroutine or Sub for short.
27. If a subroutine is just code with structure applied, then a function is just a subroutine with a little more structure applied.
28. Defining a function is no different from defining a sub, except you use the word Function instead of Sub in the procedure's declaration.
29. Like a subroutine, a function can receive values passed by the call, but unlike a sub, a function can also return a value to the call.
30. A Sub or Function comes to an end when the End Sub or End Function delimiter is encountered.
31. The If…Then…..Else construct allows the programmer to test data values and respond by running either one piece of code or another.
32. We use the statement Else to delimit lines that are dependent on the condition being tested not resolving to true.
33. ElseIf is used to define any further conditions required by the If block.

34. Techniques involved in repeating certain lines of code and over again is called looping.
35. Code found inside the Do loop delimiters will be repeated while a condition is true or until a condition becomes true.
36. Do loops are used when you need to repeat a block of code, not for a set number of times but for as many times as it takes to complete a task.
37. The Select Case construct is much neater way of packaging a series of conditional tests that direct the program flow.

Multiple Choice Questions

1. For sum of two numbers you will use :
   (a) + (b) –
   (c) *
2. For concatetaion of numbers you use :
   (a) & (b) *
   (c) ^
3. _______ is used to divide two numbers and return an integer result.:
   (a) / (b) ^
   (c) \
4. ________ is used to define any further conditions required by the If block.
   (a) If (b) ElseIf
   (c) Else
5. ________ returns the absolute value of a number. .
   (a) Abs (b) Asb
   (c) Value
6. Exp returns :
   (a) x (b) p
   (c) e
7. Current system date is returned by :
   (a) Date (b) System
   (c) Today
8. Minute returns the value between 0 and :
   (a) 1 (b) 23
   (c) 59
9. __________ returns a string that has been converted to lowercase.:
   (a) LCASE (b) CASE
   (c) UCASE

Answers

1. a	2. a	3. a	4. b	5. a
6. c	7. a	8. c	9. a	

Subjective Questions

1. How would you add script code to HTML?
2. How would you add a script to your documents?
3. Describe the various Data Types of Visual Basic?
4. How would you send the message across in VBScript?
5. Describe the various functions in VBScript?
6. How is the Nesting If used?
7. How are the For...Next loop used?
8. How are the Do While and Do Until used?
9. What do you understand by Top/Bottom checking?
10. What is COM?
11. What is DCOM? Explain.
12. What is RMI server?

13 Active Server Pages (ASP)

How Active Server Pages Really Work?

An Active Server Page is a standard HTML file that is extended with additional features. Like a standard HTML file, an Active Server Page contains HTML tags that can be interpreted and displayed by a Web browser. Anything you could normally place an HTML file – Java applets, blinking text, client-wide scripts, client-side ActiveX controls – you can place in an Active Server Page. However, an Active Server Page has three important features that make it unique.

- An Active Server Page can contain server-side scripts. By including server-side scripts in an Active Server Page, you can create Web pages with dynamic content. To take an extremely simply example, you can create a Web page that displays different messages at different times of the day.
- An Active Server Page provides several built-in objects. By using the built-in objects accessible in an Active Server Page, you can make your scripts much more powerful. Among other things, these objects enable you to both retrieve information form and send information to browsers. For example, by using the Request object, you can retrieve the information that a user has posted in an HTML form and respond to that information within a script.
- An Active Server Page can be extended with additional components. Active Server Pages comes bundled with several standard, server-side ActiveX components. These components enable you to do such things as work with databases, send email and access the file system.

These standard ActiveX components are very useful. However, you're not limited only to these components. You can buy third-party components or create additional ActiveX components of your own. There's no limit to how you can extend Active Server Pages.

These three features define an Active Server Page. An Active Server Page is a standard HTML page that has been extended with server-side scripts, objects and components. By using Active Server Pages, you can create Web sites with dynamic content. The best way to understand how Active Server Pages work is by contrasting the process of retrieving an Active Server Page with the process of accessing a normal HTML page. The process of serving an HTML page can be broken into the following steps:

1. A user enters the Internet address of an HTML file into the address bar of a Web browser and presses Enter to request a Web page (for example, http://www.aspsite.com/hello.htm).
2. The browser sends a request for the Web page to a Web server such as Internet Information Server (IIS).
3. The Web server receives the request and recognizes that the request is for an HTML file because the requested file has the extension .htm or .html.
4. The Web server retrievers the proper HTML file from disk or memory and sends the file back to the browser.
5. The HTML file is interpreted by the user's Web browser and the result are displayed in the browser window.

Of course, this process is often more complicated (for example, the contents of forms are posted and query strings are passed). But, in broad strokes, these steps outline the moment-to-moment activity of a typical Web server. A server receives requests for particular files and responds by sending the correct file, retrieving it from the hard drive or memory.

Although IIS can be used to server static HTML pages. Active Server Pages enable IIS to serve dynamic content as well. Using Active Server Pages, you can create pages with new content in response to user requests. The Web server itself becomes active in the process of creating the Web page.

It's important to understand how this process of serving an Active Server Page contrasts with the normal process of serving an HTML page, so we'll break it into steps:

1. A user enters the Internet address of an Active Server Page file into the address bar of a Web browser and presses Enter to request an Active Server Page (for example, http://www.aspsite.com/hello.asp).
2. The browser sends a request for the Active Server Page to IIS.
3. The Web server receives the request and recognizes that the request is for an Active Server Page file because the requested file has the extension .asp.
4. The Web server retrieves the proper Active Server Page file from disk or memory.
5. The Web server sends the file to a special program named ASP.dll.
6. The Active Server Page file is processed from top to bottom and any commands it encounters are executed. The result of this process is a standard HTMl file.
7. The HTML file is sent back to the browser.
8. The HTML file is interpreted by the user's Web browser and the results are displayed in the browser window.

From the perspective of the Web server, an Active Server Page is very different from a normal HTML page. A normal HTML file is sent without processing to the browser. All the commands in an Active Server Page, on the other hand, must first be executed to create an HTML page. This allows an Active Server Page to contain dynamic content.

From the perspective of the browser, on the other hand, an Active Server Page is almost exactly the same as a normal HTML page. The only difference is that an Active Server Page typically must end with the extension .asp rather than .htm or .html. When a request is made for an Active Server Page, the browser receives a normal HTML page. This enables an active Server Page to be compatible with all browsers.

Integrating Objects and Components into Active Server Pages

An Active Server Page is primarily a scripting environment. This book assumes that you will be creating Active Server Page scripts with Microsoft's VBScript. However, you can use other scripting languages with Active Server Pages such as Jscript and PerlScript as well. Any scripting language that has a scripting engine compatible with the ActiveX scripting standard can be used in an Active Server Page.

The easiest way to add a script to an Active Server Page is by using the script delimiters <% and %>. Any text enclosed within these delimiters will be processed as a script. Here's is an example:

```
<HTML>
<HEAD><TITLE>ASP Script</TITLE></HEAD>
<BODY>
This is a
<% For i=1 to 10 %>
very,
```

```
<% NEXT %>
very long sentence.
</BODY>
</HTML>
```

When this Active Server Page is displayed by a Web browser, the following sentence is displayed:

This is a very, very, very, very, very, very, very, very, very, very, very long sentence.

The script creates 11 copies of the word very by using a VBScript FOR…Next loop.

By default, an Active Server Page assumes you'll be using VBScript as your primary scripting language. This means that you don't need to do anything beyond using the <% and %> script delimiters to use this language. However, there are three ways to explicitly specify a language for use in an Active Server Page.

First, you can use the Internet Service Manager to specify a particular scripting language as the default language for all your Active Server Pages. To do this, perform the following steps:

1. Launch the Internet Server Manager from the Microsoft Internet Information Server program group on the Start menu.
2. Right-click the name of your Web site. If you haven't changed the default configuration, the name of this Web site will be default Web Site.
3. Choose Properties to open you Web site's property sheet.
4. Click the Home Directory tab.
5. Click the Configuration button. (To do this, you must have an existing application. If you don't, create one now by clicking Create.)
6. In the Application Configuration dialog box, click the App Options tab.
7. In the Default ASP Language text box, enter the name of the scripting language that you want to use as your primary scripting language; for example, enter VBScript or script.

After you have specified a particular scripting language as the default scripting language, you can use it in your Active Server Pages simply by using the delimiters <% and %>. If you plan to use Jscript for the majority of your Active Server Pages, for example, you should configure this language as your default language.

You can also specify the primary scripting language for a particular page. To do this, place the Language directive as the very first line in your Active Server Page file, as shown in this example:

```
<%@ Language=Jscript %>
<HTML>
<HEAD><TITLE>ASP Script</TITLE></HEAD>
<BODY>
This is a
<% for i=1;I<11;I++) { %>
very,
<% } %>
very long sentence.
</BODY>
</HTML>
```

The directive in the first line of this script indicates that all the scripts contained in the file should be executed as scripts created with Jscript, rather than some other scripting language. When you use this directive, be sure to include a space between the @ character and the keyword Language. Furthermore, it's very important that this directive appear before my other commands in your Active Server Page file (otherwise you'll get an error).

A third alternative exists for including scripts in your Active Server Pages. You can use Microsoft's extended <SCRIPT> HTML tag, as in this example:

```
<HTML>
<HEAD><TITLE>ASP Script</TITLE></HEAD>
<BODY>
<SCRIPT LANGUAGE="Jscript" RUNAT="server">
function sayhello()
{
response.write("Hello!")
}
</SCRIPT>
<%
sayhello()
%>
</BODY>
</HTML>
```

Here, the <SCRIPT> tags contain a Jscript function. The Language attribute of the <SCRIPT> tag specifies which scripting language to use. The Run at attribute indicates that the script should be executed on the server rather than on the client (the browser).

The function named sayhello() is defined in the first script. The second script, marked with the usual <% and %> tags, is where the Jscript function is actually called. This Active Server Page prints the text Hello! To the screen.

If you have created client-side Jscript or VBScript scripts, you should be familiar with the <SCRIPT> tag. Microsoft's extended <SCRIPT> tag can be used to specify either client-side or server-side scripts. If you neglected to include the RUNAT="server" attribute in the preceding example, the script would be treated as a client-side script. In this case, the server would ignore the script and the browser would attempt to execute the script. (It would fail miserably because the script isn't a valid client-side script).

Why would you ever want to use the <SCRIPT> tag rather than the <% and %> script delimiters? Normally, you wouldn't use the <SCRIPT> tag. However, there are two significant differences between these two ways of specifying a script.

First, scripts that are contained in the <SCRIPT> tag are executed immediately, no matter where they appear in an Active Server Page. For example, consider the following page:

```
<HTML>
<HEAD><TITLE>ASP Script</TITLE></HEAD>
<BODY>
This is the first sentence.
<SCRIPT LANGUAGE="Jscript" RUNAT="server">
response.write("this is the second sentence.")
</SCRIPT>
</BODY>
</HTML>
```

Looking the script, you might be tempted to believe that the sentence This is the first sentence and the sentence "This is the second sentence" are printed to the screen in that order. However, when the Active Server Page is displayed in a browser, the order of the two sentence is actually reversed (or worse, nothing is displayed, because an invalid HTML page is generated).

Why does this happen? Whatever is contained in the <SCRIPT> tag is executed before anything else in a page. If you sue the View Source command on your Web browser, you'll see the following results from the preceding Active Server Page:

```
This is the second sentence,<HTML>
```

```
<HEAD><TITLE>ASP Script</TITLE></HEAD>
<BODY>
This is the first sentence.
</BODY>
</HTML>
```

This behavior of the <SCRIPT> tag has two implications. First, you can place scripts contained in a <SCRIPT> tag anywhere you want in an Active Server page. Second, the <SCRIPT> tag, for most purpose, is restricted to containing functions or procedures. The output of any script that's not contained in a function or procedure is displayed immediately and results in an invalid HTML page.

The <SCRIPT> tag has one main advantage over the <% and %> script delimiters. Using the <SCRIPT> tag, you can mix multiple scripting languages within a single Active Server Page. Consider the following example:

```
<%@ Language=Jscript %>
<HTML>
<HEAD><TITLE>ASP Script</TITLE></HEAD>
<BODY>
<SCRIPT LANGUAGE="Jscript" RUNAT="server">
function sayhello()
{
response.write("Hello!")
}
</SCRIPT>
<%
for i=1 to 10
sayhello()
NEXT
%>
</BODY>
</HTML>
```

This script prints Hello! 10 times in a row. But notice how it does this. The script contained in the <% and %> delimiters is a Visual Basic script. However, this script calls a Jscript function. The Jscript function is defined within the <SCRIPT> tag. When you want to use one scripting language as your primary scripting language, but you need to call a function from another language, you can use this method. This is useful when one language has particular functions or methods that another language lacks.

To summarize, there are three methods of including a script in an Active Server Page:

- Specify a scripting language for all your Active Server Page by using the Intenret Service Manager to specify a default language.
- Specify a scripting language for a single page by using the Active Server Page directive <%@ Language="scripting language" %>.
- Mix multiple scripting language in a single Active Server Page by using the extended <SCRIPT> tag.

Before ending this section, one final Active Server Pages directive should be discussed.

By using the Active Server Pages' output directive, you can display the value of an expression. Here's an example:

```
<HTML>
<HEAD><TITLE>ASP Example</TITLE></HEAD>
<BODY>
At the tone, the time will be: <%=TIME%>
```

```
</BODY>
</HTML>
```

You use the delimiters <%= and %> to print the value of a variable, method or function. In the preceding example, the output directive is used to output the value of the VBScript TIME function.

There's another way to accomplish the same thing. Consider the following example:

```
<HTML>
<HEAD><TITLE>ASP Example</TITLE></HEAD>
<BODY>
At the tone, the time will be: <% Response.Write(TIME) %>
</BODY>
</HTML>
```

In this example, the value of the VBScript TIME function is outputted by using the Active Server Pages Response object. The Write() method of the Response object output the value of expressions to the screen.

When should you use the Response.Write() method rather than the <%= and %> output directive? It really doesn't matter. Active Server Pager internally represent the output directive as a Response.write() method call in any case. The two methods of outputting the values of expressions are completely interchangeable.

Nevertheless, there are situations in which one method of outputting the values of expressions is more convenient than the other. For example, when you need to output the value of an expression within a script, the Response.Write() method is often easier to use. On the other hand, when you want to output the value of an expression within a section of HTML code, the <%= and %> directive is often easier to use. The following illustrates both approaches:

```
<HTML>
<HEAD><TITLE>ASP Example</TITLE></HEAD>
<BODY>
<%
for i=1 to 10
myvar=myvar&"very,"
Response.Write(I&":&myvar&"<BR>)
NEXT
%>
<HR>
This is a <%=myvar%> long sentence.
</BODY>
</HTML>
```

In this example, the Response.Write() method is used within the loop to display the value of the variable named myvar as it increases in size. The <%=and %> output directive is embedded within normal HTML code. The output directive is used to display the value of myvar.

Understanding Objects and Components

Active Server Pages include several built-in objects and installable ActiveX components. These objects and components can be used to extend the power of your Active Server Pages scripts. What are the objects and components?

An object is something that typically has methods, properties or collections. An object's methods determine the things you can do with the object. An object's properties can be read or set to specify the state of the object. An object's collections constitute different sets of key and value pairs related to the object.

To take an everyday example, the book Tom Sawyer is an example of an object. The object has certain methods that determine the things you can do with it. For example, you can read the book, use it as a doorstep or even, if you're feeling particularly malicious, tear it into shreds. The object has certain properties. For example, it has a certain number of pages and a particular author. Finally, it has a collection of key and value pairs. Each page number (the key) has a corresponding page of text (the value).

An ActiveX component is similar to an Active Server Page built-in object. However, when you are using Active Server Pages, there is one important difference between a component and an object. An instance of a component must be explicitly created before it can be used.

Active Server Pages Objects

Active Server Pages include several built-in objects. These objects enable you to extend the power of your scripts. For example, by using these objects, you can gain access to browser requests and control how the server responds to these requests. The built-in objects also provide you with control over user sessions and Web-server applications.

You have already been introduced to one example of a built-in object- the Response object. You can use the Response object to send output to a browser. However, the Response object also has several other important properties, collections and methods.

The following chapter explain in detail how to use each of the built-in objects. The following list provides a quick overview of each of the built-in objects:

- The Application object – The Application object is used to store and retrieve information that can be shared among all users of an application. For example, you can use the Application object to pass information between users of your Web site.
- The Request object – The Request object can be used to access all information sent in a request from a browser to your server. You can use the Request object to retrieve the information that a user has entered into an HTML form.
- The Response object – The Response object is used to send information back to a browser. You can use the response object to send output from your scripts to a browser.
- The Server object – The Server object enables you to use various utility functions on the server. For example, you can use the Server object to control the length of time a script executes before it times out. You can also use the Server object to create instances of other objects.
- The Session object – The Session object can be used to store and retrieve information about particular user sessions. You can use the Session object to store information that persists over the course of a visit by a user to your Web site.
- The ObjectContext object – The ObjetContext object is used to control Active Server Pages transactions. The transactions are managed by the Microsoft Transaction Server (MTS).

Active Server Pages Components

Like the built-in objects discussed in the preceding section. Active Server Pages components can be used to extend the power of your scripts. Components differ from the built in objects because they're typically used for more specialized tasks. The following list provides a brief overview of some of the components bundled with Active Server Pages:

- The Ad Rotator component – The Ad Rotator component is used to display banner advertisements on the Web pages of a Web site. You can use this component to specify how frequently different banner advertisements should be displayed.
- The Browser Capabilities component – The Browser Capabilities component can be used to display different HTML content, according to the capabilities of different browsers. For example,

you can use this component to display Web pages with frames only to frames-compliant browsers.

- The Content Linking component – using the Content Linking component, you can link together several HTML pages so that they can be navigated easily. For example, you can use this component to display the pages of an online book.
- The Counters components – The Counters components can be used to keep track of the number of visitors to your Web site. You can use the Counters components to add a hit counter to a Web page.
- The Content Rotator component – The Content Rotator component enables you to rotate through HTML content on a page. For example, you can use the component to randomly display different announcements on the home page of your Web site.
- The Page Counter component – The Page Counter component is exactly like the Counter component, in that it can be used to track the number of visitors and to add a hit counter to particular Web page.
- The Permission Checker component – The Permission Checker component can be used to display links to Web pages only to those users who have permission to see them. You can use this component to create Web pages that can be viewed only by the administrators of a Web site.
- The Collaboration Data Objects – The Collaboration Data Objects (CDO) enable you to send and retrieve email from within your Active Server Pages. You can use these objects to send email to new user after the user registers at your Web site.
- The ActiveX Data Objects – The ActiveX Data Objects (ADO) enable you to retrieve and store data in a database such as Microsoft SQL Server. These objects are extremely important.

Using Components in Active Server Pages

Before you can use a component, you must first create an instance of it. Although you can automatically access the properties, collections and methods of the built-in objects on every page, before you can use a component, you must create an instance of the component with a particular scope. In the following three sections, you will learn how to create an instance of a component with page, session and application scope.

Creating a Component with Page Scope

In most cases, you'll create an instance of a component with page scope. A component with page scope is created on a single page and dies when processing on the page ends. You can use a component with page scope only on the page where it is explicitly created. To create an instance of a component with page scope, you can use either the Server.CreateObject() method or the HTML <OBJECT> tag.

Here's an example of how you can use the CreateObject method to create an instance of the Dictionary object with page scope:

```
<%
Set MyDict=Server.CreteObject("Scripting.Dictionary")
MyDict.Add "CA", "California"
MyDict.Add "MA", "Massachusetts"
MyDict.Add "MI", "Missouri"
%>
My dictionary has <%=MyDict.Count %> entries.
<BR>
The first entry in my dictionary is <%=Mydict.Item("CA")%>
```

When this script is executed, an instance of the Dictionary object is created. Next, three key and value pairs are added to the dictionary. Finally, two properties of the dictionary are displayed.

The first statement in this example uses the Server.CreateObject method to create an instance of an object with page scope. In this example, an instance of the Dictionary object is assigned to a variable named MyDict.

After an instance of an object has been created, its methods can be called. In this example, the Add method of the Dictionary object is called to add entries to the dictionary. The first Add method call is used to add the key and value pair CA and California.

After an instance of an object has been created, you can also access its properties. In this example, the Count property of the Dictionary object is read to determine the number of entries in the dictionary. The Item property is also read to return the value of a particular key.

Finally, when you're finished using an instance of an object, you can destroy it. An object created by the preceding method will automatically be destroyed after the server finished processing the Active Server Page. Therefore, there's usually no reason to explicitly destroy an object. However, you can destroy the MyDict object explicitly by assigning MyDict to a new value or by setting the variable to the value Nothing like this:

```
<% Set Mydict=Nothing %>
```

This example illustrates the most typical method of creating and using ActiveX components in Active Server Pages. You create an instance of an object by calling the Server.CreateObject() method. After you create a new object, you can call its methods and read and set its properties.

There exists a second method for creating object with page scope. Instead of using the CreateObject method, you can create an instance of an object with the Microsoft Extended HTML <OBJECT> tag. Here is an example:

```
<OBJECT RUNAT="Server" ID="MyBrow"
PROGID="Music.BrowserType"></OBJECT>
```

Because the preceding code is not a script, you should not add it to an Active Server Page within the <% and %> script delimiters. The <OBJECT> tag is an HTML tag and it should appear within the normal HTML content of your page.

This example shows how to use three properties of the HTML <OBJECT> tag. The Id attribute provides the component with a unique identifier (a name), so you can refer to it in your Active Server Pages scripts. The PROGID is used to specify the component's registered name. This is the name the server uses to identify the component when it creates an instance of it. It's the same name you would pass to the Server.CreateObject() method.

When would you use the <OBJECT> tag, instead of the CreateObject method, to create an instance of an object? The one advantage of the <OBJECT> tag is that it does not create an instance of an object until a property or method of the object is actually called. If an Active Server Page contains object that might never be called, you should use the <OBJECT> tag to save some memory.

Microsoft recommends that you create the majority of your components with page scope. By creating components with page scope, you place less of a burden on the Web server. A page scope component releases any memory and other resources it requires when the processing of the page comes to an end.

Creating Components with Session Scope

Instead of creating components that are accessible only within a single Active Server Page, you can create components that are accessible on multiple Active Server Pages. One way to do this is to create a component with session scope. A component with session scope is available to a single visitor to your Web site on every page that the visitor requests. An instance of the component is created for each visitor. The component instance does not leave memory until the visitor leaves.

Two methods are available for creating components with session scope. One method assigns a component to a session variable by using the Server.CreateObject() method, as in the following example:

```
<%
Set Session("MyBrow")=Server.CreateObject("MSWC.BrowserType")
%>
```

This script assigns an instance of the Browser Capabilities component to the session variable named MyBrow. You can place this script within any Active Server Page or the Global.asa file. Normally, it makes sense to create the object instance in the Session_OnStart subroutine of the Global.asa file because you need to create the instance only once for each visitor.

After you have created a component with session scope, you can use its properties and methods by referring to the session variable that contains the components. For example, the following script displays the value of the Frames property of the Browser Capabilities component created with the previous script:

```
<%
if session("MyBrow").Frames = True then
    response.write "your browser supports frames!"
end if
%>
```

This method of creating a component with session scope has a significant disadvantages. The Server.CreateObject() method immediately creates an instance of a component. Even if the instance of the component is never used, it still drains resources form the server. Because a new instance of a session object is created for each visitor, this can be a significant drain.

Fortunately, there's a second way to create a component with session scope. You can create a component by using he Microsoft HTML <OBJECT> tag, like this:

```
<OBJECT RUNAT="Server" SCOPR="Session" ID"MyBrow"
PROGID="MSWC.BrowserType"></OBJECT>
```

When you crate an instance of a component with session scope with the </OBJECT> tag, you should create the instance within the Global.asa file. Create the object instance outside the Application_OnStart, Application_OnEnd, Session_OnStart and Session_OnEnd subroutines.

This example uses four properties of the, OBJECT> tag. The Scope attribute indicates that the component created should have session scope. The ID attribute provides the component with a unique identifier (a name), so you can refer to it in your Active Server Pages scripts. The PROGID is used to specify the component's registered name. This is the name the server uses to identify the component when it creates an instance of it. It's the same name you would pass to the Server.CreateObject() method.

When you create an instance of a component with the HTML <OBJECT> tag, you can access any of its methods and properties just as if the object were a built-in Active Server Page object. For example, the following script displays a property of the Browser Capabilities component.

```
<%
If MyBrow.Frames = True then
    Response.write "your browser supports frames!"
End if
%>
```

When a component is created with session scope in either of the two ways just described, any of its methods, collections or properties are available on any page that a particular user requests. However, a particular instance of the component must be created for each user. A component with session scope is created relative to a particular user session.

Creating Components with Application Scope

When you create an instance of a component with application scope, you treat it as if it were a built-in object. After it is created, any methods, collections or properties of the component can be accessed by any user on any page. The component remains available until the server shuts down, the Global.asa file is modified or the application is unloaded.

You can create a component with application scope by using methods similar to those used to create a component with session scope. First, you can create a component with application scope by using the Server.CreateObject() method. Look at this example:

```
<%
Set Application("AdRot")=Server.CreateObject("MSWC.AdRotator")
%>
```

Here, an instance of the Ad Rotator component is assigned to an application variable. You could create a component in this way within any of your Active Server Pages. However, normally, you create an instance of an object with application scope only once within the Application_OnStart event of the Global.asa file. The object then becomes available to all users of your Web site.

After an instance of the Ad Rotator component has been created with application scope, you can use its properties on any Active Server Page. For example, the following script calls the Ad Rotator's getAdvertisement property to display a banner advertisement:

```
<%=Application( "adRot" ).getAdvertisement( "schedule.txt")%>
```

You can also create a component with application scope by using the Microsoft HTML <OBJECT> tag, like this:

```
<OBJECT RUNAT="Server" SCOPE="Application" ID="AdRot"
PROGID="MSWC.AdRotator"></OBJECT>
```

When you create an instance of a component with application scope with the <OBJECT> tag, you can only create the instance within the Global.asa file. Create the object instance outside the Application_OnStart, Application_InEnd, Session_OnStart and Session_OnEnd subroutines.

In this example, four properties of the <OBJECT> tag are used. The SCOPE attribute indicates that the component should have application rather than session scope. The ID attribute provides a name for the component. The PROGID attribute enables the server to identify the component.

After creating an instance of an application object in the Global.asa file with the <OBJECT> tag, you can use the component in exactly the same way as you use an Active Server Page built-in object. You can call any of the object's methods and properties on any page without creating another instance of it.

Troubleshooting Active Server Pages

Before you can use Active Server Pages, make sure that it is properly installed on your system. If you installed Internet Information Server while installing Window NT Server, Active Server Pages might not be installed. Active Server Pages is included with IIS versions 3.0 and above (the current version is version 4.0 and version 5.0 will be released with Windows 2000). To download the latest version of Active Server Pages, visit the Internet Information Server section of the Microsoft Web site as http://www.microsoft.com/iis.

After Active Server Pages is installed, you must configure IIS to use Active Server Pages. Configure at least one directory in your Web site with permission to execute an Active Server Pages file. Follow these steps:

1. Launch the Internet Service Manager from the Microsoft Internet Information Server program group on the Start menu.
2. In the left pane of the Internet Service Manager, negative to your default Web site. (If you haven't changed anything, the Web site will be named Default Web Site)
3. Choose the directory in your Web site where you want to store your Active Server Pages files by navigating to it in the Internet Service Manager.
4. Right-click the name of this directory and choose Properties.
5. Click the tab labeled Directory or Visual Directory (depending on the type of directory).
6. In the Permission section, choose either Script or Execute.

Now that you have created a directory with permissions to execute your Active Server Pages, you must store all Active Server Pages in this directory in order to use them. If the directory is a physical directory, you can simply store your pages in the directory with this name. If the directory is a virtual directory, you must determine the physical path to this directory on your hard drive. You can determine the physical path of a virtual directory by looking at the same tabbed dialog box where you just configured the permissions.

To load an Active Server Page into a browser that's located on the same machine as IIS, use the address bar of your Web browser. For example, if the Active Server Page is located in the root directory of your Web site, you can load the page by typing http://localhost/mypoage.asp into the address bar of your Web browser. This will load the Active Server Page named mypage.asp.

Testing Your Configuration

You can test whether Active Server Page is configured properly on your system by creating a simple Active Server Page. If you can display this page successfully in your Web browser, you know that everything is set up properly. To create an active Server Page, you can use any text editor. For example, you can use Notepad, which is included as an accessory with Windows NT Server. Start Notepad (or your favorite text editor) and enter the following text:

```
<HTML>
<HEAD><TITLE>ASP Page</TITLE></HEAD>
<BODY>
<%
Response.Write("Hello World")
%>
</BODY>
</HTML>
```

Save the Active Server Pages file as test.asp make sure that your text editor doesn't append an extra .txt extension to the filename. Also, make sure that you save the file in a directory with executive or Script permission.

Now, launch your favorite Web browser in order to display your new Active Server Page. If you saved your Active server Page in the root directory of your Web site, you can load the page by typing http:/ /localhost/test.asp. If you saved the Active Server Page in a subdirectory of this root directory, include the full path by entering an address such as http://localhost/mysubdir/test/asp. If you saved the page in a virtual directory, include the name of the virtual directory in the address. For example, in your virtual directory is named myvirtualdir, you should type http://localhost/myvirtualdir/test.asp.

If everything works, the Active Server Page should be displayed. The text Hello World! Should appear in your Web browser's window.

Variables

Variables are used to store information. This example demonstrates how to create a variable, assign a value to it and insert the variable value into a text.

```
<html>
<body>
<%
Dim name
Name="Bachchan"
Response.write("My name is:" & name)
%>
</body>
</html>
```

Lifetime of Variables

A variable declared outside a procedure, could be accessed and changed by any script in the ASP page where it is declared. A variable declared inside a procedure, is created and destroyed every time the procedure is executed. No scripts outside that specific procedure can access or change that variable. To make a variable accessible in several ASP pages, declare session variables or application variables.

Session Variables

Session variables stores information about one single user and are available to all pages in one application. Common information stored in session variables is username and userid. To create a session variable, store it in an Session Object.

Application Variables

Application variables are also available to all pages in one application. Application variables are used to hold information about all users in a specific application. To create an application variable, store it in an Application Object.

Procedures

ASP code can contain procedures and functions:

```
<html>
<head>
<%
sub vbproc(num1, num2)
response.write(num1*num2)
end sub
%>
</head>
<body>
The result of the calculation is: <%call vbproc(3,4)%>
</body>
</html>
```

Use the HTML <script> tag to include procedures or functions that are written in another scripting language than the default:

```
<html>
<head>
<script language-"javascript" runat="server">
function jsproc(num1,nmum2)
```

```
{
Response.Write(num1*num2)
}
</script>
</head>
<body>
The result of the calculation is: <%jsproc(3,4)%>
</body>
</html>
```

The "language" attribute in the <script> tag defines the script language for the code block inside the script element. The "runat=server" attribute in the <script> tag defines that this is a server side script. If you omit the "runat=server" attribute, the script will be executed on the browser.

ASP forms

Examples: A Simple Form (uses the GET method). This example demonstrates how to interact with the user, with the Request. QueryString command. Save this example as demo_requery.asp

```
<html>
<body>
<form action="demo_get.asp" method-"get">
Please type your first name:
<input type="text" name="fname"><br><br>
<input type="submit" value="Submit">
</form>
<% Request.QuerySTring("fname")<>""Then
Response.write("Hello"& Request.QueryString("fname")&"!")
Response.Write("<br>How are you today!")
End If
%>
<body>
</html>
```

A simple form (uses the POST method)

This example demonstrates how to interact with the user, with the Request.Form command.

Save this example as demo_simpleform.asp

```
</html>
</body>
<form action="demo_powt.asp" method="post">
Please type your firs name:
<input type="text" name="fname">
<br>
<input type="submit" value="Submit">
</form>
<%
If Request.Form("fname")<>""Then
Response.Write("Hello" & Request.Form("fname") &"!")
Response.Write ("<br>How are you today?")
End IF
%>
<body>
</html>
```

A Form with Radio buttons

This example demonstrates how to interact with the user, through radio buttons, with the Request.Form command. Save this Example as demo_radio.asp.

```
<html>
<body>
<%
dim cars
cars=Request.Form("cars")
%>
<form action="demo_radio.asp"method="post">
<p>Please select your favorite car:</p>
<input type="radio" name="cars" value="Maruti"
<%if cars="Maruti" then Response.Write("checked")%>>
Maruti 800
<br>
<input type="radio" name="cars" value="Ambassdor"
<%if cars="Ambassdor" then Response.Write("checked")%>>
Ambasdor97
<br>
<input type="submit" value="submit">
</form>
<%
if cars<>""then
Response.Write("<p>Your favorite car is:" & cars & "</p>")
End if
%>
</body>
</html>
```

Date Types

A named storage location that can contain data that can be modified during program execution. Each variable has a name that uniquely identifies it within its scope. A data type can be specified or not.

Variable names must begin with an alphabetic character, must be unique within the same scope, can't be longer than 255 characters and can't contain an embedded period or type-deceleration character.

The attribute of a variable or field that determines what kind of data it can hold. The following listing shows the supported data types, including storage sizes and ranges.

Data Type	*Storage size*	*Range*
Byte	1 byte	0 to 255
Boolean	2 bytes	True or False
Integer	2 bytes	-32,768 to 32,767
Long (long integer)	4 bytes	-2,147,648 to 2,147,483,647
Single (Single-precision floating-point)	4 bytes	-3.402823E38 to -1.401298E-45 for negative values; 1.401298E-45 to 3.402823E38 for positive values
Double (double-prevision floating-point)	8 bytes	-1.79769313486231E308 to –4.94065645841247E-324 for negative values; 4.94065645841247E-324 to 1.79769313486232E308 for positive values

Currency	8 bytes	-922,337,203,685,477,5808 to 922,337,203,685,477.5807
Decimal	14 bytes	+/-79,228,162,514,264,337,593,543,950,335 with no decimal point; +/- 7.9228162514264337593543950335 with 29 places to the right of the decimal; smallest non-zero number is +/-0.0000000000000000000000000001
Date	8 bytes	January 1, 100 to December 31, 9999
Object	4 bytes	Any Object reference
String (variable-length)	10 bytes	+ string length 0 to approximately 2 billion
String (fixed-length)	Length of string	1 to approximately 65,400
Variant (with numbers)	16 bytes	Any numeric value up to the range of a Double
Variant (with characters)	22 bytes + string length	Same range as for variable-length String
User-defined (using Type)	Number required by elements	The range of each element is the same as the range of its data type.

Declaring Variables

To declare a variable is to tell the program about it in advance. You declare a variable with the Dim statement, supplying a name for the variable:

Dim variablename [As type]

Variables declared with the Dim statement within a procedure exist only as long as the procedure is executing. When the procedure finishes, the value of the variable disappears. In addition the value of a variable in a procedure is local to that procedure – that is, you can't access a variable in one procedure from another procedure. These characteristics allows you to use the same variable names in different procedures without worrying about conflicts or accidental changes.

A variable name:

- Must begin with a letter.
- Can't contain an embedded period or embedded type-declaration character.
- Must not exceed 255 characters.
- Must be unique within the same scope, which is the range form which the variable can be referenced – a procedure, a form and so on.

The optional As type clause in the Dim statement allows you to define the data type or object type of the variable you are declaring. Data types define the type of information the variable stores. Some examples of data types include String, Integer and Currency. Variables can also contain objects from Visual Basic or other applications. Examples of Visual Basic object types, or classes, include Object, Form1 and TextBox.

There are other ways to declare variables:

- Declaring a variable in the Declaration section of a form, standard or class module, rather than within a procedure, makes the variable available to all the procedures in the module.
- Declaring a variable using the Public keyword makes it available throughout your application./
- Declaring a local variable using the Static keyword preserves its value even when a procedure ends.

Variable Naming Conventions

A convention for nothing the range of reference for the variable. For example, local variables can be prefixed with "I", which reminds you that they can referenced only within the procedure they are defined in. The variable naming conventions are always considered into two aspects. These are scope and type. The following Listing shows the variable naming conventions as follows:

Scope	*Description*	*Example*
1	Local	lnCounter
p	Private (default)	pnStatus
g	Public(global)	gnOldREcno
t	Parameter	tnREcNo

Convention for nothing the data type of a variable. Suggested prefixes for type are as follows:

Type	*Description*	*Example*
a	Array	aMonths
c	Character	cLastName
y	Currency	yCurrencyValue
d	Date	dBirthDay
t	Datetime	tLastModified
b	Double	bValue
f	Float	fInterest
l	Logical	lFlag
n	Numeric	nCounter
o	Object	oEmployee
u	Unknown	uReturn Value

Implicit Declaration

You don't have to declare a variable before using it. For example, you could write a function where you don't need to declare AnyVal before using it:

```
Function SafeSqr(num)
    AnyVal=Abs(num)
    SafeSqr=Sqr(AnyVal)
End Function
```

Visual Basic automatically creates a variable with that name, which you can use as if you had explicitly declared it. While this is convenient, it can lead to subtle errors in your code if you misspell a variable name. For example, suppose that this was the function you wrote:

```
Function SafeSqr(num)
    AnyVal=Abs(num)
    SafeSqr=Sqr(AnyVal)
End Function
```

At first glance, this looks the same. But because the AnyVal variable was misspelled on the next-to-last line, this function will always return zero. When script encounters a new name, it can't determine whether you actually meant to implicitly declare a new variable or you just misspelled on existing variable name, so it creates a new variable with that name.

Explicit Declaration

To avoid the problem of misnaming variables, you can stipulate that script always warn you whenever it encounters a name not declared explicitly as a variable.

To explicitly declare variables.

- Place this statement in the Declaration section of a class, form or standard module:

Option Explicit

-Or-

From the tools menu, choose Options click the Editor tab and check the Require Variable Declaration option. This automatically inserts the Option Explicit statement in any new modules, but not in modules already created; therefore, you must manually add Option Explicit to any existing modules within a project.

Had this statement been in effect for the form or standard module containing the SafeSqr function, Visual Basic would have recognized AnyVal and AnVal as undeclared variables and generated errors for both of them. You could then explicitly declare AnyVal:

```
Function SafeSqr(num)
    AnyVal=Abs(num)
    SafeSqr=Sqr(AnyVal)
End Function
```

Now you'd understand the problem immediately because Visual Basic display an error message for the incorrectly spelled AnVal. Because the Option Explicit statement helps you catch these kinds of errors, it's a good idea to use it with all your code.

The Option Explicit statement operators on a per-module basis; it must be placed in the Declarations section of every form, standard and class module for which you want Visual Basic to enforce explicit variable declarations. If you select Require Variable Declaration, Visual Basic inserts Option Explicit in all subsequent form, standard and class modules, but does not add it to existing code. You must manually add Option Explicit to any existing modules within a project.

Operators

An operator is one which works on data items and performs some mathematical calculation or change the data. Visual Basic provides six types of operators. We are going to discuss these operators one by one in the coming section.

Arithmetic Operators

Most programs perform arithmetic calculations. The arithmetic operators are summarized in listing. Note the use of various special symbol not used in algebra. The caret (^) indicated exponentiation and the asterisk (*) indicates multiplication. The Integer division operator (\) and the modules (Mod) operator will be discussed shortly. Most arithmetic operators are binary operators because they each operate on two operands. For example, the expression Anysum + value contains the binary operator + and the two operands Anysum and value.

Visual Basic Operator	*Arithmetic operator*	*Algebraic expression*	*Expression*
Addition	+	x + y	x + y
Subtraction	-	z-8	z-8
Multiplication	*	yb	y*b
Division(float)	/	v/u	v/u
Division(Integer)	\	none	v\u
Exponentiation	^	qp	Q ^ p
Negation	-	-e	-e
Modules	Mod	q mod r	q mod r

The modules operator, Mod, yields the Integer remainder after Integer division. Like the Integer division operator, the modules operator rounds any fractional part before performing the operation. The expression x Mod y yields the remainder after x is divided by y. A result of 0 indicates that y divides evenly into x. Thus, 20 mod 5 yields 0 and 7 Mod 4 yields 3.

The negative operator, -, changes the sign of a number from positive to negative (or from negative to positive). The expression –8 changes the sign of 8 to negative, which yields –8. The negation operator is said to a unary operator, because it operates on only one operand. The operand must appear to the right of the negation operator.

Arithmetic expressions must be written in straight-line from when entering programs into the computer. Thus, expressions such as "a raised to the power b" must be written as:

A ^ b

So that all constant, variables and operators appear in a straight line. The algebraic notation:

A^b

Is generally not acceptable to compilers, although some special-purpose software package do exist that support more natural notation for complex mathematical expressions.

Parentheses are used in expression in much the same manner as in algebraic expressions. For example, to multiple b times quantity e + n we write:

B * (e + n)

Comparison Operator

A character or symbol indicating a relationship between two or more values or expressions. These operators include less than (<), less than or equal to (<=), greater than (>), greater than or equal to (>=), not equal <>) and equal (=). Additional comparison operators include Is and Like. Note that Is and Like can't be used as comparison operators in a Select Case statement.

This example shows various uses of comparison operators, which you use to compare expressions.

```
Dim MyResult, Var1, Var2
MyResult=(45<35) ' Returns False.
MyResult=(45=45) ' Returns True.
MyResult=(4<>3) ' Returns True.
MyResult=("5" > "4") ' Returns True.
Var1="5":Var2=4 'Initialize variables.
MyResult=(Var1>var2) ' Returns True.
Var1=5:Var2=Empty
MyResult=(Var1>var2) ' Returns True.
Var1 =0: Var2=Empty
MyResult=(Var1=var2) ' Returns True.
```

Logical Operator

These operators are used to combine one or more than one relational expression. By combining logical operators with relational operators, you can create more powerful data-testing statements.

Operator	*Description*
And	Perform a logical conjunction on two expressions.
Eqv	Perform a logical equivalence on two expressions.
Imp	Perform a logical Implication on two expressions.
Not	Perform a logical negation on an expression.

OrO Perform a logical disjunction on two expressions.

Xor Perform a logical exclusion on two expressions.

And

This example uses the And operator perform a logical conjunction on two expressions.

```
Dim A, B, C, D, MYCheck
A = 10: B = 8: C = 6: D = Null 'Initialise variables.
MyCheck = A > B and B > C 'Returns True.
MyCheck = B > A and B > C 'Returns False.
MyCheck = A > B and B > D 'ReturnsNull.
MyCheck = A and B 'Returns 8 (bitwise comparison)
```

Eqv

This example uses the Eqv operator to perform a logical equivalence on two expressions.

```
Dim A, B, C, D, MYCheck
A = 10: B = 8: C = 6: D = Null 'Initialise variables.
MyCheck = A > B Eqv B > C 'Returns True.
MyCheck = B > A Eqv B > C 'Returns False.
MyCheck = A > B Eqv B > D 'ReturnsNull.
MyCheck = A EqvB 'Returns 3 (bitwise comparison)
```

Imp

This example uses the Imp operator to perform a logical Implication on two expressions.

```
Dim A, B, C, D, MYCheck
A = 10: B = 8: C = 6: D = Null 'Initialise variables.
MyCheck = A > B Imp B > C 'Returns True.
MyCheck = A > B Imp C > B 'Returns False.
MyCheck = B > A Imp C > B 'Returns True.
MyCheck = B > A Imp C > D 'Returns True.
MyCheck = C > D Imp B > A 'ReturnsNull.
MyCheck = B Imp A 'Returns -1 (bitwise comparison)
```

Not

This example uses the Not operator to perform a logical negation on an expression.

```
Dim A, B, C, D, MYCheck
A = 10: B = 8: C = 6: D = Null 'Initialise variables.
MyCheck = Not (A > B) 'Returns False.
MyCheck = Not (B > A) 'Returns True.
MyCheck = Not (C > D) 'ReturnsNull.
MyCheck = Not A 'Returns -11 (bitwise comparison)
```

Or

This example uses the Or Operator to perform a logical disjunction on two expressions.

```
Dim A, B, C, D, MYCheck
A = 10: B = 8: C = 6: D = Null 'Initialise variables.
```

```
MyCheck = A > B Or B > C 'Returns True.
MyCheck = B > A Or B > C 'Returns True.
MyCheck = A > B Or B > D 'Returns True.
MyCheck = B > D Or B > A 'ReturnsNull.
MyCheck = A Or B  'Returns 10 (bitwise comparison)
```

Xor

This example uses the Xor operator to perform a logical exclusion on two expressions.

```
Dim A, B, C, D, MYCheck
A = 10: B = 8: C = 6: D = Null 'Initialise variables.
MyCheck = A > B Xor B > C 'Returns False.
MyCheck = B > A Xor B > C 'Returns True.
MyCheck = B > A Xor C > B 'Returns False.
MyCheck = B > D Xor A > B 'ReturnsNull.
MyCheck = A Xor B  'Returns 2 (bitwise comparison)
```

ASP Reference

This section does not contain a complete reference for ASP. It provides definitions, syntax and examples of many of the keywords commonly used.

Abs

This argument is a numeric expression. The value returned will be exactly the same as that entered in the expression, unless the number entered is negative. If the number is negative, it will be returned unsigned.

Syntax

Abs (number)

Addition (+)

Sums two numbers. This operator's (+) syntax consists of an expression + expression = a sum. Strings can be concatenated using the addition operator, but the & operator is preferred for concatenation because of its better documentation.

Syntax

One expression + one expression = Result

And

Use this operator to perform a legal conjunction with two expressions. The result is True only if both expression evaluate to True. The result is false if either expression evaluates to False.

Syntax

One expression And One expression = result

Array

Use this function to returns a variant that contains an array:

Syntax

Array (elementslist)

Asc

Use this function to return the ASCII character code representing the first letter in a string.

Syntax

Asc(string)

Call

Use this statement to shift control to a new procedure. You do not have to use the Call keyword when calling a procedure.

Syntax

[Call] name [argumentlist]

Chr

Use this function to return the particular character associated with the individual ANSI character code.

Syntax

Chr (charcode)

Class (Object)

Use this object to manipulate the events of a class that you create. It will not work to make a variable of the type Class. The Class object refers to an object defined using the Class statement. You can create a specific instance of that class when you define a class using the Class statement.

Class (Statement)

Use this statement to identify a class. You can also determine the methods and properties, as well as the variables, comprising a class. You can make class members private or public.

Syntax

Class name

Clear Method

Use this method to clear the property settings of the Err object. You use Clear to clear the Err object after an error handled. A typical use would be for deferred error handling with the On Error Resume Next statements.

Syntax

Object.Clear

Color

Use these constants to apply color in applications. You don't have to define these constants to use them, as they are build into VBScript. Color constants can be used in your code to represent the hex values. The constants are as follows: vbBlack (Black), vbRed (Red), vbGreen (Green), vbYellow (Yellow), vbBlue (Blue), vbMagenta (Magenta), vbCyan (Cyan) and vbWhite) (White).

Syntax

Object.property – ColorConstant

Concatenation

Use the concatenation operator to concatenate two string expressions. An expression is converted to a string type whenever an expression is not a string.

Syntax

Result = expressiona & expressionb

Const

Use this statement to declare constants instead of literal values. Constants are public statements (p. –public page—) by default, but are always private statements within procedures. You can't define variables or user functions because they aren't constants. You can use constants similarly to variable expressions.

Syntax

[public I Private] Const anyname = expression

CreateObject

Use this object to create and return a reference to an Automation object. Automation servers provide different types of objects. Word-processing applications might provide application objects and a document object. A spreadsheet application could offer a sheet and cells.

Syntax

CreateObject(servername.typename [, location])

Date and Time

The date and time constants enable you to use days and time in applications. These constants do not need to be defined, as they are built into VBScript. You can use date and time constants in your code to represent hex values. The day of the week Constants are vbSunday (Sunday), vbMonday (Monday), vbTuesday (Tuesday), vbWednesday (Wednesday), vbThursday (Thursday) vbFriday (Friday) and vbSaturday (Saturday).

The following table describes the other rate and time constants.

Constant	*Value*	*Description*
VbUseSystem	0	Applies the date format from your computer's regional settings.
VbUseSystemDayOfWeek	0	Applies the day of the week from your system's settings for the first day of the week
VbFirstJan1	1	Applies the week in which January 1 occurs (default)
VbFirstFourDays	2	Applies the first week that has at least four days in the new year
VbFirstFullWeek	3	Applies the first full week of the year.

Example

Date

Use this function to return the current system date.

Syntax

Date

DateAdd

Use this function to return a date with a time interval. You can use the DateAdd function to add or subtract a specified time interval form a date. You can also use DateAdd to calculate a date a given number of days from yesterday or a time a specified number of minutes from now. Adding days to a specific date requires using the Day of Year ("y") function, the Day ("d") function or the WeekDay ("w") function.

Syntax

DateAdd(interval, number, date)

Values for DateAdd Function

Value	*Description*
Yyyy	Year
Q	Quarter

M	Month
Y	Day of year
D	Day
W	Weekday
Ww	Week of year
H	Hour
N	Minute
S	Second

DateDiff

Use this function to return the number of specified intervals between two dates. You could use Datediff to calculate the number of days between two dates or the number of weeks between today and the end of the year.

Syntax

DateDiff(interval, date1, date2 [,firstdayofweek[, firstweekofyear]])

DateDiff Function Syntax

Arguments	***Purpose***
Interval	Use this string expression to identify an interval to be used to calculate the differences between the first and second date. Required.
Date1, date2	Use this argument to identify dates to be used in the calculation. Required.
Firstdayofweek	Use this constant to select the day of the week. Optional.
Firstweekofyear	Use this constant to select the first week of the year. The first week is assumed to be the week in which January 1 occurs, if not specified. Optional.

DateSerial

Use this function to return a variant data type for the Date function for a particular year, month and day. To identify, a date, such as December 31, 1999 the range of numbers for each DateSerial argument should be in the accepted range for the unit; that is 1-31 for days and 1-12 for months. However, you can also specify relative dates for each argument using any numeric expression that represents some number of days, months or years before or after a certain days.

Syntax

DateSerial(year, month, day)

DateValue

Use this function to return a Variant for the Date function. The date argument is customarily a string expression signifying a date from January 1, 100 through December 31, 9999. If a date includes invalid time information, an error is produced. The short date format defined for your system is recognized by DateValue in terms of the order of moth, day and year when a date is given as a string that includes only numbers separated by valid date separators. The DateValue function can distinguish between full month names and abbreviations.

Syntax

DateValue(date)

Day

Use this function to return an integer between 1 and 31, which denotes the day of the month. The argument is any expression that could possibly represent a date.

Syntax

Day(date)

Dictionary

Use this object to maintain keys and items associated with the keys in an array. Items are linked to a unique key.

Dim

Use this statement to declare variables. The dim statement creates the variable and allocates storage space. In Visual Basic, there are different data types and correspondingly differing amounts of storage space allocated for each. In VBScript there is only the Variant type. Variables declared with Dim are available to all procedures inside the script and only within the procedure at the procedure level. You also can declare a dynamic array using the Dim statement.

Syntax

Dim varname[((subscripts])][,varname[([subscripts])]]...

Division

Use this operator to divide one number by another. The number to the left of the division symbol (/) will be divided by the number on the right side of the symbol.

Syntax

Number / Number = Result

Do..Loop

Use this statement to continue testing for a condition while the condition is True or until the condition becomes True.

Syntax

```
Do [ {While| Until} some state of affairs]
    [some words]
[Exit Do]
[ some words]
Loop
Do [ {While | Until} condition]
[statements]
[Exit Do]
[statements]
Loop
Or
Do
[statements]
[Exit Do]
[statements]
Loop [ {while | Until} condition]
```

Erase

Use this statement to re-initialise elements in fixed-size arrays and eliminate the storage space allocation for dynamic arrays. You need to know whether an array is fixed-size (ordinary) or dynamic.

Syntax

Erase thisarray

The thisarray argument is the name of the Array variable to be erased.

Err

Use this object to store run-time information. The Err object also accepts the raise and Clear methods for generating and clearing run-time errors. The Err object has global scope. The Err object's properties are set to zero or null after the On Error Resume Next statement.

Syntax

Err.Property
Err.Method

Eval

Use this function to calculate an expression and deliver the result. In VBScript, setting one variable to the value of another variable can be interpreted in two ways. Two values are compared to see if they are identical.

Syntax

[resuot =]Eval (expression)

Exit

Use this statement to exit a block of Do...Loop, For...Next, Function or Sub code.

Syntax

The following lists show the correct syntax for Exit Statements:

- Use Exit Do to exit a Do....Loop statement. Use it only inside a Do...Loop statement. Control is transferred to the statement following the Loop statement.
- Use Exit For a exit a For loop. Use only in a For...Next statement or a For...Each loop. The Exit for control statement transfers control to the statement following the Next statement. (It transfers control to the statement immediately after the next statement.) In nested structures, control is transferred to the loop, one level above the loop where the Exit...For statement appears.
- Use the Exit function to exit a function procedure. Execution will continue with the statement immediately following the statement calling the Function.
- Use Exit property to exit the property procedure where it appears. Execution will continue with the statement after the statement that originally called the procedure.
- Use Exit Sub to exit a Sub procedure. Execution continues with the statement following the calling statement.

FileSystemObject

Use this object to retrieve files on a computer's file system.

Syntax

For example, the FileSystemObject using the Drive property has the following syntax:

FileSystemObjectname.Drives
Scripting.FileSystemObject

Filter

Use this function to return a zero-based array containing a subset of a string array based on specified filter criteria. If there are no matches of Value within InputStrings, the function returns an empty array.

Syntax

Filter(InputStrings, Value[, Include[, Compare]])

For...Each

Use this block to repeat a set of statements for each element in an array or collection. The For Each block is entered if there is at least one of the elements in the group. Once the loop is begun, all the statements in the loop are executed for the first element in the group. As long as there are more elements in the group, the statements in the loop continue to execute for each element. When there are no more elements in the group, the loop is exited and execution continue with the statement following the Next statement.

Syntax

```
For Each element In group
    [statements]
    [Exit For]
    [statements]
    Next [element]
```

For...Next

Use this statement to repeat a group of statements the number of times indicated. The counter is incremented once the loop starts and all statements in the loop have executed the step. Finally, the statements in the loop execute again or the loop is exited and execution continues with the statement following the Next statement.

Syntax

```
For counter = start To end [Step step]
[statements]
[Exit For]
[statements]
Next
```

FormatCurrency

Use this function to return a value using a currency symbol as up on your system. The definition of the currency format resides in your computer's regional settings.

Syntax

```
Formatcurrency(Expression[, NumDigitsAfterDecimal
[,IncludeLeadingDigit [,UseParensForNegativeNumbers
[, GroupDigits]]]])
```

FormatDateTime

This function returns a value that has a date or time format.

Syntax

```
FormatDateTime(Date[, NamedFormat])
```

FormatNumber

Use this function to format numbers. The format for numbers in the regional settings of your computer determines how the numbers will be displayed.

Syntax

```
FormatNumber(Expression [,NumDigitsAfterDecimal
[,IncludeLeadingDigit [,UseParensForNegativeNumbers
[, GroupDigits]]]])
```

FormatPercentFunction

Use this function to reformat a number so that it is displayed as a percentage (multiplied by 100) with a trailing % character. The format for numbers in the regional settings of your computer determines how the numbers will be displayed.

Syntax

FormatPercent(Expression[, NumDigitsAfterDecimal [, IncludeLeadingDigit [,UseParentsForNegativeNumbers [, GroupDigits]]]])

Function (Statement)

Use this statement to set up a procedure that will carry out a specific task and / or do calculations when parameters are passed to it.

Syntax

```
[Public [Default] | Private] Function name [(arglist)]
[statements]
[name = expression]
[Exit Function]
[statements]
[name = expression]
End Function
```

GetObject

Use this function to use an Automation object by launching a file. The Set statement assigns an object variable to the object.

Syntax

GetObject([pathname] [, class])

Hex

Use this function to return a string that shows the hexadecimal value of a number. Before being evaluated, the number is rounded to the nearest whole number.

Syntax

Hex(number)

Hour

Use this function to return a whole number between 0 and 23 that indicates the hour of the day.

Syntax

Hour (time)

If...Then...Else

Use this statement to set up conditions to be tested so that certain functions are performed. The statements are first evaluated, the condition immediately following the If is tested and if the condition exists [is True], the statements following are then executed. In the event that the condition does not exist (is False), the ElseIf conditions are evaluated. Under the ElseIf conditions, each condition is tested to see if it exists (is true). If none of these conditions is found to be true, then the testing moves on to the Else statements. Finally, after each process is completed, in a top-down manner, the program execution continues with statements following the End if statement. Single-line statements must not contain the keyword Else.

Syntax

```
If condition Then statements [Else elsestatements]
Or, you can use the block form syntax:
If condition Then
[statements]
[ElseIf condition-n Then
[elseifstatements]]....
```

```
[Else
[eslestatements]]
End If
```

InputBox

Use this function to display a prompt in a dialog box, wait for next input or a button click and then return the value entered. When the user clicks OK or presses Enter, the InputBox function returns whatever is in the text box. If the user clicks Cancel, the function returns a zero-length string (").

Syntax

```
InputBox(prompt[, title][, default][, xpos][, ypos][, helpfile, context])
```

InStr

Use this function to return the position of the first occurrence of one string contained inside another.

Syntax

```
InStr([start, ]string1, string2[, compare])
```

Int

Use this function to return the whole portion of a number. The Int function is like Fix, except that negative numbers cause Int to return the first negative integer less than or equal to the number, whereas Fix returns the first negative integer greater than or equal to the number.

Syntax

```
Int(number)
Fix(number)
```

Is

Use this operator to compare two object reference variables. If each object reference variable refers to the same object, the result will be True. If not, the result will be False.

Syntax

```
Result = objecta Is objectb
```

IsArray

Use this variant to return a Boolean value indicating whether the object is an array variable. True is returned when the variable is an array variable and False is returned when it is not.

Syntax

```
IsArray(varname)
```

IsEmpty

Use this function when you want a returned value to show that the variable has been initialized. True is returned if the variable is not initialized and False if it is.

Syntax

```
IsEmpty(expression}
```

IsNull

Use this function when you want a returned Boolean value to show that a given expression does not contain valid data. The IsNull function returns True when the expression is Null and False if there is valid data.

Syntax

```
IsNull (expression)
```

IsNumericFunction

Use this function when you need a returned Boolean value to indicate whether an expression can be used as a number. True is returned when the whole value can be evaluated as a number and False when it cannot.

Syntax

IsNumeric(expression)

IsObjectFunction

Use this function when you need a returned Boolean value to indicate that the reference made is to an Automation object. True is returned if the reference is made to a valid object and false if it is not.

Syntax

IsObject(expression)

JoinFunction

Use this function to return a string created by joining a number of substrings contained in an array.

Syntax

Join(list[, delimiter])

Lbound

Use this function when you want to return a value for the lowest subscript defined for an Array function. The Lbound function is used with the Ubound function to determine the size of an array. Use the Ubound function to find the upper limit of an array dimension. The lower bound for any dimension is always zero.

Syntax

Lbound(arrayname[, dimension])

Lcase

Use this function to return a string that has been converted to lowercase. This can be used only to convert uppercase letters to lowercase. Lowercase letters and other characters are not changed.

Syntax

Lcase(string)

This string argument is any valid string expression. Null is returned if the string contains Null.

Left

Use this function to return a specified number of characters from the leftmost portion of a string.

Syntax

Left(string, length)

Len

Use this function to return the number of characters in a string. In addition, you can use it to obtain the number of bytes of storage required by a variable.

Syntax

Len(string I varname)

LoadPicture

Use this function to load a picture. The graphics formats acknowledged by the LoadPicture function include bitmap (.bmp) files, icon (.ico) files, run length encoded (.rle) files, metafile (.wmf) files, enhanced metafiles (.emf) , GIF (.gif) files and JPEG (.jpg) files. It is available only on 32-bit platforms.

Syntax

LoadPicture(picturename)

Log

Use this function in order to return the natural logarithm of a number. The natural logarithm is the logarithm to the base "e". The constant e is approximately 2.718282.

Syntax

Log(number)

Mid

Use this function to display or use the characters form a string that you specify.

Syntax

Mid(string, start [, length])

Minute

Use this function to display a whole number between 0 and 59 that represents the minute of the hour.

Syntax

Minute(time)

The time argument is any expression that can represent a time. If time contains Null, Null is returned.

Mod

Use this operator in order to divide two numbers and return that remains. The modulus or remainder, operator divides number1 by number2 (rounding floating-point numbers to integers) and returns only the remainder as the result.

Syntax

Result = number1 Mod number2

Month

Use this function when you need to obtain an integer between 1 and 12 that represents the number of the month of the year.

Syntax

Month(date)

MonthName

Use this function to return the name of the month in the form of a string indicating the month in question.

Syntax

MonthName(month[, abbreviate])

MsgBox

Use this function to show a dialog box containing a message specified in the code. The MsgBox function will wait for user input and return a value indicating which button the user clicked. The constants determine the options for the message box format and what is displayed on its title bar.

Syntax

MsgBox(prompt[, buttons][, title][, helpfile, context])

Not

Use this operator when you want to make a logical expression negative. The Not operator inverts the bit values of any variable and sets the associated bit in the result. You can use the Not operator in searches to exclude members of certain categories.

Syntax

Result = Not expression

Now

Use this function to return the current date and time according to your computer's setting of system date and time.

Syntax

Now

Nothing

Use this keyword in VBScript to clear from memory an object variable from any actual object. Use the Set statements to assign Nothing to an object variable.

Syntax

Set Variable = Nothing

Number

Use this property to return or set a numeric value in order to indicate an error. The Number property is the Err object's default property.

Syntax

Object.Number [= errornumber]

On Error Resume Next

Use this statement to enable error handling. Failing to use the On Error Resume Next statement will result in fatal errors. That is, execution stops when the error message to displayed.

With On Error Resume Next, you can continue execution with the statement immediately following the statement that caused the run-time error, or with the statement immediately following the On Error Resume Next statement.

Syntax

On Error Resume Next

Option Explicit

Use this option to constrain variable to be declared. The Option Explicit statement must appear in a procedure prior to other statements. When you use the Option Explicit statement, you must explicitly declare all variables using the Dim, Private, Public or ReDim statements. You will not be allowed to create variables on the fly.

Syntax

Option Explicit

Or

Use this operator to stipulate a logical either/or action that takes place on two expressions. That is, if either 1 or 2 exists, then some action takes place.

Syntax

Result = expression1 or expression2

Private

Use this statement when you want to create private variables and designate storage. Private variables can be used only in the script of declaration. Public variables can be used throughout a program. Private is also used to declare arrays.

Syntax

Private varname [([subscripts])][, varname [([subscripts])]].....

Public

Use this statement when you want to create public variables and designate storage. Private variables can be used only in the scripts of declaration. Public statement variables are available to all procedures in all scripts.

Syntax

Public varname[([subscripts])][, varname[([subscripts])]]....

Raise

Use this method when you want to raise a run-time error. Raising an error means that you reveal and display the error that interferes with the execution of the program. If you use Raise without specifying some arguments, however and the property settings of the Err object contain values that have not been cleared, those values become the values for your error.

Syntax

Object.Raise(number, source, description, helpfile, helpcontext)

Randomize

Use this statement when you want to utilize the random-number generator. The rnd function's random-number generator is used by the randomize statement to give it a new speed value. The seed value is the starting point for the number generation. If you omit number, the new seed value will be the value returned by the system timer. When Randomize is not used, the Rnd function (no arguments) seeds with the same number the first time it is called and then uses the last generated number as a seed value.

Syntax

Randomize [number]

ReDim

Use this statement to declare dynamic array variables and to allocate or change the storage space required. The ReDim statement is used to size and resize an already declared dynamic array when it has been declared using no subscripts. You also can use the ReDim statement repeatedly to change the number of elements and dimensions in an array.

Syntax

ReDim [Preserve] varname(subscripts) [,varname(subscripts)]....

Replace

Use this function to return a string with a replaced substring. The return value of the Replace function is a string that begins at the position specified by start and concludes at the end of the expression string with substitutions made. It is not a copy of the original string from beginning to end.

Syntax

Replace(expression, find, replacewith[, start[, count[, compare]]])

RGB

Use this function to manipulate color values. You can return a whole number representing an RGB color value. The RGB color values specify the relative intensities of red, green and blue causing a color to be displayed in correspondence with the relative numerical values for R, G, and B. The low-order byte contains the value for red; the middle byte contains the value for green and the high-order byte contains the value for blue.

Syntax

RGB(red, green, blue)

Right

Use this function to return the number of characters you specify from the right side of a string.

Syntax

Right(string, length)

Rnd

Use this function with the Randomize statement to initialize the random-number generator with a seed based on the system timer. The same number sequence is generated for any given initial seed because each successive call to the Rnd function uses the previous number as a seed for the next number in the sequence.

Syntax

Rnd[(number)]

Round

Use this function to returns a number rounded to the number of decimal places you specify.

Syntax

Round(expression[, numdecimalmalplaces])

ScriptEngine

Use this function to return a string that represents the scripting language in use.

Syntax

ScriptEngine

Second

Use this function similarly to the Minute and Hour functions, to return a whole number between 0 and 59, inclusive, representing the second of the minute.

Syntax

Second(time)

Select Case

Use this statement to enable a multiple choice in which the first case that matches the criterion is performed. In Select Case a number of conditions are tested for one object; whereas the If command block can have nested If statements, Else statements and Else if statements.

The Select Case statement structures test one expression, comparing the result of this expression with the values for each Case statement in the structure. A block of code is executed if there is a match. If there is no match, the execution falls to the next Case statement.

A common use of Select Case is to provide menu options. For example, you might have a menu with ten options. Rather than using a series of If statements to test for each of the ten conditions, you could more easily code the menu using the Case structure. Select Case statements can be nested. Each nested Select Case statement must have a matching End Select statement.

Syntax

```
Select Case testexpression
[Case expressionlist-n
[statements-n]] ...
[Case Else expressionlist-n
[elsestatements-n]]
```

```
End Select
```

Set

Use this statement to assign an object reference to a variable or property, or to link a procedure reference to an event. You must use the Set statement because the Dim, Private, Private, ReDim statements will only declare a variable with reference to an object.

Syntax

```
Set objectvar = {objectexpression | New classname | Nothing}
Set object.eventname = GetRef(procname)
```

Sgn

Use this function to return a whole number that shows the sign of a number.

Syntax

```
Sgn(number)
```

Sin

Use this function in order to return the sine of an angle. The Sin function returns the relationship between two sides of a right triangle. The length of the side opposite the angle divided by the length of the hypotenuse is the ratio.

Syntax

```
Sin(number)
```

Space

Use this function to return a string that displays the requested number of spaces.

Syntax

```
Space(number)
```

Sqr

Use this function to return the square root of a number.

Syntax

```
Sqr(number)
```

StrComp

Use this function when you need to see a value that shows the consequence of a string comparison.

Syntax

```
StrComp(string1, string2[, compare])
```

String

Use the function when you want to return a repeating character string of a length you indicate.

Syntax

```
Strib(number, character)
```

The character code specifies the character or string expression whose first character is used to construct the return string.

StrReverse

Use the function to return a string in reverse order of the characteristics in a given string.

Syntax

```
StrReverse(stringa)
```

Sub

Use this statement in order to identify a procedure. A Sub statement declares the name, arguments and code that form that form the body of a Sub procedure. Subs may be public or private. A public Sub statement is one in which the procedure can be used all other scripts. Private procedure are available only to the particular procedure where declared.

Syntax

[Public [Default] | Private] Sub name [(arglist)]
[statements]
[Exit Sub]
[statements]
End Sub[ByVal | ByRef] varname[( )] Example

Subtraction(-)

Use this operator in order to obtain the difference between two numbers or to designate the negative value of a numeric expression.

Syntax

Result = number1-number2
Or
-number

Tan

Use this function in order to return the tangent of an angle. The Tan function takes an angle and returns the ratio of two sides of a right triangle. The length of the side opposite the angle divided by the length of the side adjacent to the angle is the ratio.

Syntax

Tan(number)

Time

Use this function to display the current system time.

Syntax

Time

TimeSerial

Use this function to order to display or returns a date containing the time for a specific hour, minute and second.

Syntax

TimeSerial(hour, minute, second)

Time Value

Use this function when you want to return a Date including the time. You can enter valid times using either a 12-hour or 24-hour clock. Both 1:30 P.M. and 13:30 are valid. The TimeValue function doesn't return date information but an invalid date will cause an error.

Syntax

TimeValue(time)

Trim

Use this function when you want to remove leading spaces (L.Trim) trailing spaces (Rtrim) or both leading and trailing spaces (Trim) from a string.

Syntax

Ltrim(string)
Rtrim(string)
Trim(string)

Ubound

Use this function when you want to find the largest available subscript for the array in question. The Ubound function is used with the Lbound function to determine the size of an array.

Syntax

Ubound(arrayname[, dimension])

Ucase

Use this function to convert a string to uppercase. Only lowercase letters are covered to uppercase. Non-alphabetic characters remain unchanged.

Syntax

Ucase(string)

VarType

Use this function to obtain a value indicating the subtype of a variable. The VarType function doesn't return an array value alone. The array value is combined with another value that indicates a specific array type.

Syntax

VarType(varname)

WeekDay

Use this function to return an integer associated with a day of the week. The constant standing for each day of the week are as follows: vbSunday [1 Sunday], vbMonday [2 Monday], vbTuesday, [3 Tuesday], vbWednesday [4 Wednesday], vbThursday [5 Thursday], vbFriday [6 Friday], vbSaturday [7 Saturday].

Syntax

WeekDay(date, [fristdayofweek])

WeekDayName

Use this function to return a name in the form of a string showing the specified day of the week.

Syntax

WeekDayName(weekday, abbreviate, firstdayofweek)

While..Wend

Use this statement to execute code within a block of statements that continues executing as long as a given condition is True.

Syntax

While condition
[statements]
Wend

Xor

Use this operator when you want to logically exclude a value in two expressions. If one and only one of the expressions evaluates to True, the result is True. However, if either expression is Null, the result is also Null

Syntax

```
Result = expression1 Xor expression2
```

Year

Use this function to return an integer associated with the year.

Syntax

```
Year(date)
```

Introduction to Object Hierarchies?

To provide for a consistent logic in programming under the Windows environment, Microsoft has implemented object hierarchies and others have followed their lead. Object hierarchies make it easy to program other application's object from your own code because they follow consistent and predictable style.

First, what are objects? In the real world, objects are things like a bicycle or a desk or a chair. They are objects: they are things. In computer development, objects are things like a spreadsheet or a document or an employee or a shopping cart.

Most of us know what a hierarchy is. We think of company hierarchies that are mad up of directors, managers, laborers and so forth. So object hierarchies are a collection of objects that follow some logic. In the computer world, an example of an object hierarchy is the collection of objects in Excel. At the top level we have the Application object, which is Excel itself. Excel. At the top level we have the Application object, which Excel itself. Under that we have the Worksheet object. So again, what we have is a collection of object organized in a logical manner, or an object hierarchy.

Objects have properties, methods, events and collections. Properties are attributes or ways by which we describe the object. Methods are actions we take an object. Events are notifications that we receive that an action has occurred. Collections are a group of subobjects.

As mentioned earlier, a bicycle is an object and it has properties, methods, events and collections. Properties or attributes of a bike are things like its color, the tire size, whether it has thorn-proof tires and the height of the seat. Methods or actions we take on a bicycle are things like pedaling, steering or painting. Events are notifications that we receive such as the status of the bike has changed: it is now stopped, it is now moving and so on. We can then take some action based on that notification. Collections of a bike are things like spokes or brake pads. The items in a collection are objects themselves and also can have properties, methods, events and collections.

Note that a method can change a property and an event can be fired because a method was invoked or a property changed. If we paint the bike, we could change the color property.

A worksheet in Excel is an object and also has properties, methods, events and collections. A worksheet has a background color, font type and default width for the columns – these are all properties. Methods of a worksheet include copying text onto the clipboard, creating a formula field or generating a graph from spreadsheet data. Events for a worksheet include the notification that it was activated, that the contents have changed or that the worksheet has been deactivated. A worksheet also has collections like cells, columns, rows and comments. And since each of those collections contains a collection of objects, each of the items in the collection can have properties, methods, events and collections.

We access a property, a method or a collection by what is referred to as dot syntax. The dot is a period. So if we wanted to refer to the color property of your bicycle we would code:

```
If Bicycle.Color = "Red" then
    Response.Write "You have a red bike."
End if
```

On the second line in the preceding code example we are using the dot syntax to refer to the write method of the response object. The you have a red bike is referred to as a parameter, which is information you supply to a method for detail about the action you are taking.

ASP Object Model

The ASP Object model defines five primary objects:

The Request object

The Response object

The Application object

The Server object and

The Session object.

These objects are summarized in the following table.

Object	*Purpose*
Request object	Supplies you with information from your visitor.
Response object	Methods and properties for building your response to the visitor.
Application object	Deals with properties that govern a grouping of Web pages referred to as an application.
Session object	Methods and properties pertaining to a particular visitor.
Server object	Deals with the creation of server components and server settings

Through HTML we can send data to the server in the form of Post or Get methods. Retrieving this information is the responsibility of the Request object, which supplies us with simple groupings of properties called collections, enabling us to retrieve the submitted information. For example, if we have an HTML form on our server with a text field called Login, we could retrieve the value of the Login with following code:

```
Response.Write "Hello and Welcome to: " & Request.Form ("Login")
```

With the Request object, we can also retrieve information about the visitor's browser as shown in this code segment:

```
Request.ServerVariables("REMOTE_HOST")
```

This method would return the Internet name or IP address of the visitor's computer.

Getting Information from the Visitor

When visitors fill in the fields on a form you created, they usually press a SUBMIT button. Those data are sent to the server and are available to your through the Request object. You can also retrieve the visitor's cookies through the Request object. The Request object can also supply you with header information about the visitor as well as client security certificates.

Request Object Collections

The Request object contains five collections, which are discussed below and summarized in a table.

Object	*Purpose*
Form Collection	Fields submitted form a form through the Post method.
QueryString Collection	Fields submitted from a form through the Get method.
ServerVariables Collection	Header fields available through the visitor's browser when a page is requested.
Cookies Collection	Retrieves cookies from the visitor.
ClientCertificate Collection	Retrieves the client certificate fields from the visitor's browser.

Form Collection

Most Web sites contain forms. If you want to add visitors to your mailing list, you have a form for them to supply you with their email address. If you have an e-commerce site, you eventually need a form to get billing and shipping information from visitors: or you may have a site that polls visitors in current events and the visitor's opinions are retrieved through a form.

Visitors then send or submit the information on the form by clicking on a button, a graphic or just by pressing Enter. The data that visitors supplied through the fields on the form are submitted to whatever page you supply in the Form tag. For example:

<FORM ACTION="http:L//www.whatever.com/shoppingcart.asp" METHOD=GET>

would send the contents of the form to the shoppingcart.asp page.

When you place a Form tag on your Web page, you provide Action parameter that indicates how the data should be sent. You can use the Post method as in this example.

<FORM ACTION="http://whatever.com/shoppingcart.asp" METHOD=POST>

When you use the word Post method, the fields are sent as a binary stream and are not viewed as part of the link to a page. You can also use the Get method:

<FORM ACTION="http://whatever.com/shoppingcart.asp" METHOD=GET>

When the Get method is used, the number of bytes sent is limited and the fields appear as part of the link to the page.

QueryString Collection

As we know that when a Form is submitted, the fields on the Form can be submitted using the Get and the Post methods. Where the QueryString collection is really used is when you need to pass parameters through a link. Say, for example, that you had a page that listed the products you sold. When visitors click on a product, you want them to be able to see the details of the product that they selected. You could have hundreds of pages that displayed each product or you could have a single Active Server Page that could display any product. It would required just a ProductID and would then look up the details of the product in a database. To pass the ProductID on the page through a link you would be using the QueryString collection.

You would display the product name as a link to the product details page like this:

<A
HREF="http://somewhere.com/productdetails.asp?ProductID =2355">Toothpaste</A>

The visitor who click on the word Toothpaste is then linked to the productdetails.asp page. The ProductId for the toothpaste 2355 is sent to that page.

Server-Variables Collection

The ServerVariables collection retrieves numerous header fields and environmental variables. For example, you can use the ServerVariables collection to retrieve the type of form submission made, the name of the referring page, the path to the current page, the IP Address of the visitor and the type of Internet Server.

Retrieving one of the items in the ServerVariables collection in simple – just specify the name of the server variable you want to retrieve. For example, if you want to retrieve the path to the current page you would code this:

ThePath = Request.ServerVariables("PATH_INFO")

If you want to display back to the browser the name of the referring page you could code this:

Response.write Request.ServerVariables("HTTP_REFERER")

Some of the header fields and environmental variables should always be available. Those are displayed in table. Other server variables that are not always available, but can be very important are listed in following table.

Server Variable	*Purpose*
ALL_HTTP	Contains the entire list of nonstandard headers in the form of HTTP_NAME value:
ALL_RAW	Similar to the ALL_HTTP except that they appear just as they are received without the HTTP prefix and other formatting.
ALL_MD_PATH	Contains the logical metabase path.
APPL_PHYSICAL_PATH	Contains the physical metabase path.
AUTH_PASSWORD	A password entered if basic authentication is used.
AUTH_TYPE	Type of security authentication that was used.
AUTH_USER	Name of authenticated user.
CERT_COOKIE	String containing a unique identifier for the client certificate.
CERT_FLAGS	First flag is set if a client certificate is present. Second flag is set if the Certificate Authority is a trusted source.
CERT_ISSUER	Client certificate issuer.
CERT_KEYSIZE	Number of bits in the SSL security key.
CERT_SECRETKEYSIZE	Number of bits in the SSL security key for the server.
CERT_SERIALNUMBER	Serial number of the client certificate.
CERT_SERVER_ISSUER	Certificate Authority issuing the server certificate.
CERT_SERVER_SUBJECT	Subject field for the server certificate.
CERT_SUBJECT	Subject field for the client certificate.
CONTENT_LENGTH	Number of bytes reported by the client sent to the server with the request.
CONTENT_TYPE	Type of request made.
GATEWAY_INTERFACE	CGI version in the form of CGI/version number.
HTTPS	Contains the string on if this is a secure request and off if it is not.
HTTPS_KEYSIZE	Number of SSL bits used with the request.
HTPS_SECRETKEYSIZE	Number of bits used on the server end.
HTTPS_SERVER_ISSUER	Certificate Authority for the server certificate.
HTTPS_SERVER_SUBJECT	Subject field from the server certificate.
INSTANCE_ID	Web server instance identification number.
INSTANCE_META_PATH	Metabase path for this specific instance.
LOCAL_ADDR	IP Address of the server.
LOGON_USER	If the user is logged on the NT System, this field contains that user's name.
PATH_INFO	Path to the requested page beyond the domain name root.
PATH_TRANSLATED	Physical path to the request.
QUERY_STRING	Any data submitted through the Get method or after a question mark on a link.
REMOTE_ADDR	IP Address of the machine making the request.
REMOTE_HOST	Translated name of the requester if available. If not, the field contains the IP Address.
REMOTE_USER	Name of the user if sent by the visitor.

REQUEST_METHOD	Request method used, that is, Post, Get.
SCRIPT_NAME	Virtual file location of the script being requested.
SERVER_NAME	Host name of the server.
SERVER_PORT	Port used with the request.
SERVER_PORT_SECURE	If the request is being made through a secure connection, then this field contains the value 1; if not the field contains the value 0.
SERVER_PROTOCOL	Protocol being used with version number, that is, HTTP/1.1
ERVER_SOFTWARE	Name and version of the Web sever software being run by the server.
URL	Path of requested page.
HTTP_CONNECTION	Type of connection between server and client.
HTTP_REFERER	Pate that the visitor was at before entering the current page if the progression was made through a link.
HTTP_USER_AGENT	Browser type and version as well as the visitor's operating system.

To explore the values of all the server variables and the available types you could use the following code within an Active Server Page:

```
For Each TheStuff in Request.ServerVariables
    Response.Write TheStuff & " : "_
        & Request.ServerVariables(TheStuff) & "<P>"
Next
```

This code would produce a list of all the server variables with their values in the form of :

Server Variable: Value

Cookies Collection

Cookies are a way for you to store nuggets of information on the visitor's computer. You can then use your code to retrieve the values stored on the visitor's machine.

For example, you could store the visitor's user name within a cookie. Then when visitors returned to your site, you would know who they were. With cookies, you could also store visitor preferences, such as their favorite product types or their location.

But since the data are on the visitor's system, you cannot use cookies as your only way to identify visitors. The concern is that visitors could delete the cookie from their machine; or they may have a cookie-eating program installed that deletes cookies as soon as they are added; or, your visitor may be accessing your Web site from a machine that is different from where the cookie was placed.

You retrieve a cookie through the Cookies collection. If you stored a cookie named UserName on the visitor's system, you would retrieve it like this:

```
TheValue = request.cookies("UserName")
```

The variable TheValue would contain the value of the cookie UserName. If the UserName cookie did not exist, then TheValue would be set to nothing.

Cookies can also be complex. They can contain more than one value within a single cookie. So if you needed to retrieve the visitor's name and favorite product category that were within the same cookie called Preferences, you would give the code as:

```
UserName = Request.Cookies("Preferences")("UserName")
FavCategory = Request.Cookies("Preferences")("FavCategory")
```

ClientCertificate Collection

More and more being able to positively identify visitors is necessary, particularly in the area of the e-commerce. One of the ways you can do that is through client certificates.

Users can purchase a client certificate from an organization referred to as a Certificate Authority (CA), which is a company that creates certificates for individuals. The certificate is usually tied to an email address and a password.

Normally, when you work with certificates you decide What Certificate Authorities you trust. To say that you trust a Certificate Authority means that you accept certificates issued by their company. Two of the main players in the CA industry are Thawte (http://www.thawte.com) and Verisign (http://www.verisign.com).

You use the ClientCertificate collection of the Request object to query the values of the visitor's certificate. Say you wanted to retrieve the date that the certificate expires. You would code this:

```
ExpDate = Request.ClientCertificate("ValidUntil")
```

Request Object Property

The Request Object contains only one property, which is discussed below and summarized in table.

Object	*Purpose*
TotalBytes	Number of bytes sent when a page is requested.

TotalBytes Property

The TotalBytes property contains the number of bytes that are in the parameter portion of an Active Server Page call. If you had a page with the following form:

```
<FORM ACTION=".processform.asp" METHOD="POST">
Name:<BR><INPUT TYPE="text" SIZE=30 MAXLENGTH=50 NAME="Name">
<p>Email Address:<BR><INPUT TYPE="text" SIZE=30 MAXLENGTH=50 NAME="EmailAddress">
<p>Interests:<BR><SELECT NAME="Interests" SIZE=4 MULTIPLE>
      <OPTION VALUE="News">News
      <OPTION VALUE="Sports">Sports
      <OPTION VALUE="Hiking">Hiking
      <OPTION VALUE="Other">Other
</SELECT>
<p><INPUT TYPE="submit" NAME="SubmitButton" VALUE="OK">
</FORM>
```

and the visitor to enter the name Dave in the name field, nothing for the email address field, and News for the Interests, the following would be display 54:

```
Response.Write Request.TotalBytes
```

The number 54 is the count of bytes or characters in the request if you add the fields submitted with the names of the fields on the form:

Field Name: = 4 bytes
Field Name: = EmailAddress = 12 bytes
Field Name: = Interests = 9 bytes
Field Name: = SubmitButton = 12 bytes
Data Entered: = Dave = 4 bytes
Data Entered: = Email (NA)= 0 bytes
Data Entered: = News = 4 bytes
Ampersands: = 3 bytes
Equal Signs: = 4 bytes
Total Bytes: = 4 bytes

If you were to use the Get method with the form submitted, them TotalBytes property would return 0. The property is available only when a form is submitted with the Post method.

Sending Information Back to Visitors

Role of the Response object is to allow you to communicate with your dynamic Web and send the content back to the browser. You use the information collected by the visitors and other components on your system to prepare text, graphics and much more that you want sent back to the visitors. This is done through the Response object.

Response Object Collection

The Response object includes just a single collection, which is discussed below and summarized in the table below.

Object	*Purpose*
Cookies	Allows you to write cookies to the visitor's browser.

Cookies Collection

Cookies are a way for you to store nuggets of information on the visitor's computer. You can use your code to retrieve the values stored on the visitor's system at a later time. For example, you could store the most recent searches a visitor has made on your Web site in a cookie. Then, when visitors search again, you could present them with their last five search criteria; or you could store visitor's location in a cookie and when they return to your site. You could display local information for your visitors.

But since the data are on the visitor's system, you cannot use cookies as your only way to identify visitors or in situations where the cookie must be present. You should always try to provide an alternative to cookies. Remember that visitors could delete the code at anytime, they could have a program on their machine that blocks cookies or they may be visiting your site from a different system.

You place cookies on the visitor's machine by using the Cookies collection of the Response object. To write a single simple cookie to the visitor's system you would code this:

```
Response.Cookies ("NameOfCookies") = "Value"
```

The NameOfCookie is the name you want to store the cookie as on the visitor's system. The value represents the text to store in the cookie. So if you coded this:

```
Response.Cookies("SearchCriteria1") = "ASP"
```

This would add the cookie with the name SearchCriteria1 to the visitor's system. The cookie would contain the value ASP. If the cookie already exist, the old value will be overwritten.

The preceding creates a simple cookie that contains a single value. But cookies can also be complex and contain subkeys. You refer to subkeys by specifying the name of the cookie with the name of the subkey in the fork:

```
Response.Cookies ("NameOfCookie") ("NameOfKey") = "Value"
```

In this example, NameOf Cookie is the name of the main cookie and NameOfKey represents the name of the subkey being referred to. Value would be what you want to set the cookie to.

ASP Applications

If you are a programmer from another language or have just heard programmers talk, you may be familiar with the word state, which refers to the persistence of information across an application. In a state less application all the data is lost from one procedure to the next. So each call, or in the case of Active Server Pages, each page contains its own set of data.

An application that has state is one that maintains information between calls, which can be accessed by one or more procedures within an application or the same procedure more than one time.

If you have ever programmed in Visual Basic, you are probably familiar with Public or Global variables. These variables can be visible to the entire application from the start of the application until the end. In Visual Basic, you declare one like:

Public ExpName as String

You can set its value from within any procedure:

ExpName = "Amitabh"

When the procedure that sets the value of the variable of the variable ends, the variable still contains the text Amitabh, so other procedure can refer to it or the same procedure could reference the value that was last set to it. An application like this state.

Creating an Active Server Page Application

An Active Server Page application is composed of a series of related Active Server Pages and a file called global.asa. All the files in the Active Server page application must reside in the same root directory structure where the application is defined in or in the subdirectories of that root directory.

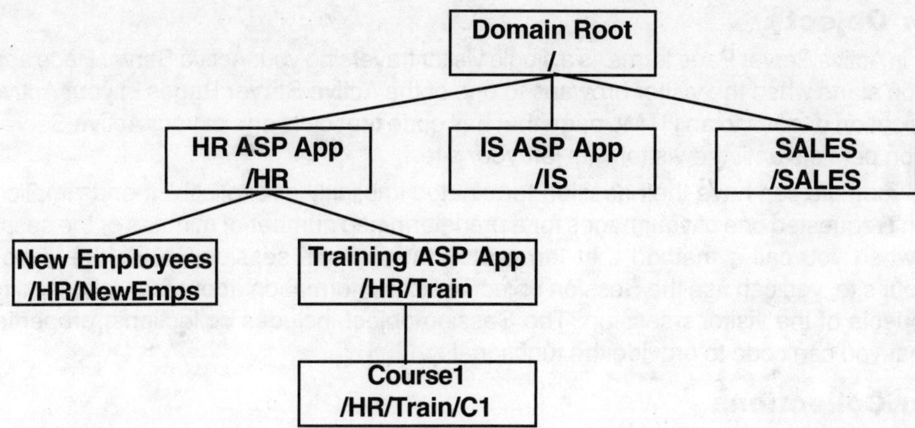

An Active Server Page applications is made up of all the files that are in the applications root or any subdirectories unless those subdirectories belong to another Active Server Page application. Using our example, we start at the top level, which isn't defined as an Active Server Page application. We then have a subdirectory called HR that is defined as an Active Server Page application, so all the Active Server Pages in that directory are part of the same HR application.

But we also said that the subdirectories are part of the application unless they are their own Active Server Page application. Under the HR directory is a directory called NewEmps. This directory is not an Active Server Page application, so any ASP pages in it belong to the HR application. If the NewEmps directory had any subdirectories, they, too, would belong to the HR application unless they were defined as their own Active Server Page application.

The second directory under HR is the Train directory. This directory is an Active Server Page application itself, so the Active Server Pages in it do not belong to the HR application. Under the Train directory is a directory called C1. Since this directory is not an Active Server Page directory, it belongs to the Train application since that is the next level up that is an Active Server Page application.

Also off the root is a directory called IS, which is defined as an Active Server Page application, so the ASP pages in that directory would belongs to IS, not HR or any other Active Server Page application.

The last directory off the root is Sales. The directory is not an Active Server Page application, so the pages in that directory are stateless. They do not belong to any Active Server Page application since they are not in a directory that is a marked as one end no levels above it are Active Server Page applications.

When you create an Active Server Page application you must be cognizant of where you are placing the ASP in relative position to what pages must be able to share state information. You normally do this by creating a directory somewhere off of the root of your Web site. You then place all the files that are part of that application into that directory.

You are then ready to mark the directory as an Active Server Page application. You do this by right-clicking on the directory within the Microsoft Management console's view of IIS and selecting Properties. After selecting Properties you should see the dialog box. Once you are in the Properties for the directory you should be in the Directory tab; if you are not, change to that tab. Then just press the Create button in the Application Settings section of this tab and enter the name that you want to give your application in the Name text box. Press OK and you now have an Active Server Page application. Now all the Active Server Pages can share information with each through the Session and the Application objects.

Session Object

A session, in Active Server Page terms, is a single visitor traversing your Active Server Page application. The session starts when the visitor browses to one of the Active Server Pages in your Active Server Page application or goes to an HTML page that has code that calls one of your Active Server Pages. The session persists until the visitor has left your site.

Visitors to your site can have their session terminated implicitly or explicitly. It ends implicitly when they haven't requested one of your pages for a predetermined number of minutes or the session ends explicitly when you call a method that terminates the visitor's session. While the visitor travels through your site, you can use the Session object to store information about the visitor and to control certain aspects of the visitor's session. The Session object includes collections, properties and a method that you can code to provide the functionality.

Session Collections

The Session object contains two collections that you can use to control data that is used in the Session-state.

Collection	*Purpose*
Contents Collection	A collection of all the variables that are available at the Session level, excluding any created with the Object tag.
StaticObjects Collection	A collection of all the objects created for the Session level that were created using the Object tag in the global.asa file.

Contents Collection

The Contents collection is a group of variables that you make available to any of the pages in your Active Server Page application through the Session object. The variables have the Session-level scope, so they are available to any ASP, but the values are not shared by all the visitors. Each visitor would have an individual copy of the variable.

StaticObjects Collection

It provides a way for you to write code that runs when a visitor starts or ends a session. It also provides a way for you to write code that run when your Active Server Page application starts and stops. But you can also create Session objects within the global.asa file that are then available, like the ones we have discussed thus far, to all the ASP in your Active Server Page application. You create these objects using an Object tag and they become available because of the StaticObjects collection.

Session Object Properties

The properties of the Session object allow you to configure and refer to aspects of the visitor's session. The properties are discussed in the following table:

Properties	*Purpose*
Codepage	Determines the character set used in the session.
LCID	Defines the Locale Identifier for the session.
SessionID	A unique identifier for the visitor's session.
TimeOut	The number of minutes between page views that cannot be exceeded or the visitor's session is assumed to have ended.

CodePage Property

The CodePage property allows you to set the character set that will be used during the session. The character set includes numbers, letters, and other characters. Different countries in the world use different character sets, so you can use this property to assist your development if your site provides international content.

The CodePage is Session property, so what you set it to is what is used by all the pages in your Active Server Page application.

LCID Property

The LCID property of the Session object allows you to set the locale that the dynamic text should be displayed as. This represents how things like time, numbers, and currency are to be displayed. Since the property is a Session property, once set, it is for all the pages in your Active Server Page application. The property is read/write and takes the following form:

```
Session.LCID = LocalID
```

where LocalID represents a number of type long that you wish to set the locale to.

SessionID Property

Initially IIS needs a way of tracking visitors as they travel through your site. The way IIS does this is through a SessionID. When you are using an Active Server Page application, IIS assigns a random number of type long to all visitors as they enter your site. Since the property is used internally by IIS to identify all visitors in their session, it is a read-only property. The property has the following syntax:

```
TheID = Session.SessionID
```

TheID is a variable into which you want to place the ID.

TimeOut Property

A session starts when visitors access one of the Active Server Pages in your Active Server Page application or when they access a page that calls code in one of your Active Server Pages. One of the ways a session can end is if visitors do not make a request for one of the pages in your Active Server Page application in a certain number of minutes.

That certain number of minutes is set through the TimeOut property. The read/write property stores the number of minutes that visitors can take between requests in your Active Server Page application without timing out.

Getting in at the Top (Server Object)

So far we have looked at the Request object that, as you have seen, allows you to get information about the visitor. We have also looked at the Response object, which we used to send information back to the visitor's browser. The Server object provides you with one property and four methods that allow you to communicate and use functionality of the Internet Information Server (IIS). You communicate with IIS in the sense that you can change how the server works with a page. The functionality of the Server Object allows you to take actions in your script for which the script itself doesn't supply the mechanism.

Visitors type in the page they want to see or they click on a link or part of the page they are on, which is an Active Server Page. The request travels through the Internet to your server. Your server retrieves the requested page and if it is an Active Server Page, it begins processing. The role of the Server object is, at this point, when the code is being processed. The Server object allows you to bring other components into the processing of the page and allows you to communicate with the server itself. The processed page is then sent to the visitor's browser.

Server Object Property

The Server object contains just one property that allows you to communicate with the Server. That property is summarized below:

Property	*Purpose*
ScriptTimeout	Seconds an Active Server Page can run before it times out.

ScriptTimeout Property

For a variety of reasons, your code can take a long time to process. The server may be very busy, a component may be stuck in a loop or your own programming logic may have the code repeating endlessly. Mainly for those reasons, you can set the maximum amount of time your script should be allowed to run before it stops and returns an error to the browser making the request.

You can set this maximum run time at the Web-browser level by viewing its properties and pressing the Configuration button on the Home directory tab. You can set this property at the Web-site level in the same way. You can also set the property at the ASP Application level by right-clicking on the application within the Microsoft Management Console view of IIS and selecting Properties.

Server Object Method

The Server object has four methods that you can use for a variety of tasks. This has been summarized in the following table:

Object	*Purpose*
CreateObject	Allows for the creation of a server component.
HTMLEncode	Encodes text into an HTML encoded format.
MapPath	Returns the physical path for a virtual or relative path.
URLEncode	Modifies the text passed so that it is encoded in URL format.

CreateObject Method

The CreateObject method allows you to instantiate a server component, which refers to an object, usually a DLL, on your server that has its own objects, collections, methods, properties, and/or events. The term instantiate means that in your code you can access the server component. You can think of these server components in three different categories. There are server components that come with the Internet Information Server. These components are specifically provided because they are designed to extend what you do with Active Server Pages. These components include the Browser Capabilities component, which provides you with information about the browser making the request; Ad Rotator component, which provides you with an easy way to include and manage banner ads on your site and the Page Counter component that can be used to provide hit information.

The second type of server component you will use are those that are on your server that provide some other additional functionality to your page but are not necessarily designed for active Server Pages. This type of server component would include Microsoft Word, which you could use to format text in the form of a Word document; a Graph Generator, which would create a graph image based on your input on the fly that you could include on your Web page; or an advanced mathematical program whose code you would use to provide some scientific calculation result on your page. The final type of server component would be the ones you create yourself. At some point, you will need to create a procedure that you can't do from within Active Server Pages and for which you can't find an existing server. On occasion you will create such a component as a workaround for some bug or deficiency within Active Server Pages. Another reason that you may create your own server components is for proposes of encapsulation, which means that you want to place code in one place and then use the code in a variety of locations; then, if the code needs to be changed, it needs to be changed in only one location.

HTML Encode Method

Say you have a Web page that displays quiz questions through an Active Server Page that tests the students' understanding of numeric comparison.

MapPath Method

The MapPath method of the Server object converts a virtual or relative path as viewed through the Web server into its real physical path. The method takes the following syntax:

ConvertedPath = Server.MapPath(PathConvert)

The PathToConvert variable is the path you wish to convert and the ConvertedPath variable is the path converted into its physical directory.

URLEncode Method

When you submit a form through the Post method, or you provide a link with a query string, the text in the query string needs to exclude certain characters since the text becomes part of the link. For example, when you want to search for the meaning of http://and you go to a search engine to look it up. When you type in that search and press the Submit button, you will see the results of your search. Part of the link to the search result is your search query. Notice that now your search string of http:// is converted to this:

http%3A%2F%2F

Response Object Methods

The method of the Response object are discussed and summarized in the following table:

Type	*Definition*
AddHeader	Allows you to add your own custom header to the response.
AppendToLog	This method provides a way for you to write information to the NT log.
BinaryWrite	This method is used to send nontextual information to the browser.
Clear	Removes any output that has been buffered.
End	Explicitly stops any processing of the code on an Active Server Page.
Flush	Sends the contents of the buffer to the browser.
Redirect	Sends the browser to a different page for the response.
Write	Sends text to the browser.

AddHeader Method

You can use the AddHeader to add your own custom header to the output of an HTML page. The method takes the following form:

Response.AddHeader NameOfHeader, ValueOfHeader

AppendToLog Method

The very handy AppnedToLog method allows you to add text to the Web Server Log for a specific request. The text can be up to 80 character and cannot contain commas. The method takes the following form:

Response.AddToLog Text2Add

Text2Add is a variable that contains the text you wish to add to the log.

BinaryWrite Method

The BinaryWrite method allows you to send nontextual, raw binary information to the browser. The method takes the following form:

Response.BinaryWrite DateToWrite

The DataToWrite variable contains the binary information that you want to send to the browser. Typically this is used to send data, such as sound or images to a browser, instead of text.

Clear Method

The Clear method empties the buffer. Since the method works with the buffer, you must have the Buffer set on or an error will occur. The method has the following syntax:

Response.Clear

End Method

The End method of the Response object tells the compiler to stop processing any remaining code. It takes the following form:

Response.End

Flush Method

One more method allows you to control this type of behavior of the output: the Flush method, which sends any buffered output and immediately sends it to the browser. Since the method clears the buffer, an error will occur if the Buffer is not on. The method has the following syntax:

Response.Flash

Redirect Method

The Redirect method sends the browser to a different page to receive a response. The method takes the following form:

Response.Redirect URL

In this case, URL is the page to which the browser is sent.

Write Method

The Write Method, which has been used frequently in the last few chapters, is probably the method you will use the most. The Write method allows to return nonbinary data to the browser. The method takes the following form:

Response.Write Text2Write

where Text2Write is the text you wish to send to the browser.

Application Object Methods

The Application object contains two methods that are used to control how the Contents collection items are referenced.

MethodPurpose

Lock Locks the application so other sessions cannot modify the variable

Unlock	Releases the lock on the application variables

Lock Method

One of the problems that can occur with a busy site or even on occasion with a site that doesn't receive much traffic, is that an application variable may be modified by more than one script at the same time. Remember that an application-level variable is available to all the sessions, so if you have hundreds of simultaneous sessions it becomes likely that more than one session may be trying to change an application variable at the same time.

Unlock Method

There are two ways that you can release a lock on the application variables. The first way is as it was earlier, by letting the code run on the page; then when the code finishes, the lock is released. But if the code takes much time to process, you may end up with a bottleneck of other sessions waiting for the application variables to be released. For what reason you should lock your application variables just prior or changing them and then release the lock just after the modification.

Objects that Extend Your ASP Capabilities

IIS comes with a variety of components that assist you in your Active Server Page development. The following table defines that:

Component	***Purpose***
Browser Capabilities	Supplies information about the browser used by the visitor and the capabilities of that browser.
Ad Rotator	Provides methods for easily managing the placement of banner ads on your site.
Page Counter	Gives you a way to display and keep track of the number of visitors to a page.
Counters	Provides a mechanism to maintain a number that is meant to be incremented through an Active Server Page application.
Content Linking	Allows you to create a series of pages that are ordered like a book and provides the navigational needs for such a style.
Content Rotator	Provides a way for you to supply random content to a page or to elements on a page.
MyInfo	Supplies an object with a variety of properties that you can use to store information about you and your company.

Browser Capabilities Component

The Browser Capabilities component provides you with information about the visitor's browser and information about their system. The following is simple call to the Browser Capabilities component:

```
Set BC = server.createobject ("MSWC.BrowserType")
Browser = BC.Browser & "" & BC.Version
FrameSupport = BC.frames
Platform = BC.platform
```

Ad rotator Component

The Ad Rotator component gives you a tool to manage the banner ads that appear on a Web page. The component will randomly display a banner ad on an Active Server Page each time the page is accessed. The component uses a separate schedule file that you create to determine how often to display one banner ad versus another banner ad. Making a call to the Banner Ad Rotator component looks like this:

```
set objAdRot = server.createobject ("MSWC.AdRotator")
Response.write
objAdRot.GetAdvertisement("./html/AdFiles/AdFile.txt")
```

Page Counter Component

The Page Counter component provides a simple way to track and display the number of hits on a Web page. The component is not installed by default with IIS. It is part of the IIS Resource Kit, which you can find at Microsoft's Web site, or if you subscribe to Microsoft Technet, it is included on a CD. You can manually install the component by placing the file from the Technet CD's location:

\IIS Resource Kit\Component\Page Counter\DLL\i386\PageCnt.dll

Counters Component

The Counters component provides a simple interface for you to store integers that you can read, write, remove and increment across all the pages in your site and even across the boundaries of Active Server Page applications. As with the last component, this component too, comes with the IIS Resource Kit. Within the resource kit it is located here:

\IIS Resource Kit\Component\Counters\DLL\i386\counters.dll

Content Linking Component

The Content Linking component provides a way for you to link pages that are to be viewed in a series, pages that have sequential meaning. For example, a book has sequential pages or a help file has a series of pages. The component is part of the standard IIS installation. The library file for the component is called nextlink.dll. The way the component works is that you supply a text file that contains all the pages that are linked together in the order in which you want them linked. In this special text file you supply not just the name of the page but the URL of the page. You don't have to name the text file anything special, it just needs to be accessible within the directory structure of the Web site where it will be used.

Content Rotator Component

The Content Rotator component provides a way for you to have different content appear on your pages whenever the page is loaded. Things such as a quote of the day, different welcome messages, tips of the day, or fortunes are examples of use of this component. This same functionality could be done by loading data from a database but this component provides an alternative mechanism that does not require a database to be loaded just to change the tip of the day.

The component is not part of the standard IIS installation, but rather is part of the IIS Resource Kit. On the Resource Kit, the library is located in the following path:

\IIS Resource Kit\Component\Content Rotator\DLL\i386\control.dll

MyInfo Component

Although the MyInfo component doesn't have the Increment and Remove methods like the Counters component, it does allow you to have strings that persist across the server. The values persist if the server goes down because they are stored in a text file outside IIS.

MyInfo component is made up of properties that you create just by setting their values and retrieve just by specifying their name. To create the component just use this line of code:

```
Set MyInfo = Server.CreateObject("MSWC.MyInfo")
```

Appliction.Contgents.Remove Method

When you create an Application variable, that variable is added to the Application Content collection. If you want to later remove that or any Application variable from the collection, use the Contents.

Remove method. Consider the following example, which adds an Application variable called "MyFavoriteGaPeopoe":

Application("MyFavoriteGaPeople") = "Terri Eden"

The Contents collection for the Application object now shows one entry. To delete that entry from the collection, use the Contents. Remove method:

Application.Contents.Remove("MyFavoriteGaPeople")

Application.Contents.RemoveAll Method

Whereas the Contents.Remove method deletes a single entry from the Application collection, the Contents.RemoveAll method deletes all the entries in the Application Content collection. The following code shows three entries added to the Contents collection for the Application object. Then the contents.RemoveAll method is used to delete the three entries from the collection.

Application("HinesvillegeGaPeople2") = "Robert Hendry"
Application("TeacherWhoChangeMylife") = "Margie Pyle
Land-Auburn 66"
Application("GoodPEople") = "Robert Futrell"
Application("GoodGirl") = "Micailah Burgress"
Application.Contents.RemoveAll()

One of the nice features of ASP is the ability to use the Application or Session objects to store global Information in an array.

Revision

1. An Active Server Page is a standard HTML file that is extended with additional features.
2. An Active Server Page can contain server-side scripts.
3. An Active Server Page provides several built-in objects.
4. An Active Server Page can be extended with additional components.
5. All the commands in an Active Server Page, must first be executed to create an HTML page. This allows an Active Server Page to contain dynamic content.
6. An Active Server Page is primarily a scripting environment.
7. By default, an Active Server Page assumes you'll be using VBScript as your primary scripting language.
8. You can also specify the primary scripting language for a particular page.
9. This behavior of the <SCRIPT> tag has two implications. First, you can place scripts contained in a <SCRIPT> tag anywhere you want in an Active Server page. Second, the <SCRIPT> tag, for most purpose, is restricted to containing functions or procedures.
10. Active Server Pages include several built-in objects and installable ActiveX components.
11. Active Server Pages include several built-in objects.
12. The Application object is used to store and retrieve information that can be shared among all users of an application.
13. The Request object can be used to access all information sent in a request from a browser to your server.
14. The Response object is used to send information back to a browser.
15. The Server object enables you to use various utility functions on the server.
16. The Session object can be used to store and retrieve information about particular user sessions.
17. The ObjetContext object is used to control Active Server Pages transactions.
18. The Ad Rotator component is used to display banner advertisements on the Web pages of a Web site.
19. The Browser Capabilities component can be used to display different HTML content, according to the capabilities of different browsers.
20. Using the Content Linking component, you can link together several HTML pages so that they can be navigated easily.

21. The Counters components can be used to keep track of the number of visitors to your Web site.
22. The Content Rotator component enables you to rotate through HTML content on a page.
23. The Page Counter component is exactly like the Counter component, in that it can be used to track the number of visitors and to add a hit counter to particular Web page.
24. The Permission Checker component can be used to display links to Web pages only to those users who have permission to see them.
25. The Collaboration Data Objects (CDO) enable you to send and retrieve email from within your Active Server Pages.
26. The Activex Data Objects (ADO) enable you to retrieve and store data in a database such as Microsoft SQL Server.
27. In most cases, you'll create an instance of a component with page scope.
28. Instead of creating components that are accessible only within a single Active Server Page, you can create components that are accessible on multiple Active Server Pages.
29. When you create an instance of a component with application scope, you treat it as if it were a built-in object.
30. You can test whether Active Server Page is configured properly on your system by creating a simple Active Server Page.
31. A variable declared outside a procedure, could be accessed and changed by any script in the ASP page where it is declared.
32. Session variables stores information about one single user and are available to all pages in one application.
33. Application variables are also available to all pages in one application.
34. Each variable has a name that uniquely identifies it within its scope.
35. To declare a variable is to tell the program about it in advance.
36. A convention for nothing the range of reference for the variable.
37. An operator is one which works on data items and performs some mathematical calculation or change the data.
38. Most programs perform arithmetic calculations.
39. A character or symbol indicating a relationship between two or more values or expressions.
40. Logical operators are used to combine one or more than one relational expression.
41. Object hierarchies make it easy to program other application's object from your own code because they follow consistent and predictable style.
42. In computer development, objects are things like a spreadsheet or a document or an employee or a shopping cart.
43. Objects have properties, methods, events and collections.
44. The ASP Object model defines five primary objects: The Request object, The Response object, The Application object, The Server object and The Session object.
45. The Request object can also supply you with header information about the visitor as well as client security certificates.
46. QueryString collection is really used is when you need to pass parameters through a link.
47. The ServerVariables collection retrieves numerous header fields and environmental variables.
48. Cookies are a way for you to store nuggets of information on the visitor's computer.
49. The TotalBytes property contains the number of bytes that are in the parameter portion of an Active Server Page call.
50. Role of the Response object is to allow you to communicate with your dynamic Web and send the content back to the browser.
51. An application that has state is one that maintains information between calls, which can be accessed by one or more procedures within an application or the same procedure more than one time.
52. A session, in Active Server Page terms, is a single visitor traversing your Active Server Page application.
53. The Contents collection is a group of variables that you make available to any of the pages in your Active Server Page application through the Session object.
54. StaticObject collection provides a way for you to write code that runs when a visitor starts or ends a session.

55. The properties of the Session object allow you to configure and refer to aspects of the visitor's session.
56. The CodePage property allows you to set the character set that will be used during the session.
57. The LCID property of the Session object allows you to set the locale that the dynamic text should be displayed as.
58. The functionality of the Server Object allows you to take actions in your script for which the script itself doesn't supply the mechanism.
59. The CreateObject method allows you to instantiate a server component, which refers to an object, usually a DLL, on your server that has its own objects, collections, methods, properties, and/or events.
60. The MapPath method of the Server object converts a virtual or relative path as viewed through the Web server into its real physical path.
61. You can use the AddHeader to add your own custom header to the output of an HTML page.
62. AppendToLog method allows you to add text to the Web Server Log for a specific request.
63. The BinaryWrite method allows you to send nontextual, raw binary information to the browser.
64. The Clear method empties the buffer.
65. The End method of the Response object tells the compiler to stop processing any remaining code.
66. Flush method, sends any buffered output and immediately sends it to the browser.
67. The Redirect method sends the browser to a different page to receive a response.
68. The Browser Capabilities component provides you with information about the visitor's browser and information about their system.
69. The Ad Rotator component gives you a tool to manage the banner ads that appear on a Web page.
70. The Page Counter component provides a simple way to track and display the number of hits on a Web page.
71. The Counters component provides a simple interface for you to store integers that you can read, write, remove and increment across all the pages in your site and even across the boundaries of Active Server Page applications.
72. The Content Linking component provides a way for you to link pages that are to be viewed in a series, pages that have sequential meaning.

Multiple Choice Questions

1. This operator's syntax consists of an expression + expression :
   (a) Addition (b) Sum
   (c) Expression
2. This statement is used to shift control to a new procedure :
   (a) Control (b) Shift
   (c) Call
3. To add 2 strings you can use :
   (a) Add (b) Concatation
   (c) Two
4. Use this function to return a date with a time interval :
   (a) DateAdd (b) Return
   (c) Date
5. Use this statement to declare variables :
   (a) DIM (b) Declare
   (c) Variable
6. Use this function to calculate an expression and deliver the result :
   (a) Calc (b) Result
   (c) Eval
7. Use this function to return a value using a currency symbol as up on your system :
   (a) Currency (b) FormatCurrency
   (c) Symbol

8. Use this function to return a string that shows the hexadecimal value of a number :
   (a) Hex (b) String
   (c) Value
9. Use this function to show a dialog box containing a message specified in the code:
   (a) Message (b) Box
   (c) MsgBox
10. Use this statement when you want to create public variables and designate storage :
   (a) Private (b) Public
   (c) Designate
11. Buffer is cleared by :
   (a) Clear method (b) Clean method
   (c) Buffer method
12. The method which allows you to add text to the Web Server Log for a specific request is called.
   (a) AddToLog (b) AppendToLog
   (c) Log
13. This property contains the number of bytes that are in the parameter portion of an Active Server Page call:
   (a) Bytes (b) Total
   (c) TotalBytes

Answers

1. a	2. c	3. b	4. a	5. a
6. c	7. b	8. a	9. c	10. b
11. a	12. b	13. c		

Subjective Questions

1. What is ASP?
2. How do Active Server Pages work?
3. What are Objects and Components?
4. How would you create components using Session Scope?
5. How would you test your configuration?
6. What are variables?
7. What are procedures?
8. What are ASP forms?
9. How would you declare variables?
10. Describe the variable naming convention.
11. What is Implicit Declaration?
12. What is Explicit Declaration?
13. What are Object Hierarchies?
14. What are various type of Collections?
15. Describe the various types of Response methods.
16. What are ASP Applications?
17. What are Server Objects?
18. Describe the following statements:

   AddHeader method — AppendToLog method
   Binary/Write method — Clear method
   End method — Flush method
   Redirect method — Write method
   Lock method — Unlock method
   Server Object method — CreateObject method
   HTML Encode method — MapPath method
   URL Encode metod

14 Index

Symbols

T

U

V

W

X

Y